Microarray Technology
Through Applications

Microarray Technology Through Applications

Francesco Falciani (Editor)

School of Biosciences
University of Birmingham
Birmingham, UK

Taylor & Francis
Taylor & Francis Group

Published by:

Taylor & Francis Group

In US: 270 Madison Avenue
         New York, N Y 10016
In UK: 2 Park Square, Milton Park
         Abingdon, OX14 4RN

ISBN: 9780415378536 or 0-4153-7853-2

Library of Congress Cataloging-in-Publication Data

Microarray technology through applications / F. Falciani (ed.).
    p. ; cm.
  Includes bibliographical references and index.
  ISBN 978-0-415-37853-6 (alk. paper)
1. DNA microarrays. I. Falciani, F. (Francesco)
  [DNLM: 1. Gene Expression Profiling. 2. Oligonucleotide Array
Sequence Analysis—methods. 3. Genomics—methods. 4. Research Design.
QU 450 M6258 2007]
  QP624.5.D726M5143 2007
572.8'636—dc22

                                              2007000530

Editor: Elizabeth Owen
Editorial Assistant: Kirsty Lyons
Production Editor: Karin Henderson
Typeset by: Phoenix Photosetting, Chatham, Kent, UK
Printed by: Cromwell Press Ltd

Printed on acid-free paper

10 9 8 7 6 5 4 3 2 1

Taylor & Francis Group, an informa business   Visit our web site at http://www.garlandscience.com

Contents

6 The analysis of cellular transcriptional response at the genome level: Two case studies with relevance to bacterial pathogenesis 125

Thomas Carzaniga, Donatella Sarti, Victor Trevino, Christopher Buckley, Mike Salmon, Shabnam Moobed, David Wild, Chrystala Constantinidou, Jon L. Hobman, Gianni Dehò, and Francesco Falciani

7 Functional annotation of microarray experiments 155

Joaquín Dopazo and Fátima Al-Shahrour

Table 0.1 Key to link cases to protocols and supplementary information

Chapter	Protocols	Additional material
Chapter 1	Protocols 1, 2 and 3	Appendix 1
Chapter 2	Protocol 5	
Chapter 3	Protocol 6	Appendix 2
Chapter 4	Protocol 8	
Chapter 5	Protocol 7	
Chapter 6	Protocols 2 and 3	Appendix 4
Chapter 7		
Chapter 8	Protocols 2 and 3	Appendix 3
Chapter 9		

Contributors

Fátima Al-Shahrour, Department of Bioinformatics, Centro de Investigación Príncipe Felipe (CIPF), Valencia, Spain

Sara Alvarez, Molecular Group Cytogenetics, Spanish National Cancer Centre (CNIO), C/ Melchor Fernández Almagro 3, E-28029, Madrid, Spain

David Blesa, Molecular Group Cytogenetics, Spanish National Cancer Centre (CNIO), C/ Melchor Fernández Almagro 3, E-28029, Madrid, Spain

Christopher Buckley, MRC Centre for Immune Regulation, Division of Immunity and Infection, The University of Birmingham, Birmingham, UK

Thomas Carzaniga, Dipartimento di Scienze Biomolecolari e Biotecnologie, Università degli Studi di Milano, Via Celoria 26, 20133 Milan, Italy

Juan C. Cigudosa, Molecular Group Cytogenetics, Spanish National Cancer Centre (CNIO), C/ Melchor Fernández Almagro 3, E-28029, Madrid, Spain

Ana Conesa, Centro de Genómica, Instituto Valenciano de Investigaciones Agrarias, Valencia, Spain

Chrystala Constantinidou, School of Biosciences, The University of Birmingham, Edgbaston, Birmingham, B15 2TT, UK

Gianni Dehò, Dipartimento di Scienze Biomolecolari e Biotecnologie, Università degli Studi di Milano, Via Celoria 26, 20133 Milan, Italy

Joaquín Dopazo, Department of Bioinformatics, Centro de Investigación Príncipe Felipe (CIPF), Valencia, Spain

Francesco Falciani, School of Biosciences, The University of Birmingham, Edgbaston, Birmingham, B15 2TT, UK

Javier Forment, IBMCP, Technical University of Valencia, Valencia, Spain

José Gadea, IBMCP, Technical University of Valencia, Valencia, Spain

Anna González-Neira, Genotyping Unit. Spanish National Cancer Centre (CNIO), C/ Melchor Fernández Almagro 3, E-28029, Madrid, Spain

Nick Haan, BlueGnome, Cambridge, UK

Jon L. Hobman, School of Biosciences, The University of Birmingham, Edgbaston, Birmingham, B15 2TT, UK

Antony Jones, Functional Genomics Laboratory, School of Biosciences, The University of Birmingham, Edgbaston, Birmingham B15 2TT, UK

Cristina Largo, Molecular Group Cytogenetics, Spanish National Cancer Centre (CNIO), C/ Melchor Fernández Almagro 3, E-28029, Madrid, Spain

Steve D. Minchin, School of Biosciences, The University of Birmingham, Edgbaston, Birmingham, B15 2TT, UK

Shabnam Moobed, Keck Graduate Institute of Applied Life Sciences, Claremont, CA 91711, USA

Mercedes Robledo, Hereditary Endocrine Cancer Group. Spanish National Cancer Centre (CNIO), C/ Melchor Fernández Almagro 3, E-28029, Madrid, Spain

Sandra Rodríguez-Perales, Molecular Group Cytogenetics, Spanish National Cancer Centre (CNIO), C/ Melchor Fernández Almagro 3, E-28029, Madrid, Spain

Mike Salmon, MRC Centre for Immune Regulation, Division of Immunity and Infection, The University of Birmingham, Birmingham, UK

Donatella Sarti, School of Biosciences, The University of Birmingham, Edgbaston, Birmingham, B15 2TT, UK

Nigel J Saunders, Bacterial Pathogenesis and Functional Genomics Group, Sir William Dunn School of Pathology, University of Oxford, South Parks Road, Oxford OX1 3RE, UK

Graham Snudden, BlueGnome, Cambridge, UK

Lori A. S. Snyder, The Sir William Dunn School of Pathology, University of Oxford, UK

Victor Trevino, School of Biosciences, The University of Birmingham, Edgbaston, Birmingham, B15 2TT, UK

Joseph T. Wade, Department of Biological Chemistry and Molecular Pharmacology, Harvard Medical School, 240 Longwood Avenue, Boston, MA 02115, USA

Jeroen van Dijk, RIKILT, Wageningen, the Netherlands

David Wild, Keck Graduate Institute of Applied Life Sciences, Claremont, CA 91711, USA and Warwick Systems Biology Centre, Coventry House, University of Warwick, Coventry CV4 7AL, UK

Donling Zheng, King's College London, Department of Life Sciences, Room 3, 132 Franklin-Wilkins Building, 150 Stamford Street, London SE1 9NN, UK

Abbreviations

ALO	allelic-specific oligonucleotide
AML	acute myeloid leukemia
BAC	bacterial artificial chromosome
BCCP	biotin carboxyl carrier protein
BSA	bovine serum albumin
CBM	carbohydrate-binding molecule
CCD	charge-coupled device
CC-RCC	clear-cell RCC
cDNA	complementary DNA
CGH	comparative genomic hybridization
ChIP	chromatin immunoprecipitation
CMOS	Complementary metal-oxide-semiconductor
CRP	cAMP receptor protein
CTAB	hexadecyltrimethylammonium bromide
CV/CD	common variant/common disease (hypothesis)
DIP-chip	DNA immunoprecipitation chip
DMD	digital micromirror device
DNA	deoxyribose nucleic acid
DOP	degenerated oligonucleotide priming
ELISA	enzyme-linked immunosorbent assay
ESE	exomic splicing enhancer
EST	expressed sequence tag
FDR	false discovery rate
FISH	fluorescence *in situ* hybridization
GalNAc	*N*-acetylgalactosamine
GC	gas chromatography
GMAT	genome-wide mapping technique
GST	glutathione-*S*-transferase
kb	kilobasepairs
LC	liquid chromatography
LC–MS	liquid chromatography–mass spectrometry
LD	linkage disequilibrium
LPS	lipopolysaccharide
MAGE	microarray gene expression
MAGE-OM	MAGE object model
MAGE-Stk	MAGE software toolkits
MALDI-TOF	matrix-assisted laser desorption ionization/time-of-flight
MAS	maskless array synthesizer
Mb	megabasepairs
MBP	maltose-binding protein
MGED	Microarray gene expression data (society)
MIAME	minimum information about a microarray experiment
MM	mismatch oliogmer
MTC	medullary thyroid cancer
Neu5Ac	*N*-acetylneuraminic acid
NMR	nuclear magnetic resonance
NHS	*N*-hydroxysuccinimide
NTA	Ni^{2+}-Nitrailotriacetate
OGT	Oxford Gene Technology
ORF	open reading frame
PBM	protein-binding microarray
PC	principal component
PCA	principal component analysis
PCR	polymerase chain reaction
PI	predictive interval
PM	perfect match
PMT	photomultiplier tube
PPI	protein/protein interaction
PVDF	polyvinyldifluoride
Q-RT-PCR	quantitative reverse-transcriptase PCR
RCC	renal cell carcinoma
RIP-chip	RNA immunoprecipitation chip
ROMA	run-off transcription microarray analysis
RT	reverse transcriptase
SAGE	serial analysis of gene expression
SAR	system acquired resistance
SDS	sodium dodecyl sulfate
SELDI	surface-enhanced laser desorption/ionization

SELEX	systematic evolution of ligands by exponential enrichment	**TDT**	transmission/disequilibrium test
SNP	single nucleotide polymorphism	**TIFF**	tagged image file format (also known as TIF)
SOM	self-organizing map	**TSA**	thymidine signal amplification
SOTA	self-organizing tree algorithm	**TTS**	triplet-forming oligonucleotide target sequence
SSC	salt–sodium citrate		
SSDNA	salmon sperm DNA	**UV**	ultraviolet
STAGE	sequence tag analysis of genomic enrichment	**VHL**	von Hippel–Lindau
TAP	tandem affinity purification	**YAC**	yeast artificial chromosome

Preface

In the last few years biology has experienced unprecedented technological and conceptual developments. This has coincided with the widespread application of functional genomic technologies that allow monitoring the expression and interaction of thousands of genes, proteins and metabolites in single experiments. Because of the relatively low cost, microarray-based platforms have made the greatest impact.

Microarray Technology Through Applications is an initial introduction to the wide spectrum of microarray technologies for the non-expert and will also be a useful tool to the expert user who is looking for better understanding of specific issues behind experimental design and data interpretation.

Expression profiling was at first the only microarray-based technology to be of general use in the scientific community but in the last few years, we have seen a proliferation of microarray-based technologies and their application to a diverse set of tasks, ranging from studying the interaction of proteins with nucleic acids to measuring the concentration of a large number of proteins in biological samples. As an increasingly large number of research groups are considering using these approaches we felt that there was the need for a book that would provide a comprehensive overview of the theoretical and practical basis of microarray technology. This book will provide such an overview by means of a comprehensive introductory chapter (Chapter 1) followed by a series of case studies that represent the main application of this technology in biology. These case studies have been written by leaders in the field and describe prototypic projects simply and rigorously, with indication of how to generalize the approach to similar studies. The book is also designed to be a useful reference in the laboratory by providing a series of protocols for manufacturing and using microarrays in a number of applications (for a key to link case studies to protocols and supplementary information see Table 0.1 on page xi). Chapter 2 describes the use of immunoprecipitation techniques used in conjunction with DNA microarrays to determine the genome-wide binding profile of DNA-associated proteins. This technique (chip-on-chip) is now becoming a crucial tool for understanding gene regulation at a genome level and has applications from bacteria to human cells. Chapter 3 describes the use of microarray technology to compare the structure of different genomes. The case study in this chapter reports the application of this technique to identify chromosomal deletions and rearrangements in the genome structure of healthy versus diseased human cells. The application of this approach to the comparison of bacterial genomes and more generally in the analysis of comparative genomics datasets is discussed in Appendix 2. In Chapter 4, the use of microarray technology for identifying single nucleotide polymorphisms (SNPs) is discussed. Chapter 5 explains a recently developed technique for the identification of direct targets of transcription factors. This technique is based on the analysis of RNA produced from an *in vitro* transcription reaction that is catalyzed by the addition of a purified transcription factor. Chapter 6 contains two case studies with relevance to bacterial pathogenesis and provides the necessary information for the design, execution and analysis of experiments with two-color and single channel arrays.

Chapter 7 describes a set of bioinformatics tools designed to facilitate the biological interpretation of the result of a microarray experiment. Such tools are of fundamental importance of functionally annotate large functional genomics datasets. Chapter 8 exemplifies the application of microarray technology in agricultural research. This case study represents a potential practical application of expression profiling that uses a combination of some of the techniques described in previous two chapters. Chapter 9 deals with an application of microarray technology in proteomics providing indications on the potential issues behind this application.

Francesco Falciani

Useful links

Resource	Site	Remarks
Construction of cDNA library		
Lambda ZAP	www.stratagene.com	Commercial Product
CloneMiner	www.invitrogen.com	Commercial Product
SMART	www.clontech.com	Commercial Product
Clone LIMS and tracking		
AlmaZen	almazen.bioalma.com	Commercial Product
CloneTracker	www.biodiscovery.com/index/clonetracker	Commercial Product
B.A.S.E.	base.thep.lu.se	OpenSource Product
EST trimming, assembly, and processing pipelines		
phred, phrap, cross-match	www.phrap.org	Trimming algorithms
lucy	www.tigr.org/software/	Trimming algorithms
CAP3	genome.cs.mtu.edu/cap/cap3.html	Trimming algorithms
CLU	compbio.pbrc.edu/pti	Trimming algorithms
d2-cluster	www.ccb.sickkids.ca/dnaClustering.html	Clustering algorithm
TGICL	www.tigr.org/tdb/tgi/software/	Clustering algorithm
ESTIMA	titan.biotec.uiuc.edu/ESTIMA/	OpenSource Product
PHOREST	www.biol.lu.se/phorest	OpenSource Product
ESTWeb	bioinfo.iq.usp.br/estweb	OpenSource Product
ESTAnnotator	genome.dkfz-heidelberg.de	OpenSource Product
PipeOnline	bioinfo.okstate.edu/pipeonline/	OpenSource Product
ESTree	www.itb.cnr.it/estree/process.php	OpenSource Product
Microarray printing		
Genomic Solutions	www.genomicsolutions.com	Commercial Product
Corning	www.corning.com	Commercial Product
ArrayIt	www.arrayit.com	Commercial Product
Annotation		
Gene Ontology	www.geneontology.org	Gene Ontology Consortium
InterPro	www.ebi.ac.uk/interpro	Protein Family Annotation
Blast2GO	www.blast2go.de	Freeware tool for GO annotation
General database systems		
MySQL	www.mysql.org	OpenSource Product
PostgreSQL	www.postgresql.org/	OpenSource Product

| Oracle | www.oracle.com/database/index.html | Commercial Product |

Plant genomics databases

PLANET	mips.gsf.de/projects/plants/ PlaNetPortal/databases.html	Network of European Plant Databases
TAIR	arabidopsis.org/	The Arabidopsis Information Resource
MAIZEGDB	www.maizegdb.org/	Maize genome database, homepage
GRAMENE	www.gramene.org/	Resource for Comparative Grass Genomics
SoyBase	soybase.agron.iastate.edu/	Soybean database project

Microarray sites and data repository

MGED	www.mged.org/	International Microarray Gene Expression Data Society
MIAME	www.mged.org/Workgroups/MIAME/ miame.html	Minimum information about a microarray experiment, published by the MGED
SMD	genome-www5.stanford.edu/index. shtml	Homepage of the Stanford microarray database of Stanford University
ArrayExpress	www.ebi.ac.uk/arrayexpress/	EBI repository
RED	red.dna.affrc.go.jp/RED/	Rice Expression Database
NASCArrays	affymetrix.arabidopsis.info/	International Affymetrix transcriptomics service

Microarray protocols

General information	www.microarrays.org	Site with information on microarrays, University of California at San Francisco
KRL	www.rbhrfcrc.qimr.edu.au/kidney/ Pages/Microarray protocol.html	Microarray protocols used by the kidney research laboratory (Australia)
Microarray protocols	research.nhgri.nih.gov/microarray/ hybridization.shtml	Microarray protocols used by the National Human Genome Research Institute

Data analysis

Bioconductor	www.bioconductor.org	Freeware Project for Genomics Data Analysis
GEPAS	gepas.bioinfo.cnio.es	Web resource
CyberT	visitor.ics.uci.edu/genex/cybert	Web resource
Expressionist	www.genedata.com	Commercial Product
Rosetta	www.rosettabio.com/products/ resolver/default.htm	Commercial Product

Institutions

| CropNet | flora.life.nottingham.ac.uk/agr/ | UK Bioinformatics Resource for Crop Plants |

ILSI	www.ilsi.org/	International Life Sciences Institute
NCBI	www.ncbi.nlm.nih.gov/	National Center for Biotechnology Information
EBI	www.ebi.org	European Bioinformatics Institute
TIGR	www.tigr.org/	The Institute for Genomic Research, homepage

Introduction to microarray technology

1

Jon L. Hobman, Antony Jones, and Chrystala Constantinidou

> 'Man is a tool-using animal...Without tools he is nothing, with tools he is all' – Thomas Carlyle (1795–1881)

> 'The mechanic who wishes to do his work well, must first sharpen his tools.' – *Analects of Confucius* 15: 9

1.1 Introduction to the technology and its applications

1.1.1 Microarrays as research tools

Large-scale DNA sequencing projects and the completion of increasing numbers of genome sequences is having a major impact on biological research. The 'post-genomic era' is characterized by exploitation of genomic DNA sequence data as a research resource, and the use of high throughput experimental methods to study organism-wide events and interactions. These technical advances combined with increasing amounts of available genomic data have started to influence the direction that biological research is taking. Until quite recently, there has been a concentration on the reductive ('bottom up') view of understanding how an organism grows, adapts to changing conditions, or interacts with other organisms. This has been achieved by research groups studying single genes or regulons, or by determining the structure and function of small numbers of proteins, or by studying interactions between small numbers of cellular components. The use of the data generated in these experiments has led us to an understanding of how many cellular components work, and how some of these components interact with each other. However, in much the same way that understanding what a component in a radio does, or what happens when that component part of that radio is damaged or removed, does not lead to an understanding of how the radio works, we are faced with similar problems in describing how organisms work by looking at their components in isolation (Lazebnik, 2002). Now, there is a momentum towards the whole organism view of biology (Twyman, 2004a), using the holistic ('top down') approach of trying to understand how an organism works in its entirety and how the networks of physical and functional interactions occur between gene promoters, proteins, and noncoding RNAs (Brasch *et al.*, 2004). Attempts to understand the whole organism have led to the emergence of systems biology as a new cross-disciplinary research area, which encompasses experimental research, systems and control theory, bioinformatics, and theoretical and computational model building and prediction.

The development of appropriate technologies (experimental research tools) that exploit genomic data is playing a major role in the development of whole organism studies, and will begin to allow us to dissect networks of gene regulation (transcriptomics), understand protein production patterns and interactions with other proteins (proteomics), study the interactions of small molecules with proteins (chemical genomics), and start to catalog the small molecules and metabolites found in cells during normal and abnormal function (metabolomics).

One of the most important research tools used in transcriptomics and proteomics studies is the 'array', which is a powerful, high-throughput, massively parallel-assay format used for studying interactions between biological molecules. Arrays are a good example of how an advance in technology has allowed researchers to test hypotheses and interrogate organisms on a scale which prior to the development of this technology would have been impossible.

1.1.2 Arrays and microarrays

The use of ordered arrangements or 'arrays' of spatially addressed molecules in parallel assays is becoming an increasingly popular technology for studying interactions between biological molecules. These assays are commonly referred to as arrays, and by extension assays that have been miniaturized to a small format are called microarrays. The biological materials used in these arrays can be nucleic acids, proteins or carbohydrates, whilst arrays of chemicals and other small molecules have also been made. These materials are most commonly deposited on the surface of a planar solid substrate in an ordered arrangement so that each positional coordinate where material has been deposited contains material that represents a single gene or protein or other molecule. The most widely used solid substrates for arrays are nylon or nitrocellulose membranes, glass slides or silicon/quartz materials.

For DNA arrays, as the numbers of different nucleic acid molecules that are printed on arrays has become larger, so that the array represents the whole genome of the organism, and each spot on the array represents a single gene, the imperative has been to miniaturize the array format. There are two drivers for miniaturization: the first is simply so that all of the features (e.g. spots of DNA, each representing one gene) could be fitted onto a conveniently sized solid substrate, and the second is to use smaller and smaller amounts of biological material on the arrays, because high throughput methods of sample preparation tend to be restricted in the amount of biological material that can be purified using them. Concomitant with miniaturization of the array there has also been a trend towards decreasing the amount of biological material (e.g. RNA) used in the array experiments, as the less material that needs to be used, the easier it is to extract and purify it, and the more economical it is in terms of reagent costs.

There are several widely established formats that are used for DNA microarrays. These array types fall into two categories: those that are constructed within laboratories, and those that are produced under industrial manufacturing conditions by commercial companies. The first type, developed by the Pat Brown lab at Stanford University (Schena *et al.*, 1995, 1996; DeRisi *et al.*, 1997; Heller *et al.*, 1997) is the so-called 'home brew' or 'roll your own' glass slide microarrays, which are produced in-house, often in a core facility. The most popular technology for printing in-house arrays appears from anecdotal evidence to be contact printing, which is used by a large number of university research laboratories, and will be covered in some detail in this section (*Figure 1A*). The second format is the manufactured array, of which the best known is the Affymetrix GeneChip™ format, which is discussed in more detail in Section 1.4 (*Figure 1B*). In addition to these two well-known array formats there are other formats offered by commercial companies, such as Agilent, Nimblegen, Oxford Gene Technology, Xeotron, Combimatrix, Febit, and Nanogen. Each of these formats is more or less related in concept to the spotted array or Affymetrix formats. Agilent technologies (http://www.chem.agilent.com/Scripts/ PCol.asp?lPage=494) have developed a method for depositing long oligonucleotides (60-mers) on to glass slides, using ink jet printing (Hughes *et al.*, 2001). This printing method is also used by Oxford Gene Technology (OGT) to create arrays (http://www.ogt.co.uk/). OGT is a company created by Professor E.M. Southern, which owns fundamental European and US patents on microarray technology. Nanogen (http://www.nanogen.com/) have developed a method of electronically addressing oligonucleotides to positions on a chip, and enhancing hybridization by using electronic pulsing (Edman *et al.*, 1997; Sosnowski *et al.*, 1997; Heller *et al.*, 1999, 2000). Nimblegen (http://www.nimblegen.com/technology/) use a proprietary maskless array synthesizer (MAS) technology, to synthesize high density arrays, using photodeposition chemistry for oligonucleotide synthesis on a solid support. This system uses a digital micromirror device (DMD) in which an array of small aluminum mirrors are used to pattern over 750 000 pixels of light. The DMD creates 'virtual masks' at specific positions on a microarray chip that protects these regions from UV light that is shone over the array surface. In positions on which the UV light shines, it deprotects the oligonucleotide strand already synthesized, allowing the addition of a new nucleotide to the lengthening oligonucleotide (Nuwaysir *et al.*, 2002; Albert *et al.*, 2003). Nimblegen uses short oligonucleotide (25-mer) technology in their arrays, and are producing high density tiling arrays for resequencing and ChIP-chip experiments (see Section 1.1.4). Xeotron (who have recently been acquired by Invitrogen http://www.invitrogen.com/content .cfm?pageid=10620) use a proprietary platform technology for synthesis of DNA microarrays. The arrays are made by *in situ* parallel combinatorial synthesis of oligonucleotides in three-dimensional nano-chambers. The process of oligonucleotide synthesis uses photogenerated acids to deprotect oligonucleotide capping, and uses digital projection photolithography to direct deprotection and parallel chemical synthesis (Gao *et al.*, 2001; Venkatasubbarao, 2004). This method has also been used to produce peptide arrays and can be used for other syntheses. Combimatrix (http://www.combimatrix.com/) uses a different technology to generate the acids used to detritylate capped oligonucleotides during *in situ* phosphoramidite synthesis. Rather than using light-directed acid generation, Combimatrix uses a specially modified 'CMOS' semiconductor to direct synthesis of DNA in response to a digital command. Each feature on the

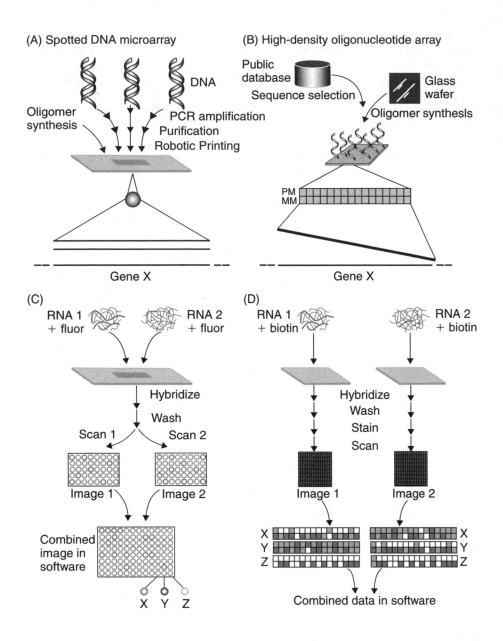

array is a microelectrode, which can selectively electrochemically generate acid, during oligonucleotide synthesis using phosphoramidite chemistry. Febit (http://www.febit.de/index.htm) market an all-in-one machine that synthesizes oligonucleotide arrays using maskless light activated synthesis of microarrays controlled by a digital projector, hybridizes the fabricated arrays, and analyzes the data. New microarray technologies such as nonplanar DNA microarrays made by companies such as Illumina (http://www.illumina.com), PharmaSeq (http://www.pharmaseq.com), and SmartBead Technologies (http://www.smartbead.com) are constantly evolving, and a recent review of the state of the art and future prospects

Figure 1.1

Expression analysis experiments using spotted glass DNA microarrays, and Affymetrix DNA microarrays. (A) Spotted glass microarrays are produced by the robotic spotting of PCR products, cDNAs, clone libraries or long oligonucleotides onto coated glass slides. Each feature (spot) on the array corresponds to a contiguous gene fragment of 40–70 nucleotides for oligonucleotide arrays, to several hundred nucleotides for PCR products. (B) Affymetrix high-density oligonucleotide arrays are manufactured using light directed *in situ* oligonucleotide synthesis. Each gene from the organism is generally represented by ten or more 25-mer oligonucleotides, which are designed to be a perfect match (PM) or a mismatch (MM) to the gene sequence. (C) For spotted arrays, gene expression profiling experiments commonly involve the conversion of RNA or mRNA to cDNA and labeling of the cDNA with a fluorescent dye for two samples. These are cohybridized to the probes on the array, which is then scanned to detect both fluorophores. The spots X, Y, and Z at the bottom of the image represent (X) increased levels of mRNA for gene X in sample 1, (Y) increased levels of mRNA for gene Y in sample 2, and (Z) similar levels of mRNA of gene Z in both samples. (D) During Affymetrix GeneChip transcription experiments, cRNA is biotinylated, and hybridized to the GeneChip. The GeneChip is then stained with avidin conjugated to a fluorophore, and scanned with a laser scanner. Results show: (X) Increased levels of expression of genes in sample 1, (Y) Increased levels of gene expression for sample 2, and (Z) similar gene expression levels for both samples. Reprinted from Harrington *et al.* (2000) *Curr Opin Microbiol* **3**: 285–291, with permission from Elsevier. (A color version of this figure is available at the book's website, www.garlandscience.com/9780415378536)

indicates that the evolution of the technology is proceeding rapidly (Venkatasubbarao, 2004).

Aside from differences in the method of array manufacture (deposition of prepared material versus *in situ* synthesis) the major differences between DNA array types is the nucleic acid material that is deposited onto the solid surface. Early-spotted DNA arrays deposited PCR products amplified from genes or open reading frames (ORFs), or spotted plasmid preparations from gene libraries, cDNAs, or expressed sequence tag clones (ESTs). As more complete genome sequences have been deciphered, complete genome sequence data is being used for the design of the materials deposited on the arrays. Primer design software can be used to design PCR primers to amplify regions from each gene from a sequenced genome for arraying (see http://colibase.bham.ac.uk/ as an example of a website that integrates genome analysis and primer design software tools), and PCR arrays have the advantage that they will represent both the sense and antisense strands of DNA. Single stranded oligonucleotide arrays by their nature can only be sense or antisense arrays, so their design requires careful thought because transcriptomics experiments use labeled complementary DNA (cDNA) made from mRNA to hybridize onto the array, so sense oligonucleotide arrays will hybridize to these cDNAs, but antisense ones will not. Genome sequence data and bioinformatics techniques have been used in the rational design and chemical synthesis of long oligonucleotides (40–100-mer) to represent genes on arrays, so that each has a matched melting temperature and length, which is claimed to improve the reliability of hybridization signals. Both of these two 'longmer' methods rely on a single, long nucleotide fragment to represent each gene or ORF on the array. The alternative strategy employed for DNA arrays (such as Affymetrix and Nimblegen arrays) has been the use of multiple short oligonucleotides (generally 25-mer) on an array to represent a gene. The use of high density oligonucleotide arrays such as these to 'tile' across a genome so that intergenic regions as well as ORFs and genes are represented by multiple oligonucleotides, has led to greater flexibility in the experiments that can be performed on these arrays, which is detailed below.

1.1.3 Principles of DNA array technology

All arrays are used as a tool to determine interactions between molecules immobilized to the solid surface, and molecules that are in a complex mixture in a solution, which is in contact with the array (*Figure 1.1*). Those interactions that occur are then detected and quantified. All DNA arrays harness the ability of nucleic acids with complementary sequences to hybridize to each other under suitable conditions. In DNA arrays, one nucleic acid is immobilized on a solid surface, and the nucleic acid in the hybridization solution is labeled with a radioactive isotope, chemical or dye molecule that can be detected quantitatively. Throughout this chapter we will be adopting the convention that the nucleic acid material tethered to the solid surface of the microarray is termed the 'probe', and the nucleic acid that is labeled by a reporter molecule such as a dye or isotope will be termed the 'target' (Phimister, 1999). The theory behind this nomenclature is that the 'probe' nucleic acids immobilized to the array surface are used to interrogate the complex mixture of labeled nucleic acids (e.g. cDNAs in transcriptomics experiments) for nucleic acids in the hybridization solution that are complementary to, and will associate with, the immobilized (tethered) probe. Unfortunately, this widely adopted nomenclature for description of probe and target in microarrays is the opposite of that commonly adopted (in our laboratory at least) for Southern/northern blotting, where a radioactively labeled DNA or cDNA probe would be used to hybridize to an immobilized nucleic acid on a nylon membrane. Clearly, this difference in nomenclature can lead to some confusion.

Key to DNA microarray technology is nucleic acid hybridization, which occurs where single stranded (denatured) nucleic acids are incubated together under conditions that promote the formation of base paired duplex molecules by C:G or A:T base pairing. The double stranded nucleic acid hybrids are therefore composed of sequences that are complementary to each other. Conditions that favor hybridization between nucleic acids can be promoted by manipulating time, temperature and ionic strength of the hybridization buffer (stringency) and will be affected by the concentration and complexity of the sample (Young and Anderson, 1985; Stoughton, 2005). Hybridizations are commonly conducted at low stringency (high salt concentration) often in the presence of formamide, because formamide allows the hybridization to be carried out at relatively low temperatures (42°C). The stability of the DNA hybrids is dependent on whether there are any mismatches in the nucleic acid duplex, with stability decreasing as the number of mismatches increases. The stability of the nucleic acid duplex determines how strongly attached the target DNA is to the immobilized probe nucleic acid, and how easy or difficult it is to wash away labeled target DNA from nonspecific binding to the attached probes. Washes to destabilize and consequently remove mismatched hybrids are conducted under increasing stringency (for practical purposes this is generally decreasing salt concentration and/or increasing temperature), so that nonspecific hybridization is minimized.

Solution nucleic acid hybridization was used widely in the 1960s and 1970s to determine DNA homology, and investigate nucleic acid structure (Anderson, 1999). The colony blot (Grunstein and Hogness, 1975), Southern blot (Southern, 1975), and dot blot were methods that were primarily used to identify identical or closely related DNA sequences by hybridization, and immobilized DNA onto a solid (though permeable) surface such as a nitrocellulose or later a nylon membrane, and then hybridized radioactively labeled denatured nucleic acids to the immobilized and denatured nucleic acids. The DNA macroarray is essentially a direct development from the dot blot and Southern blot. The Panorama® macroarrays marketed by Sigma-Genosys consist of PCR products immobilized on nylon membranes; where each PCR product represents all, or part, of a single gene. In transcriptomics experiments these immobilized PCR products are hybridized with radioactively labeled cDNA products, washed, and the resulting membrane scanned using a phosphor-imager system. For macroarrays, comparisons between two samples have to be conducted using two separate hybridizations, because the 'test' and 'control' samples are labeled with the same reporter molecule. Aside from miniaturization, the key difference between a standard glass slide DNA microarray experiment and experiments conducted using systems such as the Panorama® macroarray, is that the glass slide arrays use competitive hybridization between cDNAs derived from two sources, and the immobilized probe DNA. In these experiments, each cDNA from a different source (e.g. wild-type and mutant) is labeled with a different fluorescent dye. Competitive hybridization has a number of advantages: it allows for a direct comparison between two samples, e.g. a wild-type (control) and a mutant (test) on the same array, and very importantly from the point of view of data reliability and reproducibility, competitive hybridization overcomes irregularities in probe spot properties or local hybridization conditions on the array, which may adversely affect hybridization signal intensities. In competitive hybridization, the signal from each spot on the array is therefore a ratio of signals from both the control and test samples, where the control and test have undergone matched conditions. This means that even if the probe deposition, or hybridization conditions are irregular, accurate signal intensity ratios can be obtained (Stoughton, 2005).

Although the protein (Cutler, 2003), carbohydrate (Wang, 2003), and small molecule (Spring, 2005) microarray formats are different from DNA microarrays, and interactions between biological molecules will be different, the principle of how these other arrays work is essentially identical to the concept of DNA arrays in terms of the idea of using the immobilized biological molecule (antibodies, proteins, small molecules or carbohydrate) 'probes' to interrogate a complex mixture of labeled 'target' biological molecules for those that will bind to, or associate with them (see *Figure 1.2*).

1.1.4 Microarrays as tools for biological research applications

The major modes of use of DNA microarrays are:

- expression profiling,
- pathogen detection and characterization,

Figure 1.2

Microarray-based technologies used for the study of biological interactions. Modified from Shin *et al.* (2005) *Chemistry, a European Journal* **11**: 2894–2901, with permission from Wiley-VCH.

- comparative genome hybridization (CGH),
- genotyping,
- whole genome resequencing,
- determining protein DNA interactions (ChIP-chip) (Stears *et al.*, 2003; Buck and Lieb, 2004; Stoughton, 2005) (see *Figure 1.3* for an overview of some of the applications of DNA microarrays).

Several other DNA array-based applications are developing, which include:

- regulatory RNA studies (Wassarman *et al.*, 2001; Zhang *et al.*, 2003),
- alternative splicing and RNA binding protein studies,
- methylome analysis (reviewed in Mockler and Ecker, 2005).

DNA microarrays are probably best known as a research tool to study whole organism/tissue genome-wide transcriptional profiling, where the transcription profiles of all of the genes on a genome, from a subset of genes from the genome, or from a particular tissue, can be simultaneously assayed. The nucleic acid that is being assayed in transcriptional profiling by the arrays is messenger RNA, which is generally converted to cDNA

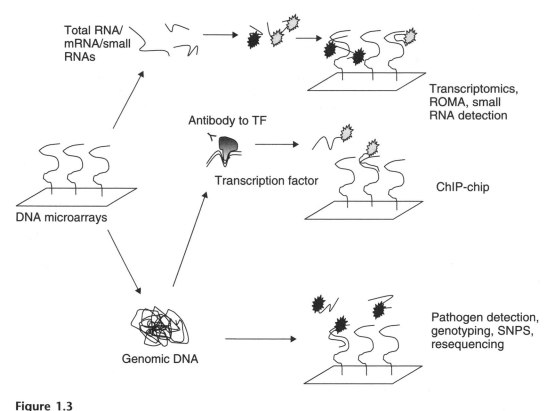

Total RNA/ mRNA/small RNAs

Transcriptomics, ROMA, small RNA detection

Antibody to TF

Transcription factor

ChIP-chip

DNA microarrays

Genomic DNA

Pathogen detection, genotyping, SNPS, resequencing

Figure 1.3

Examples of applications for DNA microarrays.

before hybridization onto the array. It is common in these types of experiments to compare the transcription profiles of two or more individuals or cultures of cells, in order to determine the transcriptional differences between them. In their simplest form (sometimes referred to as type I experiments (DeRisi *et al.*, 1997; Yang and Speed, 2002) DNA array transcriptomics experiments compare, for example, a wild-type with a mutant, a healthy with a diseased individual, or unstressed with stressed cells. More complicated transcriptomics experiments, which use an invariant control, such as genomic DNA, or pooled RNA for one channel in the experiment (sometimes referred to as type II experiments (DeRisi *et al.*, 1997; Yang and Speed, 2002)), can be used to compare multiple individuals, conditions or treatments. The number of published *in vivo* transcriptomics papers in which microarrays have been used is growing at an extremely rapid rate, for both prokaryotic and eukaryotic systems (Stoughton, 2005), with thousands of papers now published using this technology. One variation of *in vivo* transcriptional profiling is the *in vitro* transcriptional profiling method: run-off transcription microarray analysis (ROMA), (Cao *et al.*, 2002; Zheng *et al.*, 2004, and this book), which uses microarray technology to profile the abundance of run-off transcripts generated *in vitro* using DNA template, purified RNA polymerase, a regulatory protein and nucleotides.

There are a number of array-based methods that use DNA arrays to interrogate the DNA content of cells, tissues or other samples. These methods rely on the high density of probes contained on a DNA array to multiplex hybridizations, in order to either assay for a large number of DNA sequences (detection of the presence or absence of DNA sequences), or to use the high density of probes to increase resolution in locating the presence of a hybridizing piece of DNA (mapping onto the chromosome). Comparative genome hybridization (CGH) is a method that has been extensively used to detect the absence or presence of particular genes or chromosomes, or variations in gene copy number in eukaryotes, as these gene deletions or duplications are often associated with diseases such as cancer, and with developmental abnormalities, such as Down's syndrome. CGH uses cohybridization onto metaphase chromosomes of labeled total genomic DNA from a 'test' and 'reference' population of cells to localize and quantitatively measure DNA copy number differences between these populations, and associate the copy number aberrations with the disease phenotype (Kallioniemi et al., 1992). One disadvantage of CGH is that the use of metaphase chromosomes results in a relatively low resolution of detection using CGH (Pinkel et al., 1998). Array CGH is a further refinement of the technique that allows high resolution of detection of where deletions or gene duplications occur on the chromosome, by using high density or tiling DNA arrays (Pinkel et al., 1998; Albertson et al., 2000; Dunham et al., 2002; Ishkanian et al., 2004). Similarly, the use of microarrays in genotyping and detection of single nucleotide polymorphisms (SNPs) is becoming widely used, because arrays offer rapid, parallel allele discrimination (Fan et al., 2000; Hirschhorn et al., 2000; Syvänen, 2001; Lindroos et al., 2002; Kennedy et al., 2003; Matsuzaki et al., 2004). A further development of DNA rather than RNA-based uses for arrays is in the detection and characterization of pathogenic microorganisms. Use of DNA arrays for detection is a technology that has become popular because of the ability of DNA arrays to be both flexible and multiplexing (Call et al., 2003; Korczak et al., 2005). DNA microarrays have also been used to compare pathogenic and non-pathogenic variants of related bacterial species using comparative genomic hybridization (Behr et al., 1999; Salma et al., 2000; Schoolnik, 2002; Call, 2005), and for resequencing of pathogen strains, detailed below.

A fundamental goal of understanding how organisms regulate gene expression is the study of how regulatory proteins interact with DNA, and influence transcription from promoters within their cognate regulon. The site specificity of regulator interaction with genomic DNA, and what influences their binding to genomic DNA *in vivo* is fundamental to studies on the control of gene expression. This can be problematic when mutations or a deletion of the gene encoding the regulator, or overexpression of regulatory proteins is lethal to the host cells (Lieb et al., 2001). Many of these studies on transcription factor interactions with DNA have been conducted *in vitro*, using purified proteins and DNA in gel shift assays, DNAase I footprinting assays or SELEX (systematic evolution of ligands by exponential enrichment (Gold et al., 1997)) and other methods. These methods have been limited by the need to overproduce and purify regulatory proteins for assays, and there are indications that *in vitro* binding of a protein to DNA is not always an accurate predictor of a regulator's binding sites *in vivo* (Lieb et *al.*, 2001; Buck and Lieb, 2004). Although computational approaches to determining putative regulator binding sites on a genome have been used quite extensively (Van Helden *et al.*, 1998; Hughes *et al.*, 2000; Stormo, 2000; Tan *et al.*, 2001; Van Helden, 2003), they require experimental confirmation. This can be achieved using gel shift assays, footprinting, run-off transcription assays or by the use of transcriptomics, but with multiple transcription factors regulating the expression of a single gene or operon, regulation of gene expression by one regulator may be masked by the effects of other regulators. Scale also plays an important role. In the case of some putative transcription factor binding sites there could be many hundreds or thousands of potential binding sites for that regulator throughout a genome. There are therefore many strong arguments for using *in vivo* selection of a transcriptional regulator. Genomic SELEX (Gold *et al.*, 1997) is one method that has been used to enrich for regulatory protein binding sites. Another is chromatin immunoprecipitation (ChIP) (reviewed in Kuo and Allis, 1999), which in its original form was used to study histone protein interactions with the chromosome in eukaryotes, and uses formaldehyde to cross-link proteins to the genomic DNA of the organism *in vivo*, followed by immunoprecipitation of protein–DNA complexes using antibodies (Cosma *et al.*, 1999; Krebs *et al.*, 1999). ChIP has been used for mapping interactions between histone proteins and target genes (Kuo and Allis, 1999), or for identifying binding sites for regulators using ChIP and quantitative PCR (Buck and Lieb, 2004). Combining ChIP with high density whole genome microarrays (ChIP-chip) has allowed researchers to map the *in vivo* genomic interactions of DNA-binding regulatory proteins in eukaryotes (Ren *et al.*, 2000; Iyer *et al.*, 2001; Lieb *et al.*, 2001; Horak *et al.*, 2002; Nagy *et al.*, 2003), and is now being used to study transcriptional regulation in prokaryotes (Grainger *et al.*, 2004; Herring *et al.*, 2005).

A burgeoning use of high density microarrays is for genomic resequencing, (reviewed in Mockler and Ecker, 2005), a method which relies upon the decreased stability, and therefore decreased signal generated by mismatched DNA hybrids, to identify base sequence differences between the array, which has been designed to the genome of one organism, and the genomic DNA of the tested organism. The principles of this technology, and examples of the early use of this technology have been discussed previously (Khrapko *et al.*, 1989; Southern *et al.*, 1992; Southern, 1996; Hacia, 1999), but essentially the methods use the power of microarrays to multiplex the interrogation of nucleic acid sequence by hybridization, in gain of signal or loss of signal assays. Resequencing arrays require extremely high resolution (1 bp) ultra high density array design (Hacia, 1999; Mockler and Ecker, 2005) and allow the complete resequencing of an organism in a single hybridization (Wong *et al.*, 2004). High density high resolution arrays have been used to resequence multiple *Bacillus anthracis* strains (Zwick *et al.*, 2004). But, because resequencing arrays can only compare the genome of the test organism with a known reference sequence, they are not useful in identifying novel sequences or rearrangements of sequences (Mockler and Ecker, 2005). This is an important point, and shows one of the weaknesses of DNA arrays, because DNA arrays will only tell you if DNA homologous to the probe sequences on the array is present, not if any novel genetic material is present in the strain.

Clearly, DNA arrays can be used for a wide variety of experimental techniques, and the higher the resolution of the DNA array in terms of genome coverage, and the higher the number of overlapping short probes on the array, the more experimental techniques can be used on that single array platform.

1.2 The design of a microarray

There are different formats of microarray available, either as commercially obtainable complete kits (such as Affymetrix, or pre-printed oligonucleotides on glass slides) or completely 'home made' versions, with every variation in between. The format of the nucleic acid used in the array is also variable. Historically the first microarrays (Affymetrix always being the exception) were spotted with PCR products of varying size, using primers designed to hybridize towards the 3' and 5' end of each gene of interest. Over the years the design of the PCR primers has become more sophisticated, aiming to generate PCR amplicons of the same size representing each gene (e.g. Sigma-Genosys http://www.sigma-genosys.com/gea.asp). There has been a definite shift towards long oligonucleotide design for arrays, however. Companies such as Ocimum (http://www.ocimumbio.com/web/default.asp, formerly the genomics business of MWG biotech), Operon (http://www.operon.com/), Sigma-Genosys and Illumina (http://www.illumina.com/) all offer oligonucleotides (usually one oligonucleotide representing one gene and varying in size between 20 and 70 nucleotides) representing either entire genomes or specific gene groups (apoptosis, cytokines etc.). Or they can design and synthesize microarray oligonucleotide sets for a customer. For the in-house manufacture of DNA arrays the oligonucleotide option offers an easier, faster and even cheaper alternative to PCR products. Long oligonucleotides have the added advantages that they obviate quality control failed PCR reactions and PCR product concentration problems associated with PCR product arrays.

Different companies use different 'optimal' criteria for the design of their oligonucleotides. As a result the commercially available sets vary in their length, location within the target gene, sensitivity, and specificity. The quality of the long oligonucleotide design is directly linked to the quality of the genome sequence from which it was designed, and the accuracy and effectiveness of the annotation information, especially regarding gene definition. As new genes are identified, probes need to be redesigned and the oligonucleotide sets need to evolve to account for this. It is important that the design is optimized to avoid false-positive data. Imbeaud and Auffray (2005) propose that time is better spent creating gene lists coupled with probe sequence information and annotation, ranked according to accuracy and confidence, rather than invalidating false-positive data arising from bad array design.

1.3 Glass slide DNA microarrays

Self-spotted glass slide microarrays are now widely used in research because production of them is relatively cheap and there is great flexibility about what can be printed on the glass slide array. It is therefore cost effective for a single facility to be able to print arrays for a large number of individual projects or research groups using a single printing machine.

1.3.1 The technology

Array technology has emerged from two distinct areas of research, and interpretations of the history of the development of arrays have proved controversial (Schena *et al.*, 1998; and response to this: Ekins and Chu, 1999). The first area of research that has impacted on the development of arrays in the form commonly recognized now as the 'classic' two-color array is the 'ambient analyte theory' of why miniaturized and highly parallel ligand-binding assays are more sensitive than any other ligand-binding assay (Ekins *et al.*, 1989a, 1989b). This idea was applied to immuno-diagnostics and involved the use of solid substrates (plastic) to immobilize antibodies, and the use of fluorescent labels to quantify 'ratiometric' inter-actions between antibodies, using confocal laser microscopy to detect the abundance of each fluorescent dye on an immobilized antibody microspot (Ekins *et al.*, 1989a, 1989b, 1990; Ekins and Chu, 1999). This system contains many elements that are found in arrays used in the present. The second key area of research impacting on microarray development has been nucleic acid hybridization technology, and the use of covalent attachment to glass slides of oligonucleotides and nucleic acids for hybridization purposes (Khrapko *et al.*, 1989; Southern *et al.*, 1992), and in parallel to this, the development of *in situ* combinatorial synthesis of peptide and oligonu-cleotide arrays (Fodor *et al.*, 1991). All of the elements of the technology used for glass slide DNA microarray transcriptomics experiments as used at present were first published by Dr Patrick Brown's research group (Schena *et al.*, 1995, 1996; DeRisi *et al.*, 1997). These publications described the use of different fluorescent dye reporter molecules for each nucleic acid sample, and printing of arrays from nucleic acid solutions using contact printing technology. The concept of a dual-labeled microarray transcriptomics experiment is very simple, as it relies on a competitive hybridization between the two samples of cDNAs made from cellular RNA to the tethered 'probes' on the array. Because the dyes used to label the cDNA fluoresce at different wavelengths, the abundance of one transcript compared with the abundance of the other transcript can be visualized as a color, thus giving a comparative result (reviewed in Freeman *et al.*, 2000).

1.3.2 Microarray fabrication

There are two major technologies used for printing glass slide microarrays: **noncontact** deposition methods, and **contact** deposition methods.

Noncontact printing technologies use ink-jet printing technology to deposit small droplets of liquid onto a solid surface, and are used either to print presynthesized oligonucleotides or other biological materials, or for *in situ* synthesis of oligonucleotides on a solid surface (Case-Green *et al.*, 1999; Theriault *et al.*, 1999; Holloway *et al.*, 2002). The technology of printing using ink-jet methods has centered on piezoelectric technologies, or thermal, acoustic or continuous flow technologies (Theriault *et al.*, 1999) developed by commercial organizations, of which Agilent is perhaps the most well known. Recently, a design for open source ink-jet array printers has become freely available (Lausted *et al.*, 2004).

Contact printing uses physical deposition of small volumes of liquid from a metal pin onto the slide surface. This can be from solid pins, where the spotting solution adheres to the outside of the pin, or split pins that draw spotting solution into an internal reservoir by capillary action, and deposit small volumes of spotting buffer during contact with the slide surface. Contact printing marries the principles of how sixteenth and seventeenth century ink pens worked, with twentieth and twenty-first century precision engineering and robotics. The printing process itself is dependent on extreme accuracy of positioning of the pins in the *X*, *Y*, and *Z* axes, reproducible control of the speed of movement of the print head, and the ability of the print pins to take up nucleic acids in solution by capillary action, and deposit highly reproducible spots of the solution onto the glass surface. Glass microarrays can be printed using robotic spotting machines which are available from several suppliers (or even using a machine that can be constructed according to plans available from the Pat Brown laboratory website (http://cmgm.stanford.edu/pbrown/mguide/index.html) (Castellino, 1997).

Glass microarrays have historically been printed with DNA preparations of plasmid/cosmid/BAC gene libraries, probes made by PCR from genomic DNA, with cDNAs from gene expression libraries, or with long oligonucleotides that have been designed from complete and annotated genome sequences. The most common sizes of probe are 100–500 bp for DNA and 40–70-mer for long oligonucleotides. In these cases one PCR product/cDNA or oligonucleotide has generally been used to represent one gene or ORF from the organism. The chemically synthesized oligonucleotide, purified PCR product, or other nucleic acid material is then deposited onto the solid substrate.

The technical details of printing glass slide DNA microarrays, the pin types used, and spotting buffers are discussed in Appendix 1.

1.3.3 A typical glass slide microarray transcriptomics experiment

Glass slide microarrays are widely used in genome-wide expression profiling (transcriptomics) experiments. A general description of a common type of DNA microarray transcriptomics experiment can be broken down into seven phases:

(i)   array printing,
(ii) experimental design,
(iii) sample preparation,
(iv) sample labeling,
(v)   hybridization,
(vi) data/image acquisition,
(vii) data analysis.

1.3.4 Array printing

In array printing, glass microarray slides, which are identical in dimensions to microscope slides, are commonly used. The microarray slide is uniformly coated with a chemical compound that will interact with and immobilize nucleic acids, irreversibly binding them to the surface. Nucleic acids (clone library, cDNAs, PCR products, long oligonucleotides) are deposited on the slide by contact printing, and treated with UV light or baking at 80°C to crosslink them to the slide surface. The printed slides are stored desiccated at room temperature, in the dark, until required for experimentation.

1.3.5 Experimental design

The design of a microarray experiment is highly important, and if the experiment is incorrectly designed, the information produced from it could be, at worst, useless. Clearly knowing what questions you wish to answer, and what hypotheses you wish to generate or test are extremely important (Stekel, 2003a, 2003b), as are the statistical implications of replication, power analysis, whether comparisons made on the microarray are to be direct or indirect (by making comparisons within or between slides, the so-called type I or type II experiments), and whether experimental design needs to be single or multi-factorial (Kerr and Churchill, 2001; Churchill, 2002; Yang and Speed, 2002; Stekel, 2003a, 2003b).

1.3.6 Sample preparation

In sample preparation, RNA from the host organism is isolated, converted to cDNA and labeled with dyes before hybridization to the array. Of critical importance to any successful transcriptomics experiment is the quality, quantity, and integrity of the total or messenger RNA (mRNA) that is used. There are numerous methods for RNA isolation, including the use of Trizol®, and other phenol-based methods, as well as RNA isolation kits that are commercially available from companies such as Ambion (http://www.ambion.com/), Qiagen (www.qiagen.com/), and Promega (http://www.promega.com/). These kits are designed for rapid and reliable RNA extraction, and are widely used because they are simple, reliable, reproducible, and are manufactured under stringent quality control. Because of the instability of RNA in general, and the short half-life of most bacterial RNAs in particular, of equal importance to the choice of RNA extraction methodology is how to 'freeze' (prevent further production, and degradation of) the RNA in the cell, prior to the extraction procedure. RNAprotect™ (Ambion) and RNAlater™ (Qiagen) are commercially available solutions that perfuse cells, and 'freeze' the cellular RNA profile. Other researchers use 5% phenol in ethanol which achieves the same result, prior to RNA extraction and purification. Total RNA is commonly extracted from cells, but it has to be borne in mind that mRNA is only a small fraction of total RNA, so quite large quantities of total RNA (commonly between 10 and 20 µg) are commonly extracted in order to obtain sufficient mRNA for a transcriptomics experiment. However, extraction of total RNA has advantages, because simple monitoring for any degradation of the ribosomal RNA caused by RNAases, indicates whether the RNA quality has been compromised. There are a number of methods used to monitor RNA quality and quantity, of which agarose gel electrophoresis is the simplest method. Recently, instrumentation which uses very small amounts of sample for analysis of quantity and quality have been developed. UV microspectrophotometers, such as the Nanodrop™ (http://www.nanodrop.com/) are used to measure RNA concentration, and the Agilent™ 2100 Bioanalyser (http://www.chem.agilent.com/Scripts/PDS.asp?lPage=51) uses an RNA electrophoresis 'chip' which combines both quantification and qualitative measures of RNA integrity in the same assay, using minimal amounts of RNA.

1.3.7 Sample labeling, hybridization, and detection

In sample labeling, there are two widely used methods of labeling cDNA made from total or messenger RNA. These methods are called **direct** or **indirect** labeling. Both methods convert purified and quantified total RNA or mRNA into cDNA by reverse transcriptase (RT) using either oligo dT priming to the 3' polyadenylation site of eukaryotic mRNA, or random hexamer/nonamer priming to prokaryotic RNA (which lacks the polyadenylation site that eukaryotic mRNA has). The difference between **direct** and **indirect** labeling lies in how the cyanine dyes (CyDyes™ (Cy3 or Cy5)) or other fluorescent dyes such as the Alexa™ dyes are incorporated into the cDNA. In **direct** labeling a dye conjugated nucleotide is incorporated directly into the cDNA by the reverse transcriptase enzyme (RT). In the **indirect** labeling method, the RT enzyme incorporates an amino-allyl UTP into the cDNA instead of dTTP. The amino-allyl labeled cDNA is then reacted under alkaline conditions with NHS esters of CyDyes in a direct chemical coupling reaction. There are advantages and disadvantages to using both methods. The **direct** incorporation method is simple and rapid, but it is generally agreed that there is a bias of incorporation in the direct labeling reaction, so that cDNA is labeled at higher efficiency by Cy3 than by Cy5. This problem has been overcome to a certain extent by increasing dNTP concentrations in the Cy5 labeling reactions of some commercially available labeling systems, such as the Promega Chipshot™ kit, which prevents 'choking' of the RT enzyme by Cy5 (http://www.promega.com/faq/chipshot.html). The other major strategy employed to overcome dye incorporation bias in direct labeling is by performing dye swap experiments (reviewed in Dobbin et al., 2003), where, for example, in one experiment the control sample is labeled with Cy3, and the test sample is labeled with Cy5, and then a duplicate experiment is performed where the control sample is labeled with Cy5, and the test sample is labeled with Cy3. However, other evidence suggests that using direct labeling methods the dye incorporation bias may be cDNA sequence specific (Dombkowski et al., 2004) which is more difficult to overcome. Although the **indirect** methods of labeling results in less biased dye incorporation when compared with **direct** labeling, the disadvantage of this method is that the protocols are more complicated and time consuming than direct labeling.

There are other technologies available that have been designed primarily to increase sensitivity/decrease the amount of RNA needed in hybridization. These fall into two categories: **sample amplification** and **signal amplification**. In the first category, linear RNA amplification by T7 polymerase *in vitro* transcription (Van Gelder *et al.*, 1990; Phillips and Eberwine, 1996; Scherer *et al.*, 2003; Speiss *et al.*, 2003) has been used for some considerable time to amplify RNA from limited samples. Other methods such as template-switching PCR have also been developed to address the need to reduce the amount of RNA required in microarray hybridization experiments, and to amplify it in a less labor-intensive manner than the Eberwine method (Petalidis *et al.*, 2003). **Signal amplification** technologies such as dendrimer (or 3DNA) technology marketed by Genisphere® (Stears *et al.*, 2000; and http://www.genisphere.com/ array_detection_home.html) have been used to boost signals from small amounts of RNA/cDNA available in some transcriptomics experiments, and this technique has been compared with both direct and indirect labeling (Manduchi *et al.*, 2002). This signal amplification is achieved by using a three-dimensional structure of oligonucleotides (dendrimer) to which several hundred CyDye molecules are attached, as the method of labeling cDNA. Other methods such as tyramide signal amplification (TSA; http://probes.invitrogen.com/handbook/sections/0602.html) (Karsten *et al.*, 2002) and rolling circle amplification (Nallur *et al.*, 2001) are methods that are being investigated to boost signals from microarray experiments.

In sample hybridization, for anyone familiar with Southern/northern blotting techniques and fluorescence *in situ* hybridization (FISH) (Anderson, 1999), most of the techniques and solutions used in microarray hybridizations will be very familiar. For a microarray experiment, some glass slide formats require a 'pre-soak' in a solution of sodium borate, which is used to reduce autofluorescence on the slide surface, prior to the slide being prehybridized, commonly in a solution that will contain salt–sodium citrate (SSC), sodium dodecyl sulfate (SDS), bovine serum albumin (BSA) and a DNA blocker (often salmon sperm DNA (SSDNA)). The purpose of both BSA and SSDNA is to 'block' unreacted chemical groups on the slide surface, in order to reduce nonspecific hybridization. After prehybridization, slides are hybridized under high salt conditions that promote inter-molecular pairing between the tethered DNA 'probe' and the two samples of labeled cDNA (made from total RNA, or mRNA), one labeled with the CyDye™ Cy3, the other labeled with Cy5. After increasingly stringent washes (decreasing concentrations of SSC) to remove CyDye-labeled target cDNA which is binding nonspecifically to the slide surface, or hybridizing with mismatches to the surface immobilized 'probes', the slide is dried prior to scanning. These steps are summarized in *Figure 1.1C*.

1.3.8 Technical challenges

There are a number of technical challenges in obtaining high quality, repro-ducible data from glass slide microarray experiments. These challenges are found throughout each of the seven steps of the microarray experimental process detailed in Section 1.3.3, and key to data reproducibility is the use of reproducible experimental techniques.

From the experimental point of view, the technical challenges of producing high quality data start with the slide surface, material to be spotted, and quality of the printed spots. It is important that there should be replicates of each spot on the microarray and there should be multiple independent repli-cations of the experiment, which improves the statistical reliability of the data. In terms of the transcriptomics experiment, selection and reproducibility of growth conditions, and care in extraction and purification of RNA are extremely important, as is obtaining sufficient RNA for the experiment. Experiments conducted with RNA that is of less than the highest quality will be compromised. Throughout the labeling process, production of cDNA and dye incorporation into the cDNA should be monitored, and the amounts of labeled cDNA to be hybridized on the slide carefully balanced. Reproducible washing of the slides after hybridization is important in maintaining reproducibility of data, and many laboratories use slide hybridization and washing machines to standardize the hybridization and washing conditions. Once the raw data has been generated, data acquisition and analysis play an extremely important role in the quality of the results, and subsequent conclusions drawn from those results. Many of these technical aspects of microarraying, and guides to troubleshooting are available (Bowtell and Sambrook, 2003; http://www.corning.com/Lifesciences/technical_information/techDocs/troubleshootingUltraGAPS_ProntoReagents.asp?region=ge&language=en).

1.4 Affymetrix microarrays

Affymetrix GeneChips® (www.affymetrix.com/) are a widely used DNA microarray format in which the microarrays are manufactured at a central facility using a process of photolithography (Fodor *et al.*, 1991, 1993; Pease *et al.*, 1994) rather than spotting of the nucleic acid which is commonly seen in glass array formats (see Section 1.4.2). Methods and kits used in labeling, hybridization, and detection, as well as the hardware used to perform hybridization and slide scanning of Affymetrix arrays are 'off the shelf' from Affymetrix, which leads to greater standardization of procedures from laboratory to laboratory, and should allow easier comparison of results between laboratories and subsequent metadata analysis. By comparison, in glass slide microarrays, there are a multiplicity of probe formats, printing technologies, substrates for printing arrays, and attachment chemistries. Additionally, different labeling methods, dyes used in detection, and different hybridization and array washing methodologies are used by different laboratories.

The Affymetrix microarray platform is conceptually and technologically different to glass slide microarrays at a number of levels (*Figure 1.1B* and see also Section 1.3), but the underlying principles of how Affymetrix microarrays and glass slide microarrays work, remains the same. They both use labeled target nucleic acids to interrogate nucleic acid probes attached to a solid support (Lander, 1999).

1.4.1 The technology

Affymetrix technology is based on oligonucleotides designed from DNA sequences, but instead of a single long oligonucleotide as is commonly used for glass slides (40–70-mer), Affymetrix arrays use a large number (generally between 5 and 20 depending on the ORF size) 25-mer oligonucleotides to represent a gene or ORF. Affymetrix have also produced high density arrays for *Escherichia coli* that contain one 25-mer oligonucleotide probe, on average, per 30 bases across the genome, with oligonucleotides spanning intergenic sequences at high density (Selinger *et al.*, 2000). These high density arrays are sometimes known as 'tiled' arrays, as they replicate the idea of tile coverage of a floor or wall, with only small gaps (by analogy, the grouting) between the tiles. The 25-mer oligonucleotides used in Affymetrix arrays are designed to correspond to different regions of the whole gene, and are designed to be perfect matches (PMs) to the genomic sequence (*Figure 1.1B*). However, a mismatch oligonucleotide (MM) which generally contains a nucleotide mismatch to the genome sequence in the center of the oligonucleotide at position 13 is also included as a control. The logic for doing this is that the MM oligonucleotide provides a signal that should be equivalent to background, nonspecific hybridization, and can be used to subtract background hybridization signals from the array data. This is especially important with low abundance transcripts in a complex mixture, where the signal intensity for a particular array feature may be close to the background intensity. The added advantage of using this system is that the signal seen for each gene or ORF is a composite, so it does not rely on the signal from one oligonucleotide or PCR product, but on signals from each of the oligonucleotides representing the gene or ORF. This minimizes the possibility that either false-positive and false-negative signals could skew data sets in a one oligonucleotide or PCR product per gene array, and has proven useful in determining 'signal' from 'noise' in GeneChip® experiments.

1.4.2 Microarray fabrication

Affymetrix arrays are produced in a central facility under environmental conditions closely resembling those in which computer silicon chips are manufactured, using procedures that also mirror those used to make silicon chips. Affymetrix array fabrication involves synthesis of the oligomers directly on the solid substrate (a quartz wafer) using combinatorial chemistry, and an adaptation of photolithography masking techniques. In the manufacturing process, nucleotides are added one by one to a growing oligonucleotide chain, with all of the features being synthesized in parallel (Fodor *et al.*, 1991; Pease *et al.*, 1994), where up to 400 Genechip® arrays are manufactured on a single quartz wafer. This wafer is then cut up into single Genechips®, which are mounted in carrier cartridges ready for hybridization experiments.

Photolithographic masks are used to either block or allow light through over different sections of the wafer on which the array will be fabricated (see *Figure 1.4*). Synthesis of the oligonucleotides is only activated when UV light shines on a particular feature area of the wafer, which is modified with photolabile protecting groups. Illumination of the protecting groups yields reactive hydroxyl groups, which are reacted with phosphoramidites which themselves are protected at the 5'-hydroxyl end with a photolabile group. Coupling occurs at sites exposed to light. After capping and oxidation, the substrate is rinsed, and illuminated through a second mask where it deprotects additional hydroxyl groups on the surface bound oligonucleotides, these hydroxyl groups are able to participate in chemical coupling with photolabile 5'-protected deoxynucleoside phosphoramidite monomers that are added to the chip, once light is shone onto the substrate. Once the phosphoramidites have coupled, and unbound phosphoramidite has been

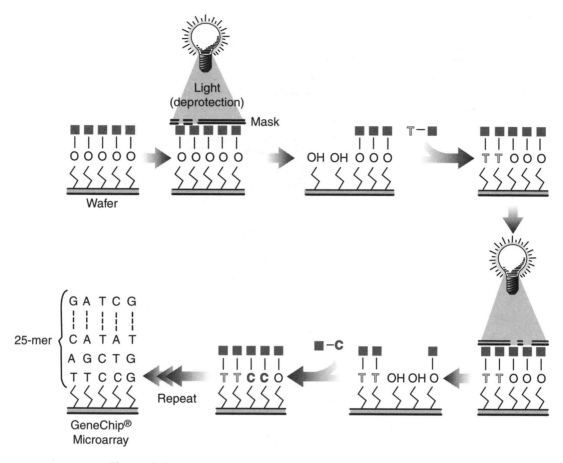

Figure 1.4

The Affymetrix photolithography method of microarray chip synthesis. Reproduced with permission from Affymetrix.

washed away, a different mask is placed over the wafer, and a new round of deprotection, followed by chemical coupling of a different phospho-ramidite monomer occurs. In this way the different oligonucleotide features are built up on the wafer (Pease *et al.*, 1994).

1.4.3 Sample labeling, hybridization, and detection

In Affymetrix microarray gene expression profiling experiments (transcrip-tomics) using eukaryotic RNA, a one-cycle or two-cycle reverse transcriptase mediated linear amplification methodology is used to convert mRNA into cDNA. This RT reaction is primed by a T7-oligo(dT) primer. In a subsequent *in vitro* transcription process the cDNA produced in the linear amplification stage is used as a template for production of an antisense cRNA that is labeled with biotin. This is achieved by incorporation of a biotinylated ribonucleotide analog (a pseudouridine base) into the cRNA during the *in vitro* transcription. Fragmented biotin labeled target cDNA is then hybridized on a Genechip® for 16 h, washed, and then stained first with streptavidin, then with a biotinylated antistreptavidin antibody and then with streptavidin–phycoerythrin. The arrays are washed after each stage and the fluorescence emission signal produced by the bound phycoerythrin when it is excited by an argon laser is measured at 570 nm using an Affymetrix scanner. This is in contrast to the dual fluorescent dye labeling methodology used in spotted microarrays, which compare the signals of two differently labeled cDNAs in a competitive hybridization. (See *Figure 1.1C* and *D* for an overview of the process.)

For prokaryotic RNA labeling for Affymetrix arrays, a random priming and end-labeling methodology is used, because prokaryotic mRNA lacks a poly-A tail. In this methodology cDNA is synthesized from RNA by RT, using random hexamer oligonucleotide primers. The cDNA products are then fragmented using DNase I and labeled with biotin–dideoxyuracil triphosphate (bio-ddUTP) using terminal transferase. Hybridization, washing, and staining are carried out as described above for the eukaryotic arrays, using preprogrammed wash cycles on the Affymetrix fluidics workstation (Affymetrix, 2004). It should be noted that different Affymetrix arrays require different wash programs.

1.4.4 Applications

The experimental uses of Affymetrix microarrays include: expression analysis studies, through transcription profiling; studies on the regulation of transcription through studies on DNA methylation; determining sites of transcription factor binding in the genome; studying origins of replication.

Affymetrix microarrays have been very widely used in genome-wide transcriptional profiling (Lockhart *et al.*, 1996), and have been used for SNP analysis (Wang *et al.*, 1998; Huber *et al.*, 2001). Affymetrix microarrays have also been used for detection and resequencing of DNA (Lipshutz and Fodor, 1994; Hacia, 1999; Vahey *et al.*, 1999), CGH (Primig *et al.*, 2000; Harding *et al.*, 2002) and for studying the effects of small RNAs on gene regulation and expression (Wassarman *et al.*, 2001; Schmid *et al.*, 2003; Zhang *et al.*, 2003).

Protein–DNA interaction studies have also been successfully performed using Affymetrix oligonucleotide arrays, but this has required the enzymatic synthesis of a second DNA strand complementary to the oligonucleotide sequence in order that the DNA-binding protein could actually interact with the immobilized probe on the array (Bulyk *et al.*, 1999), and Affymetrix arrays have been used for many of the applications of arrays detailed in Section 1.1 (and see *Figure 1.3*), particularly the whole genome/tiling arrays that have been developed (Kapranov *et al.*, 2002). The wide variety of experimental purposes for which Affymetrix and the conceptually similar Nimblegen arrays have been used demonstrates the utility of tiling arrays as research tools. The Affymetrix website (www.affymetrix.com) contains a great deal of very useful information about the array format, technology, and applications and has a database of over 3300 publications that have used Affymetrix technology.

1.5 Microarray platforms for protein studies and other applications

The general principle of using massively parallel assays for studies on the interactions between biological molecules is being used in protein/peptide (Reimer *et al.*, 2002) and carbohydrate/glycan arrays (Feizi *et al.*, 2003; Wang, 2003), and is also being used in chemical genomics studies (Spring, 2005). Protein and other arrays are attractive research tools because, although transcriptomics will tell us about the transcriptional responses of a cell to different conditions, transcriptomics only tells us a part of the whole story about cellular responses. The wider argument for the importance of studying the proteome of a cell is that proteins are the molecules that make the cell work, and that protein function is dependent on its physical structure, any post-transcriptional modifications, localization, and the interactions in which the protein participate. Transcriptomics is appropriate for studying how transcription of genes and operons occurs at the genome-wide scale, but will only tell us that. Transcriptomics will not tell us about any of the key features of proteins produced from the mRNA, such as function and interactions with other biological molecules, or even the abundance of a protein, because transcript abundance does not necessarily reflect protein abundance (Twyman, 2004a). This is because regulation of transcription is just one step in the process of protein production, and one of many mechanisms that govern the level of a particular protein in the cell (Hepburne-Scott and Herick, 2005), especially as rates of RNA degradation and protein turnover are rarely the same. In addition, the multiplicity of post-translational modifications of proteins that can occur (Mann and Jensen, 2003) can generate diversity in protein products encoded by a single gene, with estimates that the number of protein species in humans are likely to be at least an order of magnitude higher than the number of genes, because of the diversity of alternative gene splicing events and post-translational modifications.

The other information molecules in the cell, apart from the nucleic acids and proteins, are the carbohydrates. These molecules are also now being exploited in arrays because of their structural diversity, and because they are displayed on cell surfaces and exposed regions of macromolecules (Wang, 2003). Carbohydrates are being used to measure binding affinities of proteins and whole cells to them.

1.5.1 Protein arrays

Protein arrays have been in the shadows cast by DNA arrays for some time, because of the greater technical difficulties of working with proteins, but protein arrays are now emerging as an extremely powerful research tool (reviewed in Twyman, 2004b). Protein/peptide arrays are attractive research tools because they enable researchers to study multiple protein interaction events and activity networks simultaneously, as DNA microarrays do for nucleic acids (Walter *et al.*, 2000; Cutler, 2003). Protein arrays have the potential to be used to detect and monitor the expression levels of proteins, as well as to investigate protein functions and interactions. The potential and actual uses of protein arrays are therefore very wide, especially in the study of 'interactomics' where they can be used for protein–antibody, protein–protein, protein–substrate, protein–drug, enzyme–substrate or multiple analyte diagnostic tests (http://www.functionalgenomics.org.uk/ sections/resources/protein_arrays.htm), and in principle can measure binding constants and catalytic activities for all proteins in the cell simultaneously. Protein arrays also offer the opportunity to study the protein modifications that pharmaceutical compounds may take part in, and study post-translational modifications of proteins (Stears *et al.*, 2003).

Protein arrays in use today are direct developments of the methodology and experimental techniques used by Ekins and co-workers, who developed the 'ambient analyte theory' of why miniaturized and highly parallel ligand-binding assays are more sensitive than any other ligand-binding assay, during the late 1980s (Ekins *et al.*, 1989a, 1989b, 1990; Templin *et al.*, 2002). Ekins and colleagues used microspotted proteins and peptides in ligand-binding assays, detecting binding by fluorescent dye labeling of proteins and scanning of the solid substrate using confocal microscopy. Peptide arrays developed from technologies designed for combinatorial chemical synthesis of different classes of compound, whereas protein arrays require material purified from cells as the probes (Reimer *et al.*, 2002). Protein arrays can be divided into two groups: protein expression arrays and protein function arrays, which will be discussed in detail below.

1.5.2 Protein expression arrays (capture arrays) and protein function arrays

Protein expression arrays or **capture arrays** are also known as **protein profiling arrays**, and are directly equivalent in their experimental function to DNA microarrays used for transcriptomics studies, even though the interactions between proteins detected on a protein capture array are different to those between nucleic acids on a DNA array. The primary purpose of a protein expression array is to simultaneously quantify the levels of a number of proteins produced by a cell, in a highly parallel assay. However, unlike DNA arrays that are spotted with nucleic acids designed to hybridize to denatured nucleic acids in the hybridization solution, protein expression arrays are commonly composed of antibodies immobilized onto the slide surface (De Wildt *et al.*, 2000) and can be made with monoclonal or polyclonal antibodies, antibody fragments or synthetic polypeptide ligands (Mitchell, 2002). Alternatively the probes used on the array can be peptides, alternative protein scaffolds or nucleic acid aptamers (oligonucleotides that bind specifically to proteins) (http://www.functionalgenomics.org.uk/ sections/resources/protein_arrays.htm). In the most directly equivalent protein array experiment to the standard DNA array transcriptomics experiment, a capture array could be used to compare the fluorescently labeled protein content of a healthy and diseased individual, in a two-color comparative protein experiment (see *Figure 1.5*).

The number of features on a DNA array is largely determined by the technology available to produce arrays with a high density of features, because (assuming money is not a limiting factor) production of nucleic acid probes by PCR or synthesis of oligonucleotides is straightforward.

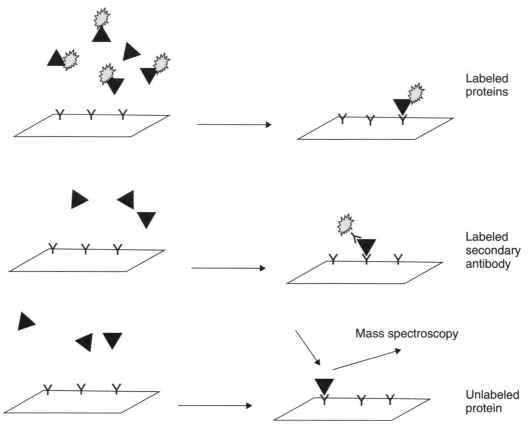

Figure 1.5

Common detection methods used to detect protein binding to features on protein capture arrays. Modified from Twyman (2004b) *Principles of Proteomics* p. 198, with permission from BIOS Scientific publishers.

However, the size of a protein array will be not be determined by technical issues such as array spotting, because the technology to spot high density arrays is the same as that for DNA arrays. Instead, the size of a protein array is determined by the availability of purified antibodies or other protein probes, and their specificity (Mitchell, 2002; Cutler, 2003; Barry and Soloviev, 2004; Hepburne-Scott and Herick, 2005; Poetz *et al.*, 2005). This limitation seems to be one of the major bottlenecks in protein array production. Nonetheless, progress is being made, and there are some commercially available protein expression arrays that are now available, as well as arrays produced by research groups. Whatman Schleicher and Schuell are marketing the serum biomarker chip (http://www.whatman.com/products/?pageID=7.60.337) which is an antibody capture array that can be used to profile the relative abundance of 120 human serum proteins.

Expression/capture (antibody) arrays do have some disadvantages, of which the main one is that the array will only tell you if the protein is being expressed, not whether it is functionally active, or what proteins or molecules it interacts with.

Protein function arrays are designed to study the biological activity of a number of proteins, or interactions between proteins and other proteins, DNA, substrates, and small molecules, in a parallel assay format. Commercial protein function arrays are now starting to become available to the wider research community, and the technology will mature so that it becomes more widely used. At the time of writing this chapter, the Invitrogen Yeast ProtoArray™ protein/protein interaction (PPI) microarray, based on work conducted in Michael Snyder's laboratory at Yale University (Zhu *et al.*, 2001) is released to the market, as well as a smaller human Protoarray (http://www.invitrogen.com). Sigma-Aldrich is now marketing the Panorama™ human p53 microarray (http://www.sigmaaldrich.com) which are streptavidin-coated glass slides, to which are immobilized p53 proteins via biotin and a proprietary tag biotin carboxyl carrier protein (BCCP) (Boutell *et al.*, 2004). These arrays have been used to study DNA–protein interactions, and protein–protein interactions. Protagen (http://www.protagen.de/) has used UNIclone® technology, licensed from the Max-Planck-Society in Munich to make UNIchips® containing 2500 different human proteins, which are being used for testing of specificity of antibody binding, and defining markers for cancer and autoimmune disease.

1.5.3 The technology

Protein microarrays have developed along similar lines to DNA microarrays, in that they are either constructed from large 'biological' molecules (full protein/antibody) or small (peptide, aptamer) 'synthetic' molecules. Protein arrays have been made on solid planar surfaces or on beads. Probably because proteins are more diverse molecules than nucleic acids, several different solutions to the immobilization of proteins on arrays have been used. These include immobilization of the proteins to the array surface via chemical (glass arrays), hydrophobic (hydrogels) or electrostatic (nitrocellulose) interactions (reviewed in Walter *et al.*, 2000). Other substrates have been used, which include silica, polyvinyldifluoride (PVDF), and polystyrene. In all cases retention of functionality has always been an overriding concern.

Chemical synthesis of peptides for arrays either on solid supports, for subsequent spotting, or *in situ* (Emili and Cagney, 2000; Reineke *et al.*, 2001; Frank, 2002; Houseman *et al.*, 2002), and light-directed synthesis of *in situ* peptide arrays (Fodor *et al.*, 1991; Gao *et al.*, 2003) have been used, because combinatorial chemical synthesis offers simplicity and controllable immobilization to the solid surface, and because chemical synthesis of peptides avoids the difficulties and time/cost constraints encountered in purifying large numbers of functionally active proteins.

As with all biological interaction assays, using arrays for proteomics studies relies upon using the correct conditions for protein–protein, protein–antibody or protein–substrate interactions. This means that the ligand-binding interactions that occur on a protein array have to be conducted under conditions that favor the binding of the target molecules to the immobilized protein probe, followed by a washing step that removes targets that bind in a nonspecific manner to the probes. Different proteins may interact with each other under different conditions, and also require different binding and washing conditions.

1.5.4 Microarray fabrication

The technology of producing protein/peptide microarrays is very similar to that of DNA microarrays (Lueking et al., 1999; MacBeath and Schreiber, 2000; MacBeath 2002). Protein and peptide arrays can be spotted/inkjet printed, or peptide arrays can be synthesized in situ on a solid substrate. The key issue in protein/peptide arrays is the functionality of the probe molecules. For protein arrays, this is clearly dependent upon the proteins being purified sufficiently in order that contaminating proteins do not compromise the specificity, that the purified protein is biologically active, and that the immobilization strategy does not compromise the functionality of the protein. To be functional, proteins must be correctly folded, so a number of different slide chemistries have been used to immobilize proteins either by chemical interactions, hydrophobic interactions or electrostatic interactions.

Early work on protein arrays tended to work in the microtiter plate format, but this has moved to a format that has echoed very closely the early technologies employed in DNA microarrays. For example, Lueking and co-workers (Lueking et al., 1999) first spotted and inkjet printed proteins from an expression library on to PVDF filters, which were screened using monoclonal antibodies. Haab et al. (2001) demonstrated printing of antibody or antigen probe arrays onto slides coated with poly-L-lysine and labeling of the target proteins with CyDye™ NHS esters. Subsequent arrays are becoming larger and more sophisticated. Another approach to immobilization has concentrated on using the affinity tags used in protein purification to immobilize proteins on to a solid surface. Zhu et al. (2001) used Saccharomyces cerevisiae genes to make fusion proteins tagged with both hexahistidine and glutathione-S-transferase (GST) tags. These could be purified using affinity columns and then be immobilized onto solid surfaces coated with nickel sulfate and/or glutathione. Ofir et al. (2005) have produced proteins fused to the carbohydrate-binding module (CBM). The proteins are purified by affinity chromatography using cellulose, and immobilized to cellulose, whilst the BCCP domain has been used to purify and immobilize proteins (Boutell et al., 2004).

1.5.5 Sample labeling, hybridization, and detection

Detection of binding onto the protein array can be achieved by a number of different methods. These include fluorescence labeling of the protein 'targets' using Cy3 and Cy5, fluorescein or Texas Red. Monofunctional NHS esters of CyDyes can be used to label amine residues in antibodies and other proteins, or maleimide esters of CyDyes can be used to label free thiol groups. Detection of binding to the array can then be quantified using a standard DNA microarray scanner. Another method of detection is using enzyme-linked immunosorbant assays (ELISA) technology, where detection of binding between the probe and the target molecule is detected by the use of a secondary labeled antibody directed to the target protein.

An interesting alternative method of labeling is the Universal Labeling System™ (ULS) developed by Kreatech (http://www.kreatech.com/) to label proteins using the stable binding properties of platinum complexes to N and S side chains of cysteine, methionine, and histidine. This technology is applicable to DNA and RNA as well as proteins, by binding to the N7 position of guanidines, and provides researchers with a rapid, one-step nonenzymatic labeling technology, which can use a number of different reporter molecules, such as CyDyes, haptens or enzymes.

Other methods for detection rely on the enzymatic function of, for example, a kinase target protein phosphorylating the immobilized probe using ^{33}P-γ-ATP as the substrate (Zhu and Snyder, 2001; Reimer *et al.*, 2002). Unlabeled proteins can be directly detected using matrix-assisted laser desorption ionization/time-of-flight (MALDI/TOF) spectroscopy, surface enhanced laser desorption/ionization (SELDI) spectroscopy, electrospray tandem mass spectrometry or surface plasmon resonance (SPR) technology (Emili and Cagney, 2000; Mitchell, 2002). Hybridization conditions for protein arrays are most similar to those used in ELISA assays, but still rely upon the general principle that the array is incubated with the solution containing the protein(s) under conditions that promote binding between the immobilized probe and the target molecules, is washed under conditions that remove weakly binding target molecules, and binding/interactions are then detected and quantified using appropriate scanners.

1.5.6 Applications

There are four major areas where protein arrays can potentially be used, or are being used:

- proteomics,
- functional analysis,
- diagnostics,
- isolation of phage display library peptides.

So, the uses of protein arrays are very wide. Schweitzer and colleagues summarized protein–protein interaction, antibody-specificity profiling, immune profiling and protein–small molecule interaction studies, as major uses of protein arrays, but also identified uses such as phosphoinositide probing, use of the arrays as enzyme substrates, uses in DNA binding and protein interaction domain screening (reviewed in Schweitzer *et al.*, 2003). (See *Figure 1.6* for a summary of applications for protein arrays.)

In proteomics studies capture arrays may provide a viable alternative to two-dimensional gel electrophoresis and mass spectrometry in the identification of *de novo* protein synthesis, in experiments comparing protein expression between two different cells. Whole-organism, as well as production of organ- or disease-specific protein arrays, is probably some time away because of the large numbers of protein probes that would be required to profile expression levels.

In functional analysis studies, protein arrays can be used to map protein–protein interaction, and study antibody specificity, enzyme activity, ligand

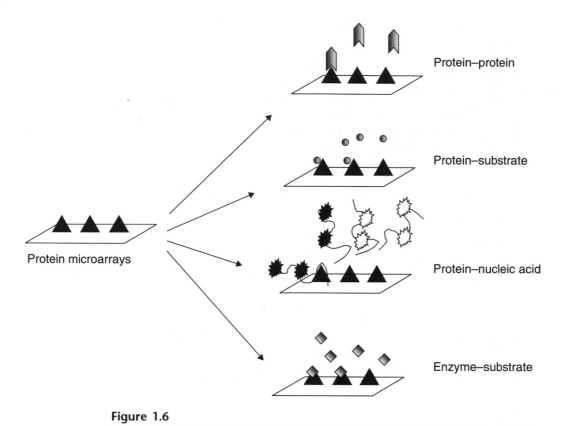

Figure 1.6

Examples of applications for protein microarrays. Modified from Twyman (2004b) *Principles of Proteomics*, p. 203, with permission from BIOS Scientific publishers.

binding, and in epitope mapping studies. In diagnostics, protein arrays can be used to multiplex assays such as those used to detect the presence of disease specific antibodies and antigens from blood or sera samples, or to profile samples for new disease markers. There are specialized protein arrays, such as the lectin array produced by Zheng and colleagues, which can recognize and bind to specific structural epitopes of carbohydrates (Zheng *et al.*, 2005). Protein arrays can also be used to isolate (or 'pan') for individual clones from phage display libraries, which interact with the probes spotted onto the array (http://www.functionalgenomics.org.uk/sections/resources/protein_arrays.htm).

1.5.7 Technical challenges

In 2002 Mitchell summarized some of the many challenges to building a viable protein microarray chip. These were:

(i) using a slide surface chemistry that allowed diverse proteins to retain their secondary and tertiary structure, and therefore activity on the microarray;

(ii) identifying and isolating an agent with which to capture the proteins of interest (antibodies, other proteins, or other molecules);

(iii) accurately measuring the degree of protein binding with a system that is both sensitive and with a wide (or at least suitable) dynamic range;

(iv) being able (if necessary) to identify the protein adhering to the microarray (Mitchell, 2002).

Protein array technology is highly dependent on being able to immobilize proteins that retain their functionality, and there are considerable technical challenges in successfully immobilizing fully functional proteins, which retain their complex three-dimensional structure and activity whilst being tightly bound to the slide surface. Although denatured proteins are useful in screening for antibody cross-reactivity, or selecting ligand-binding proteins (http://www.functionalgenomics.org.uk/sections/resources/protein_arrays.htm), functional arrays require functional proteins. Another challenge is that as each protein–protein interaction is unique, there is no standard set of conditions (as there are in hybridizing and washing nucleic acid hybrids) that can be used in protein arrays.

Unlike DNA or oligonucleotide microarrays where the nucleic acid probes to be spotted onto the array can be easily produced by *in vitro* methods such as PCR, or by chemical synthesis, a major technological hurdle to be overcome for the production of protein arrays is the provision of sufficient quantities of purified protein probes to spot on the array. Peptides can be chemically synthesized, but the expression of proteins still relies upon a biological element in their production, whether in *in vitro* cell-free translation systems, or, more conventionally, in cell-based expression systems or from recombinant clones. The overexpression and purification of single proteins for structural or functional studies has been a labor intensive, time consuming and at times frustrating bottleneck in biochemical, structural, and molecular biology research for some considerable time. The development of several different systems in the past decade that tag proteins with a short, amino acid sequence (such as FLAG, C-myc, BCCP, CBD, and hexahistidine tags) that allows any protein so tagged to be purified by an affinity column in a one-step process, has been a major step forward in simplifying the purification of proteins. Using high throughput cloning and overexpression methods, with simple protein purification strategies that rely on an added tag, rather than the intrinsic qualities of the protein, to automation and robotics will aid the high throughput overexpression and purification of proteins. Automation, such as the Expressionfactory™ from Nextgen Sciences (http://www.nextgensciences.com/product._gtpf_ef.htm) will simplify the high throughput cloning, expression, and purification of recombinant proteins, as will the use of phage display libraries to generate antibody and other protein diversity. Some of the other technological solutions to high throughput protein production have been discussed in some detail in a recent article by Sakanyan (2005).

Aside from producing sufficient numbers of proteins, of sufficient purity, there are several other technical challenges associated with protein microarrays. The first is simply that proteins are far harder to work with than nucleic acids, and there are potential problems with lack of specificity, particularly with antibodies (Mitchell, 2002). The second is that unlike nucleic acids, proteins cannot be amplified *in vitro* by a process like PCR, or linear amplification (Van Gelder *et al.*, 1990; Phillips and Eberwine, 1996) which may make detection of poorly expressed proteins difficult, and protein concentrations in a single biological sample are more variable than mRNA. The operational range of a protein chip may have to be of the order of 10^{14} (compared with 10^4 for mRNA detection in DNA arrays) in order to accurately detect the levels of a particular protein (Mitchell, 2002). Detection technologies and other methods to address this difficulty in detecting low abundance proteins are being developed (Shao *et al.*, 2003), but in terms of detection methods currently being used, there are two major areas of concern: these are whether labeling proteins with fluorescent dyes will affect their binding abilities on the array; and, if the alternative detection strategy of using a labeled secondary antibody is used, whether the antibodies used are specific enough. Because directly labeled proteins extracted from biological systems often occur in complexes, there are some concerns that a strong signal may be due either to a high concentration of one protein binding to the array feature (spot), or may be due to signals from a co-immunoprecipitated complex. This raises questions about whether protein expression/capture arrays are in fact quantitative (Barry and Soloviev, 2004), and whether multiplexing protein expression arrays can be refined to give quantitative protein expression profiling, although signal amplification, multicolor detection and competitive displacement approaches appear to offer some promise in quantitative protein profiling in a multiplex format. One approach to protein expression profiling, which addresses the specificity issues, is the use of sandwich immunoassays in an array format. This method has the advantage that it uses two antibodies that bind to separate epitopes on the target protein. One of the antibodies is immobilized on the array surface, whilst the other antibody is labeled and used to identify the protein captured by the first antibody.

The challenge of retention of protein function, or at least its ability to interact correctly with other proteins or substrates may be an issue when these proteins are spotted on microarrays. A simple comparison between proteins and DNA demonstrates this point. DNA (unlike RNA) is actually quite stable, tolerating drying and immobilization to solid substrates, whilst retaining its ability to hybridize with other nucleic acids, or act as a substrate for interactions with DNA-binding proteins, once rehydrated. Proteins are by their very nature less stable than DNA, and present challenges to the researcher who wishes to immobilize them on an array, and retain their biological activity. Clearly, any antibody or other protein immobilized on a microarray needs to retain functional activity, ideally for weeks or months after spotting. As all proteins are by their nature different to each other (pI, solubility, hydrophobicity, etc.) retention of protein function whilst immobilized may pose technical difficulties. By contrast, nucleic acids are made up of combinations of four nucleotides, rather than the 20 amino acids that may be present in a protein, and therefore have similar chemical characteristics. A number of different approaches have been used to attempt to retain protein function, commonly tagging of the protein with a biotin or hexahistidine tag at the N- or C-terminal of the protein has been a popular methodology (Boutell *et al.*, 2004), which is crucial to the functionality of the probes on the array.

A detailed discussion of the theory and practice of protein arrays, and the wider context of proteomics within which protein arrays falls is dealt with in some detail in the recent book by Twyman (2004a, 2004b).

1.5.8 Carbohydrate/glycan arrays

Carbohydrate/glycan arrays are not as widely used as DNA or protein arrays, but have many potential and actual uses. The importance of carbohydrates in a wide variety of cell recognition processes, such as host–pathogen interactions, and the recruitment of neutrophils to sites of cellular damage, is now widely recognized (Kiessling and Cairo, 2002). In particular, interactions between carbohydrates and proteins are involved in cell–cell adhesion, recognition and mediation of immune system function, metastasis, and intracellular traffic (Drickamer and Taylor, 2002; Mahal, 2004; Shin et al., 2005). Within cells, carbohydrates exist in diverse forms: attached to proteins, as glycoproteins; attached to lipids as found in microbial lipopolysaccharides, or simply as polysaccharides, such as bacterial capsular polysaccharides (Kiessling and Cairo, 2002). As approximately 50% of all eukaryotic proteins and some prokaryotic proteins are glycosylated, it is clear that glycans are important molecules in information transfer (Schwartz et al., 2003; Blixt et al., 2004).

There are several major complications in working with carbohydrates, compared with protein or DNA. The first is the fact that whilst nucleic acids and polypeptides are linear assemblies, carbohydrates can be much more complicated molecules, with branched, three-dimensional structures. Whilst this attribute, and the wide variety of monosaccharide building blocks for carbohydrates, as well as a variety of different linkages, and lengths of linkages (Werz and Seeberger, 2005) means that carbohydrate/glycan molecules are diverse, and therefore have a large capacity for carrying information, it also means that their synthesis and/or purification is difficult. Analysis of protein–glycan interactions has also been hampered by the complicated nature of analysis methods (Schwartz et al., 2003).

1.5.9 The technology

The material used for carbohydrate/glycan arrays can be produced in several different ways. Isolation of carbohydrates from natural sources requires multiple and varied purification steps (Wang, 2003), as does enzymatic synthesis. Chemical synthesis of carbohydrates/glycans requires multiple selective protection and deprotection steps during synthesis of oligosaccharides, because of the often three-dimensional nature of the branched linkages in oligosaccharides (Feizi et al., 2003). However, the development of automated synthesis methods (Houseman and Mrksich, 2002) has allowed researchers to make arrays of naturally occurring carbohydrates, as well as chemically modified oligosaccharides (Werz and Seeberger, 2005).

Because carbohydrates have a low mass and are mainly hydrophilic, it is difficult to directly immobilize them onto solid matrices by noncovalent bonding (Wang, 2003). Immobilization of carbohydrate probes has been made, however, by spotting and immobilization onto nitrocellulose membranes (Wang *et al.*, 2002), by linkage of oligosaccharides to lipids, which were immobilized to nitrocellulose (Fukui *et al.*, 2002), or by immobilization to hydrophobic polystyrene slides (Willats *et al.*, 2002; reviewed in Wang, 2003). Schwartz and co-workers have been able to immobilize carbohydrates by covalent attachment via a flexible linker molecule (Schwartz *et al.*, 2003).

1.5.10 Labeling and detection

Carbohydrate arrays are being used for a diverse set of purposes. This means that labeling techniques have to be suited to what the array is detecting. Immunostaining of the carbohydrate arrays (Fukui *et al.*, 2002) or detection of binding using labeled secondary antibodies (Wang *et al.*, 2002; Guo *et al.*, 2004) have been used to detect binding to carbohydrate arrays, as has the fluorescent labeling of bacteria using a cell-permeable nucleic acid stain (Disney and Seeberger, 2004). Houseman and Mrksich (2002) constructed a carbohydrate array that was used to probe rhodamine-labeled lectin targets. Binding of the lectins to the carbohydrate probes on the array was visualized by confocal fluorescence microscopy. Labeling of free carbonyl groups on glycoproteins and carbohydrates can also be achieved, using CyDye™ hydrazides, whilst protein targets of the oligosaccharide probes can be labeled as discussed in the protein array section of this chapter.

1.5.11 Applications

Like protein arrays, carbohydrate/glycan arrays are being used both as 'capture' arrays to detect the expression of a particular biological molecule such as an antibody, and as assays for functional interaction studies, such as for studying the specificity of antibodies. (See *Figure 1.7* for an overview of some of the applications of carbohydrate arrays.) An example of carbohydrate arrays which have been used in both the capture and functional role is one where Thirumalapura and co-workers have constructed lipopolysaccharide (LPS) microarrays, made with LPS isolated from *Escherichia coli* O111, *E. coli* O157, *Francisella tularensis*, and *Salmonella typhimurium*. They have used these arrays both to monitor the specificity of monoclonal antibodies, and to detect antibodies to *F. tularensis* in canine blood samples (Thirumalapura *et al.*, 2005). Other recent examples of the use of glycan arrays in the capture mode are those such as the detection of bacterial toxins using immobilized *N*-acetyl galactosamine (GalNAc) and *N*-acetyl-neuraminic acid (Neu5Ac) derivatives (Ngundi *et al.*, 2005), and the use of lectin arrays to profile cell surface carbohydrate expression (mannose, galactose, and GlcNAc expression) in mammalian cells (Zheng *et al.*, 2005). Other experimental uses of these arrays have been the determination of glycan-dependent binding profiles of HIV gp120 glycoprotein (Adams *et al.*, 2004), the determination of kinase activity in whole blood samples (Diks *et al.*, 2004), and Disney and Seeberger (2004) have used carbohydrate microarrays to study interactions between cells and carbohydrates, and to detect pathogens. Carbohydrate/polysaccharide arrays have also been used in studies on heparin–protein interactions and carbohydrate–nucleic acid interactions (reviewed in Werz and Seeberger, 2005).

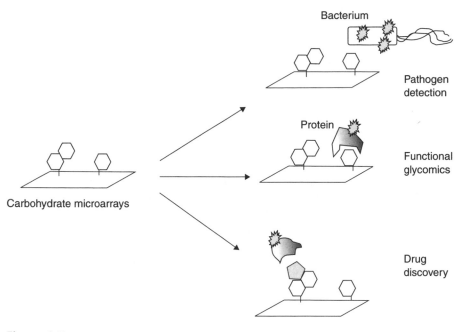

Figure 1.7

Examples of some applications of carbohydrate arrays. Modified from Shin *et al.* (2005) *Chemistry, a European Journal* **11**: 2894–2901, with permission from Wiley-VCH.

1.5.12 Technical challenges

At present there are a number of technical challenges that carbohydrate arrays are facing. The diversity of carbohydrate molecules and their three-dimensional structures mean that there is great diversity amongst the carbohydrates (e.g. cellulose and chitin). Chemical synthesis of substituted sugars *in situ* on carbohydrate arrays, enzymatic synthesis, and purification of naturally occurring carbohydrates can therefore be time consuming and difficult, limiting the number of carbohydrate ions in the array. Therefore the establishment of arrays that are representative of the glycome will require a considerable research effort (Feizi *et al.*, 2003).

Key questions on the use of carbohydrate arrays in the capture mode revolve around specificity and sensitivity, which for arrays used to detect bacteria by their interaction with carbohydrate probes, was summarized by Mahal (2004) as:

- Can carbohydrate arrays be used to accurately type the wide variety of known bacteria?
- As detection limits for bacteria are fairly high are they less sensitive than PCR and ELISA methods?

As carbohydrate arrays become more widely used, other questions and answers to these questions will be forthcoming. Clearly, however, these specialist arrays will be of considerable use as research tools in the study of cell–cell interactions.

1.5.13 Other array formats

The use of **small molecule arrays** in finding compounds that interact with proteins is of considerable interest to the pharmaceutical industry, because the majority of clinically used small molecules target receptor proteins or enzymes, as their normal function is to bind to small molecules as substrates or signals (Spring, 2005). Technology based on *in situ* synthesis of small molecules to solid supports, or synthesis of small molecules followed by immobilization onto a solid support, labeling of particular protein targets, and detection of interactions with the small molecules on the array clearly echoes DNA, protein, and carbohydrate array technology. (See *Figure 1.8* for an overview of some of the uses of small molecule arrays.) This allows large panels of synthetic and naturally occurring small molecules to be assayed simultaneously for interactions with these compounds. Potential uses of small molecule arrays include protein–small molecule interaction detection and identification of riboswitches (small molecules/metabolites that bind to mRNA, which modulate translation of proteins) (Spring, 2005).

1.6 Data/image acquisition

The aim of any microarray experiment is to produce high quality data that can be reproducibly compared between different laboratories. The difficulties in achieving this aim are integral throughout the design and experimental procedures, and have been discussed throughout this chapter. It is, however, the final digital image that determines the quality of data representing the samples hybridized to the microarray slide. Good quality data derived from this image strongly affects downstream analysis, and data

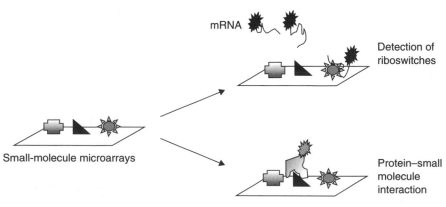

Figure 1.8

Examples of some applications of small molecule arrays.

quality is dependent on both the process and the hardware used for the image acquisition, and the process and software used for its subsequent analysis.

The final digital microarray image that is outputted from the scanner/imager is a 16-bit TIFF (tagged image file format) file made up of a matrix of pixels with 65 536 (2^{16}) levels of gray, determining the dynamic range, and the saturation level of the data. Each TIFF image representing each fluorescent sample channel is on average approximately 30 Mb in size. The image itself consists of two areas, the foreground (or spots) and the background. The spots on the array image have a diameter of 50–100 µm and each spot represents a single gene. The background in the image is defined as all of the rest of the area on the image between the spots.

The average DNA microarray slide, and therefore image derived from that slide, contains tens of thousands of 50–100 µm spots per square centimeter, so the demands of microarray image acquisition have pushed imaging technology to its limits. This is because to succesfully image each feature (spot) on the array a resolution power is required that can distinguish these spots from the background, and accurately measure the signal generated in each of two channels by each of those spots. Imaging technology has also had to deal with the problem of the relative abundance of gene products, with the majority of the known genes expressed at high levels, but with many of the probably more biologically relevant genes expressed at very low levels. High sensitivity in imaging is therefore also required, so that transcripts with a relative abundance of between one copy in 100 000 and one copy in 500 000 (three to ten copies per cell) and with a linear dynamic range of five orders of magnitude, can be accurately detected (Schena, 2000; Luo and Geschwind, 2001). Imaging technology has delivered scanners and imagers that have the sensitivity, accuracy, and efficiency required to perform this task. Different design approaches from different companies with different technological platforms or with different consumers in mind has led to a wide range of microarray scanners with a variety of different features and capabilities being available. Instruments that can accommodate different substrates such as glass, plastic, ceramic, and metal, and are also able to detect visible light, phosphorescence, chemifluorescence, and radiative isotopes as well as fluorescence, are on the market.

In general, the choice of microarray scanner is mainly between two kinds of instrument: those that are generically known as **scanners** and **imagers**. The fundamental difference between the two types of instrument is the way they excite, collect, and detect the light from the labeled sample on the microarray slide.

Scanners use narrowband illumination (such as lasers) to excite the fluorophores (*Figure 1.9A*). The fluorescence emission signal resulting from this focused excitation beam on a single pixel is captured with a photomultiplier tube (PMT) detector. These point-by-point signals are then transmitted to a computer which assembles them to the final digital image.

Imagers, on the other hand, use wideband illumination such as from a xenon lamp to illuminate the slide surface, obtaining images as snapshots covering large surface areas of the slide (*Figure 1.9B*). The resulting signal is captured using a charge-coupled device (CCD), an electronic sensor that detects and records light patterns.

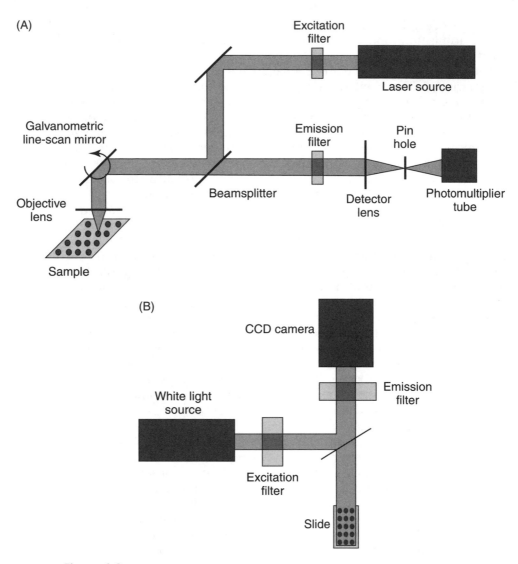

Figure 1.9

Components of (A) scanner and (B) imager systems. Reprinted with permission from Stephanie Weiss, *Choosing Components for a Microarray Scanner* http://sales.hamamatsu.com/assets/applications/Combined/genescanner.pdf

   DNA samples can be labeled with a variety of labels other than radionucleotides or fluorophores, although fluorescence labeling of cDNA, DNA or RNA samples has until now been the most commonly used labeling technology. Although very widely used, there are a number of inherent problems with fluorescent dyes, such as photobleaching (photochemical damage and reduced emission signal, particularly from Cy5) and bleed through (cross-contamination of the detected emission signal between the different simultaneously used fluorophores) which are primarily associated with scanners rather than imagers. In scanners the pixel size is determined by how far the scanning beam moves each time it collects a data point, and the size of the laser beam determines the signal area from which data is collected. Therefore the distance between the data points defines the pixel resolution with some of the new scanners reaching resolutions of about 5 μm. The ability to focus the excitation beam on a single pixel also allows the scanner to focus more energy to excite the fluorophores and can collect more light in less time. Although advantageous for detecting low intensity signals, this property can also lead to problems with saturation, which in turn can lead to photobleaching and a nonlinear emission response.

By comparison, imagers have detectors that are less sensitive than PMTs, and an excitation source that is less efficient, thus requiring longer periods to collect the emission signal and external amplification to reach the sensitivities that PMT scanners can achieve. However, as they can collect many more photons per pixel without reaching saturation, imagers are believed to provide more detailed images with less 'noise' (unwanted background signal). In general, the use of fluorophores contributes to high signal to noise ratios, as all organic materials fluoresce, making it very difficult for the scanner to distinguish between real signal and noise.

Microarray readers are improving in their sensitivity, speed, and accuracy. The result of this improvement is that they are more versatile, and are able to read an ever broader range of slides, substrates, and labeling molecules. These microarray readers are becoming at the same time bigger, incorporating robotics, which allow higher sample slide throughputs, thus appealing to core microarray facilities, or smaller (as standalone units) appealing to individual researchers. A word of caution though, as studies suggest that part of the problem in comparing data obtained from different laboratories is due to the use of the diverse types of scanning instruments and scanning procedures utilized by the researcher (Ramdas *et al.*, 2001; Bengtsson *et al.*, 2004; Lyng *et al.*, 2004). A number of websites are available providing comparative information on available scanners, for example http://www.images.technologynetworks.net/resources/mascanner.pdf. Articles in *The Scientist* that have presented reviews on the various instruments available from the various manufacturers along with new innovations and future designs can be found in *The Scientist* (http://www.the-scientist.com) within their 'Tools and Technology' series (e.g. Cortese, 2001; Hitt, 2004).

1.6.1 Image analysis

Image processing fundamentally consists of three main parts (see Yang *et al.*, 2000 for detailed description):

(i)   *addressing/gridding* – defined as the process during which each spot is assigned coordinates;
(ii) *segmentation* – this is the classification of pixels as either foreground or background;
(iii) *measurement* – this is the process through which the intensity of each spot is determined, foreground and background intensities for each spot are calculated and quality measurements take place.

1.6.2 Addressing or gridding

The basic geometry and structure of the microarray image is determined by the predefined parameters of the printing process. Therefore, the separation between and within rows and grids, the number of spots and the relative position of the array in the images is known, prior to gridding. It is therefore, the aim of the gridding process to match the known, predefined layout with the actual scanned image data. Theoretically, a straightforward process, which only begins to run into trouble when the image contains uneven grid positions, curving within a grid, varying spot size, and distance between spots. All are problems usually arising during the printing process and have to be overcome by the gridding process. A number of algorithms have been written to aid automatic gridding despite the above problems, addressing issues such as rotation of array in image, skewness of the array, and weak signals at the boundary of the grid. User intervention can result in enhanced efficiency, but with the tradeoff of slower analysis and more variability between users. Most software systems provide for both manual and automatic gridding procedures and these are discussed by Schadt *et al.* (1999), Steinfath *et al.* (2001), Jain *et al.* (2002) and Yang *et al.* (2002).

1.6.3 Segmentation

During the segmentation process pixels are defined as either foreground, that is belonging to the spot, or background, belonging to the space between spots. Most software systems available use one of four methods to achieve this separation: fixed circle, adaptive circle (Buhler *et al.*, 2000) adaptive shape (Buckley, 2002) or histogram segmentation (Chen *et al.*, 1997).

Fixed circle segmentation is historically the first and simplest method used. During this process a circle with a predetermined fixed diameter is fitted to all spots in the image. Although it is easy to implement, it does not deal very well with spots of varying shape and size. During adaptive circle segmentation though, the circle diameter is estimated for each spot individually. The algorithm detects the edges of spots by looking for sharp changes in the slope of pixel intensity vectors. Although this approach is an improvement on fixed circle segmentation, it still has some difficulties dealing with noncircular shapes. Adaptive shape segmentation on the other hand goes one step further fitting any irregular shape around what the algorithm defines as a spot. It uses the known geometry of the array to define starting points or 'seeds' within which the spot is expected to be (Adams and Bishof, 1994; Yang *et al.*, 2002). The spot grows outward from the 'seed' based on the difference between a pixel's value, and the progressive mean of values from the neighboring pixels. But even this method has difficulties dealing with highly irregular shapes. The final method of spot identification is histogram segmentation (Chen *et al.*, 1997). This method uses a target area that is chosen to be larger that any other spot, which is expected to contain both foreground and background pixels. Foreground and background intensity are determined from the histogram of pixel values for pixels within this area, with background for example being defined as pixels with mean intensity between the 5th and 20th percentile

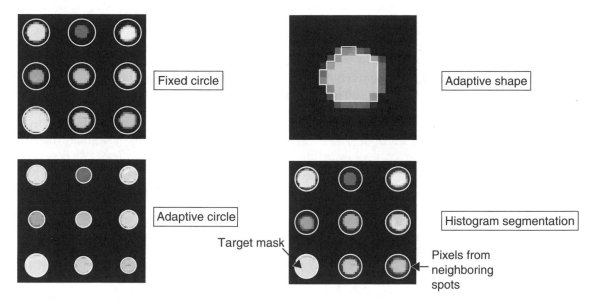

Figure 1.10

Segmentation strategies used by software analyzing microarray scanner image data. (A color version of this figure is available at the book's website, www.garlandscience.com/9780415378536)

and foreground as those pixels with mean intensity between the 80th and 95th percentile. This method, however, may not work well when a large target area is set to compensate for variation in spot size. Each of the four segmentation methods is shown in *Figure 1.10*.

1.6.4 Measurements: spot intensity and background intensity

The total fluorescence of the spot on the array is proportional (as long as fluorescence intensity is within the linear dynamic range) to the total amount of hybridization on that spot. Therefore, spot intensity is represented by the sum of pixel intensities within the spot area. The mean or median pixel value are computed as calculations during the analyis procedure.

It is generally accepted that nonspecific hybridization and other chemicals on the glass slide surface contribute to the intensity determined for each spot, and data can only be reliably extracted from the microarray image if fluorescence from regions not occupied by DNA is different from regions occupied by DNA. There are several different methods for calculating the background intensity, and each gives rise to surprisingly different results (Ahmed *et al.*, 2004; Bengtsson and Bengtsson, 2006). The most common methods that are used are: local background (Axon Instruments Inc., 1999; Eisen and Brown, 1999; Medigue *et al.*, 1999), morphological opening (Soille, 1999; Beare and Buckley, 2004), and constant background calculations. The local background method focuses on small regions surrounding the spot area, calculating the median of the pixel intensities in this region. Although most software packages use this approach there are variations on the shape and size of the background regions used, as well as proximity of pixels to the spot area (some packages disregard pixels immediately surrounding the spots as this results in background estimates that are less sensitive to the quality of the segmentation procedure).

During the morphological opening method an image is created which is an estimate of the background for the entire slide. For each individual spot the background is estimated by sampling this background image at the nominal center of the spot. The background image itself is created by segmenting the entire microarray image into elements with a side length at least twice as long as the spot separation distance, calculating within these sections the local minimum and maximum peaks of signal intensity and using these values as filters, thus removing all the spots. Although this method does tend to produce lower background estimates it is said to be less variable (Yang *et al.*, 2002). The constant background method is a global procedure during which a constant background value is subtracted from all spots. There are various ways to calculate this background value, i.e. using the third percentile of all the spot foreground values, or using a set of negative control spots. Care should be taken with the constant background method, as some evidence suggests that nonspecific binding of fluorescent dyes to the negative control spots is lower than their binding to the glass slide.

1.6.5 Quality measures

Quality measures are used by the software as indicators on how good and reliable are the foreground and background measurements. Some of the most commonly used methods are measurements of the variability of pixel intensity values within each spot, *b*-values representing the fraction of background intensities less than the median foreground intensity, signal-to-background ratios and measurements relative to spot morphology, such as spot size and circularity measure. Based on these measurements different software will automatically flag the various spots accordingly, although at the same time the software will allow manual flagging as well.

The choice of method used at every step is a decision that can have major knock-on effects in data interpretation for the entire experiment. It has been stated that choice of background correction method has a larger impact on the data than the segmentation method used, and background adjustment itself has, in turn, a greater impact on low intensity spots (Yang *et al.*, 2000). Users should be aware of the advantages and limitations of each method, and make informed decisions most suited to their circumstances.

1.6.6 Data storage

Microarray experiments result in large and complex sets of data. Handling, processing, storing, and accessing these data is proving to be an overwhelming task to researchers belonging to the 'average' molecular biology laboratory. Researchers have been becoming aware of the enormity of the task over the last few years, but it has been the tendency for each individual laboratory/group/collaboration to provide solutions tailored to their needs, thus creating additional problems of efficient access and exchange of data between laboratories.

The data storage 'problem' is a multilayered one, requiring both 'local' and 'public' solutions that are both comparable and easily accessible. There are five fundamental areas that need to be addressed and they are in a way part of a cascade. If the first step is badly done the whole series is irreversibly affected.

(i) *Large quantity of data.* The sheer quantity of data is overwhelming, especially as in some central laboratories, multiple experiments are carried out, each with a large number of variables that are unique to that experiment. Each of these variables needs to be logged and saved in an efficient, self evident, reproducible, easily accessible and safe manner. Hardware and software that can cope with the load, multiple users and provide safe back-ups are necessary.

(ii) *Data access and exchange.* The efficiency of recording how the above data was generated will dictate how easy data access and exchange with other groups is.

(iii) *Platform compatibility.* Microarray data is generated by different groups using different microarray platforms. Validation and comparison of such data (e.g. cDNA and short and long oligonucleotide DNA arrays) is paramount. Software that allows the integrated statistical analysis of multiple data sets is required.

(iv) *Annotation data.* A parallel and compatible system is required which would link expression data to up-to-date annotation data. This would allow the identification of significant changes in gene expression and extraction of biological pathways.

(v) *Assessing quality of data.* The ability to assess and validate microarray data needs to be addressed, especially when different microarray technologies are involved.

Until recently, the majority of publicly available microarray data tended to be scattered around the internet, mainly as supplementary data to a published article, or on laboratory websites. Calls for making it easier for interested investigators to find relevant data has resulted in databases that allow users to query the data that has already been uniformly processed and filtered. However, these databases have tended to have diverse purposes, being platform specific, organism specific or project specific. This has meant that there is no direct access to the experimental information in a standard format that would allow researchers to judge the quality of the data, and if need be to repeat the experiment or to re-analyze the data.

The Microarray Gene Expression Data (MGED) Society rose to the challenge trying to address all of these related data problems. The MGED Society was formed in November 1999, as an international organization of biologists, computer scientists, and data analysts, who at the time were carrying out and analyzing the first generation of microarray experiments. Its aim is to facilitate the sharing of microarray data generation by functional genomics and proteomics data (Ball *et al.*, 2002; MGED, 2002). The initial establishment of e-mail discussion groups and regular meetings amongst the founders and other users of the technology has resulted in the creation of MIAME and MAGE.

MIAME – Minimum Information About a Microarray Experiment – is a document which outlines the minimum information that should be reported about a microarray experiment to enable its unambiguous interpretation and reproduction (Brazma *et al.*, 2001; Whetzel *et al.*, 2006).

MAGE – MicroArray Gene Expression – consists of three parts that together aim to create a common data format that would facilitate sharing data between different projects (described in detail in Spellman *et al.*, 2002). MAGE-OM is an object model that represents a MIAME compliant conceptualization of a microarray experiment. It is independent of experimental platforms, or image analysis and normalization methods used. MAGE-ML is a by-product of the MAGE-OM, and is a mark-up language that translates the OM to an XML, and therefore easier to read, based database. Finally MAGE-Stk is a set of software toolkits enabling users to create MAGE-ML. A number of journals, including *Nature*, are already requiring papers submitted to them that contain microarray data, to be MIAME compliant.

In practice, what effect does MIAME have on the decisions an average molecular laboratory needs to make upon embarking on the treacherous path of microarrays? For many, especially if the bioinformatics and technological support is there, a local MIAME-supportive database (such as BASE, or others created by the various consortia (for example BµG@Sbase at http://www.bugs.sgul.ac.uk/bugsbase) would allow the gradual reporting of information associated with each experiment as it is generated in the laboratory by the various members of the group. As in most databases, accounts allowing different levels of access and rights, should give a flexibility and ease of use. Once data is published the group would need to transfer it to a MIAME compliant public database such as ArrayExpress (www.ebi.ac.uk/arrayexpress) and GEO (www.ncbi.nlm.nih.gov/geo). Other databases are being developed that will eventually be MIAME compliant. ArrayExpress and GEO also allow the direct deposition of data to their websites, thus facilitating groups that do not have access to local databases. What is important across the entire process is for all groups to adhere to the same standards and ontologies, using compatible tools such as those defined by MGED, thus making crosstalk and cross-comparison possible.

Acknowledgments

J.L.H. and C.C. acknowledge funding from the U.K. Biotechnology and Biological Sciences Research Council (BBSRC) grant EGA16107. The Functional Genomics Laboratory at the School of Biosciences was funded by a BBSRC/Wellcome Trust Joint Infrastructure Fund grant JIF13209.

References

Adams EW, Ratner DM, Bokesch HR, McMahon JB, O'Keefe BR, Seeberger PH (2004) Oligosaccharide and glycoprotein microarrays as tools in HIV glycobiology: Glycan-dependent gp120/protein interactions. *Chem Biol* **11**: 875–881.

Adams R, Bishof L (1994) Seeded region growing. *IEEE Trans Pattern Anal Machine Intel* **16**: 641–647.

Affymetrix (2004) Gene expression analysis technical manual (http://www.affymetrix.com/support/technical/manual/expression_manual.affx).

Ahmed AA, Vias M, Iyer NG, Caldas C, Brenton JD (2004) Microarray segmentation methods significantly influence data precision. *Nucleic Acids Res* **32**: e50.

Albert TJ, Norton J, Ott M, Richmond T, Nuwaysir K, Nuwaysir EF, Stengele KP, Green RD (2003) Light-directed 5'→3' synthesis of complex oligonucleotide microarrays. *Nucleic Acids Res* **31**: e35.

Albertson DG, Ylstra B, Segraves R, Collins C, Dairkee SH, Kowbel D, Kuo W-L, Gray JW, Pinkel D (2000) Quantitative mapping of amplicon structure by array CGH identifies CYP24 as a candidate oncogene. *Nat Genet* **25**: 144–146.

Anderson MLM (1999) Types of hybridization and uses of each method. In: *Nucleic Acid Hybridization* Bios Scientific Publishers, Oxford, pp. 11–16.

Axon Instruments Inc. (1999) *GenePix 4000: A User's Guide*. Axon Instruments Inc., Union City, CA (Taken over by Molecular Devices Corporation, Sunnyvale, CA).

Ball CA, Sherlock G, Parkinson H *et al.* (2002) Microarray Gene Expression Data (MGED) Society standards for microarray data. *Science* **298**: 539.

Barry R, Soloviev M (2004) Quantitative protein profiling using antibody arrays. *Proteomics* **4**: 3717–3726.

Beare R, Buckley M (2004) *Spot: cDNA Microarray Image Analysis – User's Guide*. CSIRO Mathematical and Information Sciences (http://www.cmis.csiro.au/IAP/Spot/doc/Spot.pdf).

Behr MA, Wilson MA, Gill WP, Salamon H, Schoolniik GK, Rane S, Small M (1999) Comparative genomics of BCG vaccine by whole-genome DNA microarray. *Science* **284**: 1520–1523.

Bengtsson A, Bengtsson H (2006) Microarray image analysis: background estimation using quantile and morphological filters. *BMC Bioinformatics* **7**: 96.

Bengtsson H, Jonsson G, Vallon-Christersson J (2004) Calibration and assessment of channel-specific biases in microarray data with extended dynamical range. *BMC Bioinformatics* **5**: 177.

Blixt O, Head S, Mondala T *et al.* (2004) Printed covalent glycan array for ligand profiling of diverse glycan binding proteins. *Proc Natl Acad Sci USA* **101**: 17033–17038.

Boutell JM, Hart DJ, Godber BL, Kozlowski RZ, Blackburn JM (2004) Functional protein microarrays for parallel characterisation of p53 mutants. *Proteomics* **4**: 1950–1958.

Bowtell D, Sambrook J (eds) (2003) *DNA Microarrays: A Molecular Cloning Manual*. Cold Spring Harbor Laboratory Press, Cold Spring Harbor, NY.

Brasch MA, Hartley JL, Vidal M (2004) ORFeome cloning and systems biology: standardized mass production of the parts from the parts-list. *Genome Res* **14**: 2001–2009.

Brazma A, Hingamp P, Quackenbush J *et al.* (2001) Minimum information about a microarray experiment (MIAME)-toward standards for microarray data. *Nat Genet* **29**: 365–371.

Buck MJ, Lieb JD (2004) ChIP-chip: considerations for the design, analysis, and applications of genome-wide chromatin immunoprecipitation experiments. *Genomics* **83**: 349–360.

Buckley MJ (2002) *The Spot User's Guide*. CSIRO Mathematical and Information Sciences (http://www.cmis.csiro.au/IAP/Spot/spotmanual.htm).

Buhler J, Ideker T, Haynor D (2000) *Dapple: Improved Techniques for Finding Spots on DNA Microarrays*. Technical report UWTR 2000-08-05. University of Washington.

Bulyk ML, Gentalen E, Lockhart DJ, Church GM (1999) Quantifying DNA–protein interactions by double-stranded DNA arrays. *Nat Biotechnol* **17**: 573–577.

Call DR (2005) Challenges and opportunities for pathogen detection using DNA microarrays. *Crit Rev Microbiol* **31**: 91–99.

Call DR, Borucki MK, Loge FK (2003) Detection of bacterial pathogens in environmental samples using DNA microarrays. *J Microbiol Methods* **53**: 235–243.

Cao M, Kobel PA, Morshedi MM, Wu MF, Paddon C, Helmann JD (2002) Defining the *Bacillus subtilis* σ^W regulon: a comparative analysis of promoter consensus search, run-off transcription/macroarray analysis (ROMA), and transcriptional profiling approaches. *J Mol Biol* **316**: 443–457.

Case-Green S, Pritchard C, Southern EM (1999) Use of oligonucleotide arrays in enzymatic assays: assay optimization. In: *DNA Microarrays – A Practical Approach* (ed. M. Schena). Oxford University Press, Oxford, pp. 61–76.

Castellino AM (1997) When the chips are down. *Genome Res* **7**: 943–946.

Chen Y, Dougherty ER, Bittner ML (1997) Ratio-based decisions and the quantitative analysis of cDNA microarray images. *J Biomed Optics* **2**: 364–374.

Churchill GA (2002) Fundamentals of experimental design for cDNA microarrays. *Nat Genet* **32**: 490–495.

Cortese JD (2001) Microarray readers: Pushing the envelope. *Scientist* **15**: 36.

Cosma MP, Tanaka T, Nasmyth K (1999) Ordered recruitment of transcription and chromatin remodeling factors to a cell cycle- and developmentally regulated promoter. *Cell* **97**: 299–311.

Cutler P (2003) Protein arrays: The current state-of-the-art. *Proteomics* **3**: 3–18.

DeRisi JL, Iyer WR, Brown PO (1997) Exploring the metabolic and genetic control of gene expression on a genomic scale. *Science* **278**: 680–686.

De Wildt RM, Mundy CR, Gorick BD, Tomlinson IM (2000) Antibody arrays for high-throughput screening of antibody–antigen interactions. *Nat Biotechnol* **18**: 989–994.

Diks SH, Kok K, O'Toole T, Hommes DW, van Dijken P, Joore J, Peppelenbosch MP (2004) Kinoem profiling for studying lipopolysaccharide signal transduction in human peripheral blood mononuclear cells. *J Biol Chem* **279**: 49206–49213.

Disney MD, Seeberger PH (2004) The use of carbohydrate microarrays to study carbo-hydrate–cell interactions and to detect pathogens. *Chem Biol* **11**: 1701–1707.

Dobbin K, Shih JH, Simon R (2003) Questions and answers on design of dual-label microarrays for identifying differentially expressed genes. *J Natl Cancer Inst* **95**: 1362–1369.

Dombkowski AA, Thibodeau BJ, Starcevic SL, Novak RF (2004) Gene-specific dye bias in microarray reference designs. *FEBS Lett* **560**: 120–124.

Drickamer K, Taylor ME (2002) Glycan arrays for functional genomics [Review]. *Genome Biol* **3**: 1034.1–1034.4.

Dunham MJ, Badrane H, Ferea T, Adams J, Brown PO, Rowenzweig F, Botstein D *et al.* (2002) Characteristic genome rearrangements in experimental evolution of *Saccharomyces cerevisiae*. *Proc Natl Acad Sci USA* **99**: 16144–16149.

Edman CF, Raymond DE, Wu DJ, Tu E, Sosnowski RG, Butler WF, Nerenberg M, Heller MJ (1997) Electric field directed nucleic acid hybridization on microchips. *Nucleic Acids Res* **25**: 4907–4914.

Eisen MB, Brown PO (1999) *ScanAlyze*, (available at http://rana/Stanford.Edu/software).

Ekins R, Chu FW (1999) Microarrays: their origins and applications. *Trends Biotechnol* **17**: 217–218.

Ekins RP, Chu F, Biggart E (1989a) Development of microspot multi-analyte ratio-metric immunoassay using dual fluorescent-labelled antibodies. *Anal Chim Acta* **227**: 73–96.

Ekins RP, Chu F, Micallef J (1989b) High specific activity chemiluminescent and fluorescent markers: their potential application to high sensitivity and 'multi analyte' immunoassays. *J Biolumin Chemilumin* **4**: 59–78.

Ekins RP, Chu F, Biggart E (1990) Multispot, multianalyte, immunoassay. *Ann Biol Clin* **48**: 655–666.

Emili AQ, Cagney G (2000) Large-scale functional analysis using peptide or protein arrays. *Nat Biotechnol* **18**: 293–397.

Fan J-B, Chen X, Halushka MK, Lipshutz RJ, Lockhart DJ, Chakravarti A (2000) Parallel genotyping of SNPs using generic high-density oligonucleotide tag arrays. *Genome Res* **10**: 853–860.

Feizi T, Fazio F, Chai W, Wong C-H (2003) Carbohydrate microarrays – a new set of technologies at the frontiers of glycomics. *Curr Opin Struct Biol* **13**: 637–645.

Fodor SPA, Read JL, Pirrung MC, Stryer L, Lu AT, Solas D (1991) Light-directed, spatially addressable parallel chemical synthesis. *Science* **251**: 767–773.

Fodor SP, Rava RP, Huang XC, Pease AC, Holmes CP, Adams CL (1993) Multiplexed biochemical assays with biological chips. *Nature* **364**: 555–556.

Frank R (2002) The SPOT-synthesis technique – Synthetic peptide arrays on membrane supports – principles and applications. *J Immunol Methods* **267**: 13–26.

Freeman WM, Robertson DJ, Vrana KE (2000) Fundamentals of DNA hybridization arrays for gene expression analysis. *Biotechniques* **29**: 1042–1055.

Fukui S, Feizi T, Galustian C, Lawson AM, Chai W (2002) Oligosaccharide microarrays for high-throughput detection and specificity assignments of carbohydrate–protein interactions. *Nat Biotechnol* **20**: 1011–1017.

Gao X, LeProust E, Zhang H, Srivannavit O, Gulari E, Yu P, Mishiguchi C, Xiang Q, Zhou X (2001) A flexible light-directed DNA chip synthesis gated by deprotection using solution photogenerated acids. *Nucleic Acids Res* **29**: 4744–4750.

Gao X, Zhou X, Gulari E (2003) Light directed massively parallel on-chip synthesis of peptide arrays with t-Boc chemistry. *Proteomics* **3**: 2135–2141.

Gold L, Brown D, He Y-Y, Shtatland T, Singer BS, Wu Y (1997) From oligonucleotide shapes to genomic SELEX: novel biological regulatory loops. *Proc Natl Acad Sci USA* **94**: 59–64.

Grainger DC, Overton TW, Hobman JL, Constantinidou C, Tamai E, Wade JT, Struhl K, Reppas N, Church G, Busby SJW (2004). Genomic studies with *Escherichia coli* MelR protein: applications of chromatin immunoprecipitation and microarrays. *J Bacteriol* **186**: 6938–6943.

Grunstein M, Hogness DS (1975) Colony hybridization: A method for the isolation of cloned cDNAs that contain a specific gene. *Proc Natl Acad Sci USA* **72**: 3961–3965.

Guo Y, Feinberg H, Conroy E, Mitchell DA, Alvarez R, Blixt O, Taylor ME, Weis WI, Drickamer K (2004) Structural basis of distinct ligand-binding and targeting properties of the receptors DC-SIGN and DC-SIGNR. *Nat Struct Mol Biol* **11**: 591–597.

Haab BB, Dunham MJ, Brown PO (2001) Protein microarrays for highly parallel detection and quantitation of specific proteins and antibodies in complex solutions [Research]. *Genome Biol* **2**: 0004.1–0004.13.

Hacia JG (1999) Resequencing and mutational analysis using oligonucleotide microarrays. *Nat Genet* **21**: 42–47.

Harding MA, Arden KC, Gildea JW, Gildea JJ, Perlman EJ, Viars C, Theodorescu D (2002) Functional genomic comparison of lineage-related human bladder cancer cell lines with differing tumorigenic and metastatic potentials by spectral karyotyping, comparative genomic hybridization, and a novel method of positional expression profiling. *Cancer Res* **62**: 6981–6989.

Harrington CA, Rosenow C, Retief J (2000) Monitoring gene expression using DNA arrays. *Curr Opin Microbiol* **3**: 285–291.

Heller MJ, Tu E, Holmsen A, Sosnowski RG, O'Connell J (1999) Active microelectronic arrays for DNA hybridization analysis. In: *DNA Microarrays – A Practical Approach* (ed. M. Schena). Oxford University Press, Oxford, pp. 167–185.

Heller MJ, Forster AH, Tu E (2000) Active microeletronic chip devices which utilize controlled electrophoretic fields for multiplex DNA hybridization and other genomic applications. *Electrophoresis* **21**: 157–164.

Heller RA, Schena M, Shalon D, Bedilion T, Gilmore J, Woolley DE, Davis RW (1997) Discovery and analysis of inflammatory disease-related genes using cDNA microarrays. *Proc Natl Acad Sci USA* **94**: 2150–2155.

Hepburne-Scott H, Herick K (2005) Protein function arrays vs. protein expression arrays – An introduction. *Origins* **18**: 4–6 (www.sigmaaldrich.com).

Herring CD, Raffaelle M, Allen TE, Kanin EI, Landick R, Ansari AZ, Palsson BØ (2005) Immobilization of *Escherichia coli* RNA polymerase and location of binding sites by use of chromatin immunoprecipitation and microarrays. *J Bacteriol* **187**: 6166–6174.

Hirschhorn JN, Sklar P, Lindblad-Toh K, Lim Y-M, Ruiz-Gutierrez M, Bolk S, Langhorst B, Schaffner S, Winchester E, Lander ES (2000) SBE-TAGS: An array based method for efficient single-nucleotide polymorphism genotyping. *Proc Natl Acad Sci USA* **97**: 12164–12169.

Hitt E (2004) Advances in Microarray Readers. *Scientist* **18**: 42.

Holloway AJ, van Laar RK, Tothill RW, Bowtell DDR (2002) Options available–from start to finish–for obtaining data from DNA microarrays II. *Nat Genet* **32**: 481–489.

Horak CE, Mahajan MC, Luscombe NM, Gerstein M, Weissman SM, Snyder M (2002) GATA-1 binding sites mapped in the beta-globin locus by using mammalian chip-chip analysis. *Proc Natl Acad Sci USA* **99**: 2924–2929.

Houseman BT, Mrksich M (2002) Carbohydrate arrays for the evaluation of protein binding and enzymatic modification. *Chem Biol* **9**: 443–454.

Houseman BT, Huh JH, Kron SJ, Mrksich M (2002) Peptide chips for the quantitative evaluation of protein kinase activity. *Nat Biotechnol* **20**: 270–274.

Huber M, Losert D, Hiller R, Harwanegg C, Mueller MW, Schmidt WM (2001) Detection of single base alterations in genomic DNA by solid phase polymerase chain reaction on oligonucleotide microarrays. *Anal Biochem* **299**: 24–30.

Hughes JD, Estep PW, Tavazoie S, Church GW (2000) Computational identification of *Cis*-regulatory elements associated with groups of functionally related genes in *Saccharomyces cerevisiae*. *J Mol Biol* **296**: 1205–1214.

Hughes TR, Mao M, Jones AR *et al.* (2001) Expression profiling using microarrays fabricated by an ink-jet oligonucleotide synthesizer. *Nat Biotechnol* **19**: 342–347.

Imbeaud S, Auffray C (2005) 'The 39 steps' in gene expression profiling, critical issues and proposed best practices for microarray experiments. *Drug Discovery today* **10**: 1175–1182.

Ishkanian AS, Malloff CA, Watson SK *et al.* (2004) A tiling resolution DNA microarray with complete coverage of the human genome. *Nat Genet* **36**: 299–303.

Iyer VR, Horak CE, Scarfe CS, Botstein D, Snyder M, Brown PO (2001) Genomic binding sites of the yeast cell-cycle transcription factors SBF, and MBF. *Nature* **409**: 533–538.

Jain AN, Tokuyasu TA, Snjiders AM, Segraves R, Albertson DG, Pinkel D (2002) Fully automatic quantification of microarray image data. *Genome Res* **12**: 325–332.

Kallioniemi A, Kallioniemi OP, Sudar D, Rutovitz D, Gray JW, Waldman F, Pinkel D (1992). Comparative genomic hybridization for molecular cytogenetic analysis of solid tumors. *Science* **258**: 818–821.

Kapranov P, Cawley SE, Drenkow J, Bekiranov S, Strausberg RL, Fodor SP, Gingeras TR (2002) Large-scale transcriptional activity in chromosomes 21 and 22. *Science* **296**: 916–919.

Karsten SL, Van Deerlin VMD, Sabatti C, Gill LH, Geschwind DH (2002) An evaluation of tyramide signal amplification and archived fixed and frozen tissue in microarray gene expression analysis. *Nucleic Acids Res* **30**: e4.

Kennedy GC, Matsuzaki H, Dong S *et al.* (2003) Large-scale genotyping of complex DNA. *Nat Biotechnol* **21**: 1233–1237.

Kerr MK, Churchill GA (2001) Statistical design and the analysis of gene expression microarray data. *Genet Res* **77**: 123–128.

Khrapko KR, Lysov YuP, Khorlyn AA, Shick VV, Florentiev VL, Mirazabekov AD (1989) An oligonucleotide hybridization approach to DNA sequencing. *FEBS Lett* **256**: 118–122.

Kiessling LL, Cairo CW (2002) Hitting the sweet spot. *Nat Biotechnol* **20**: 234–235.

Korczak B, Frey J, Schrenzel J, Pluschke G, Pfister R, Ehricht R, Kuhnert P (2005) Use of diagnostic microarrays for determination of virulence gene patterns of *Escherichia coli* K1, a major cause of neonatal meningitis. *J Clin Microbiol* **43**: 1024–1031.

Krebs JE, Kuo M-H, Allis CD, Peterson CL (1999) Cell cycle-regulated histone acetylation required for expression of the yeast HO gene. *Genes Dev* **13**: 1412–1421.

Kuo M-H, Allis CD (1999) *in vivo* cross-linking and immunoprecipitation for studying dynamic protein:DNA associations in a chromatin environment. *Methods* **19**: 425–433.

Lander ES (1999) Array of hope. *Nat Genet* **21(Suppl)**: 3–4.

Lausted C, Dahl T, Warren C, King K, Smith K, Johnson M, Saleem R, Aitchison J, Hood L, Lasky SR (2004) POSaM: a fast, flexible, open-source, inkjet oligonucleotide synthesizer and microarrayer. *Genome Biol* **5**: R58.

Lazebnik Y (2002) Can a biologist fix a radio? – Or, what I learned while studying apoptosis. *Cancer Cell* **2**: 179–182.

Lieb JD, Liu X, Botstein D, Brown PO (2001) Promoter-specific binding of Rap1 revealed by genome-wide maps of protein–DNA association. *Nat Genet* **28**: 327–334.

Lindroos K, Sigurdsson S, Johansson K, Rönnblom L, Syvänen A-C (2002) Multiplex SNP genotyping in pooled DNA samples by a four-colour microarray system. *Nucleic Acids Res* **30**: 14 e70.

Lipshutz RF, Fodor SPA (1994) Advanced DNA sequencing technologies. *Curr Opin Struct Biol* **4**: 376–380.

Lockhart DJ, Dong H, Byrne MC *et al.* (1996) Expression monitoring by hybridization to high-density oligonucleotide arrays. *Nat Biotechnol* **14**: 1675–1680.

Lueking A, Horn M, Eickhoff H, Büssow K, Lehrach H, Walter G (1999) Protein microarrays for gene expression and antibody screening. *Anal Biochem* **270**: 103–111.

Luo Z, Geschwind DH (2001) Microarray applications in neuroscience. *Neurobiol Disease* **8**:183–193.

Lyng H, Badiee A, Svendsrud DH, Hovig E, Myklebost O, Stokke T (2004) Profound influence of microarray scanner characteristics on gene expression ratios: analysis and procedure for correction BMC. *Genomics* **5**: 10.

MacBeath G (2002) Protein microarrays and proteomics. *Nat Genet* **32**: 526–532.

MacBeath G, Schreiber SL (2000) Printing proteins as microarrays for high-throughput function determination. *Science* **289**: 1760–1763.

Mahal LK (2004) Catching bacteria with sugar. *Chem Biol* **11**: 1602–1604.

Manduchi E, Scearce LM, Brestelli JE, Grant GR, Kaestner KH, Stoeckert CJ (2002) Comparison of different labeling methods for two-channel high-density microarray experiments. *Physiol Genomics* **10**: 169–179.

Mann M, Jensen ON (2003) Proteomic analysis of post-translational modifications. *Nat Biotechnol* **21**: 255-261.

Matsuzaki H, Loi H, Dong S *et al.* (2004) Parallel genotyping of over 10,000 SNPs using a one-primer assay on a high density oligonucleotide array. *Genome Res* **14**: 414–425.

Medigue C, Rechenmann F, Danchin A, Viari A (1999) Imagene: an integrated computer environment for sequence annotation and analysis. *Bioinformatics* **15**: 2–15.

Microarray Gene Expression Data (MGED) (2002) A guide to microarray experiments – an open letter to the scientific journals. *Lancet* **360**: 1019.

Mitchell P (2002) A perspective on protein microarrays. *Nat Biotechnol* **20**: 225–229.

Mockler TC, Ecker JR (2005) Applications of DNA tiling arrays for whole-genome analysis. *Genomics* **85**: 1–15.

Nagy PL, Cleary ML, Brown PO, Lieb JB (2003) Genomewide demarcation of RNA polymeraser II transcription units revealed by physical fractionation of chromatin. *Proc Natl Acad Sci USA* **100**: 6364–6369.

Nallur G, Luo C, Fang L, Cooley S, Dave V, Lambert J, Kukanskis K, Kingsmore S, Lasken R, Schweitzer B (2001) Signal amplification by rolling circle amplification on DNA microarrays. *Nucleic Acids Res* **29**: e118.

Ngundi MM, Taitt CR, McMurray SA, Kahne D, Ligler FS (2005) Detection of bacterial toxins with monosaccharide arrays. *Biosens Bioelectron* **22**: 124–130.

Nuwaysir EF, Huang W, Albert TJ *et al.* (2002) Gene expression analysis using oligonucleotide arrays produced by maskless photolithography. *Genome Res* **12**: 1749–1755.

Ofir K, Berdichevsky Y, Benhar I, Azriel-Rosenfeld R, Lamed R, Barak Y, Bayer EA, Morag E (2005) Versatile protein microarray based on carbohydrate-binding modules. *Proteomics* **5**: 1806–1814.

Pease AC, Solas D, Sullivan EJ, Cronin MT, Holmes CP, Fodor SPA (1994) Light-directed oligonucleotide arrays for rapid DNA sequence analysis. *Proc Natl Acad Sci USA* **91**: 5022–5026.

Petalidis L, Bhattacharyya S, Morris GA, Collins VP, Freeman TC, Lyons PA (2003) Global amplification of mRNA by template-switching PCR: linearity and application to microarray analysis. *Nucleic Acids Res* **31**: e142.

Phillips J, Eberwine JH (1996) Antisense RNA amplification: A linear amplification method for analyzing the mRNA population from single living cells. *Methods* **10**: 283–288.

Phimister B (1999) Going global. *Nat Genet* **21(suppl 1-1)**.

Pinkel D, Segraves R, Sudar D *et al.* (1998). High resolution analysis of DNA copy number variation using comparative genomic hybridization to microarrays. *Nat Genet* **20**: 207–211.

Poetz O, Ostendorp R, Brocks B, Schwenk JM, Stoll D, Joos TO, Templin MF (2005) Protein microarrays for antibody profiling: specificity and affinity determination on a chip. *Proteomics* **5**: 2402–2411.

Primig M, Williams RM, Winzeler EA, Tevzadze GG, Conway AR, Hwang SY, Davis RW, Esposito RE (2000) The core meiotic transcriptome in budding yeasts. *Nat Genet* **26**: 415–423.

Ramdas L, Wang J, Hu L, Cogdell D, Taylor E, Zhang W (2001) Comparative evaluation of laser-based microarray scanners. *Biotechniques* **31**: 546–550.

Reimer U, Reineke U, Schneider-Mergener J (2002) Peptide arrays: from macro to micro. *Curr Opin Biotechnol* **13**: 315–320.

Reineke U, Volkmer-Engert R, Schneider-Mergener J (2001) Applications of peptide arrays prepared by SPOT-technology. *Curr Opin Biotechnol* **12**: 59–64.

Ren B, Robert F, Wyrick JJ *et al.* (2000) Genome-wide location and function of DNA binding proteins. *Science* **290**: 2306–2309.

Sakanyan V (2005) High-throughput and multiplexed protein array technology: protein–DNA and protein–protein interactions. *J Chromatogr B* **815**: 77–95.

Salma N, Guillemin K, McDaniel TK, Sherlock G, Tompkins L, Falkow S (2000) A whole-genome microarray reveals genetic diversity among *Helicobacter pylori* strains. *Proc Natl Acad Sci USA* **97**: 14668–14673.

Schadt EE, Li C, Ellis B, Wong WH (1999) *Feature Extraction And Normalization Algorithms for High-density Oligonucleotide Gene Expression Array Data*. Technical Report 303. Department of Statistics UCLA.

Schena M (2000) *Microarray Biochip Technology*. Eaton Publishing, Natick, MA.

Schena M, Shalon D, Davis RW, Brown PO (1995) Quantitative monitoring of gene expression patterns with a complementary DNA microarray. *Science* **270**: 467–470.

Schena M, Shalon D, Heller R, Chai A, Brown PO, Davis RW (1996) Parallel human genome analysis: Microarray-based expression monitoring of 1000 genes. *Proc Natl Acad Sci USA* **93**: 10614–10619.

Schena M, Heller RA, Theriault TP, Konrad K, Lachenmeier E, Davis RW (1998) Microarrays: biotechnology's discovery platform for functional genomics. *Trends Biotechnol* **16**: 301–306.

Scherer A, Krause A, Walker JR, Sutton SE, Seron D, Raulf F, Cooke MP (2003) Optimized protocol for linear RNA amplification and application to gene expression profiling of human renal biopsies. *Biotechniques* **34**: 546–550, 552–554, 556.

Schmid M, Uhlenhaut NH, Godard F, Demar M, Bressan R, Weigel D, Lohmann JU (2003) Dissection of floral induction pathways using global expression analysis. *Development* **130**: 6001–6012.

Schoolnik GK (2002) Functional and comparative genomics of pathogenic bacteria. *Curr Opin Microbiol* **5**: 20–26.

Schwartz M, Spector L, Gargir A, Shtevi A, Gortler M, Alstock RT, Dukler AA, Dotan N (2003) A new kind of carbohydrate array, its use for profiling antiglycan antibodies, and the discovery of a novel human cellulose-binding antibody. *Glycobiology* **13**: 749–754.

Schweitzer B, Predki P, Snyder M (2003) Microarrays to characterize protein interactions on a whole-proteome scale. *Proteomics* **3**: 2190–2199.

Selinger DW, Cheung KJ, Mei R, Johansson EM, Richmond CS, Blattner FS, Lockhart DJ, Church GM (2000) RNA expression analysis using a 30 base pair resolution *Escherichia coli* genome array. *Nat Biotechnol* **18**: 1262–1268.

Shao W, Zhou Z, Laroche I, Lu H, Zong Q, Patel DD, Kingsmore S, Piccoli S (2003) Optimization of rolling-circle amplified protein microarrays for multiplexed protein profiling. *J Biomed Biotechnol* **5**: 299–307.

Shin I, Park S, Lee M-R (2005) Carbohydrate microarrays: An advanced technology for functional studies of glycans. *Chem Eur J* **11**: 2894–2901.

Soille P (1999) *Morphological Image Analysis: Principles and Applications*. Springer, New York.

Sosnowski RG, Tu E, Butler WF, O'Connell JP, Heller MJ (1997) Rapid determination of single base mismatch mutations in DNA hybrids by direct electric field control. *Proc Natl Acad Sci USA* **94**: 1119–1123.

Southern EM (1975) Detection of specific sequences among DNA fragments separated by gel electrophoresis. *J Mol Biol* **98**: 503–517.

Southern EM (1996) DNA chips: analyzing sequence by hybridization to oligonucleotides on a large scale. *Trends Genet* **12**: 110–115.

Southern EM, Maskos U, Elder JK (1992) Analyzing and comparing nucleic acid sequences by hybridization to arrays of oligonucleotides: evaluation using experimental models. *Genomics* **13**: 1008–1017.

Speiss AN, Mueller N, Ivell R (2003) Amplified RNA degradation in T7 – amplification methods results in biased microarray hybridizations. *BMC Genomics* **4**: 44.

Spellman PT, Miller M, Stewart J *et al.* (2002) Design and implementation of microarray gene expression markup language (MAGE-ML) [Research]. *Genome Biol* **3**: 0046.

Spring DR (2005) Chemical genetics to chemical genomics: small molecules offer big insights. *Chem Soc Rev* **34**: 472–482.

Stears RL, Getts RC, Gullans SR (2000) A novel, sensitive detection system for high-density microarrays using dendrimer technology. *Physio Genomics* **3**: 93–99.

Stears RL, Martinsky T, Schena M (2003) Trends in microarray analysis. *Nat Med* **9**: 140–145.

Steinfath M, Wruck W, Seidel H, Lehrach H, Radelof U, O'Brien J (2001) Automated image analysis for array hybridization experiments. *Bioinformatics* **17**: 634–641.

Stekel D (2003a) Microarrays: Making them and using them. In: *Microarray Bioinformatics*. Cambridge University Press, Cambridge, pp. 1–18.

Stekel D (2003b) Experimental design. In: *Microarray Bioinformatics*. Cambridge University Press, Cambridge, pp. 211–230.

Stormo GD (2000) DNA binding sites, representation and discovery. *Bioinformatics* **16**: 16–23.

Stoughton RB (2005) Applications of DNA microarrays in biology. *Annu Rev Biochem* **74**: 53–82.

Syvänen A-C (2001) Accessing genetic variation: genotyping single nucleotide polymorphisms. *Nat Rev Genet* **2**: 930–942.

Tan K, Moreno-Hagelsieb G, Collado-Vides J, Stormo GD (2001) A comparative genomics approach to prediction of new members of regulons. *Genome Res* **11**: 566–584.

Templin MF, Stoll D, Schrenk M, Traub PC, Vöhringer CF, Joos TO (2002). Protein microarray technology. *Trends Biotechnol* **20**: 160–166.

Theriault TP, Winder SC, Gamble RC (1999) Application of ink-jet printing technology to the manufacture of molecular arrays. In: *DNA Microarrays – A Practical Approach* (ed. M. Schena). Oxford University Press, Oxford, pp. 101–120.

Thirumalapura NR, Morton RJ, Ramachandran A, Malayer JR (2005) Lipopolysaccharide microarrays for the detection of antibodies. *J Immunol Methods* **298**: 73–81.

Twyman RM (2004a) From genomics to proteomics. In: *Principles of Proteomics*. BIOS Scientific Publishers, Oxford, pp.1–22.

Twyman RM (2004b) Protein chips and functional proteomics. In: *Principles of Proteomics*. BIOS Scientific Publishers, Oxford, pp. 193-215.

Vahey M, Nau ME, Barrick S *et al.* (1999) Performance of the Affymetrix GeneChip HIV PRT 440 platform for antiretroviral drug resistance genotyping of human immunodeficiency virus type 1 clades and viral isolates with length polymorphisms. *J Clin Microbiol* **37**: 2533–2537.

Van Gelder RN, von Zastrow ME, Yool A, Dement WC, Barchas JD, Eberwine JH (1990) Amplified RNA synthesized from limited quantities of heterogeneous cDNA. *Proc Natl Acad Sci USA* **87**: 1663–1667.

Van Helden J (2003) Regulatory sequence analysis tools. *Nucleic Acids Res* **31**: 3593–3596.

Van Helden J, André B, Collado-Vides J (1998) Extracting regulatory sites from the upstream region of yeast genes by computational analysis of oligonucleotide frequencies. *J Mol Biol* **281**: 827–842.

Venkatasubbarao S (2004) Microarrays – status and prospects. *Trends Biotechnol* **22**: 630–637.

Walter G, Büssow K, Cahill D, Lueking A, Lehrach H (2000) Protein arrays for gene expression and molecular interaction screening. *Curr Opin Microbiol* **3**: 298–302.

Wang D (2003) Carbohydrate microarrays. *Proteomics* **3**: 2167–2175.

Wang D, Liu S, Trummer BJ, Deng C, Wang A (2002) Carbohydrate microarrays for the recognition of cross-reactive molecular markers of microbes and host cells. *Nat Biotechnol* **20**: 275–281.

Wang DG, Fan JB, Siao CJ *et al.* (1998). Large-scale identification, mapping, and genotyping of single-nucleotide polymorphisms in the human genome. *Science* **280**: 1077–1082.

Wassarman KM, Repoila F, Rosenow C, Storz G, Gottesman S (2001) Identification of novel small RNAs using comparative genomics and microarrays. *Genes Dev* **15**: 1637–1651.

Werz DB, Seeberger PH (2005) Carbohydrates as the next frontier in pharmaceutical research. *Chem Eur J* **11**: 3195–3206.

Whetzel PL, Parkinson H, Causton HC *et al.* (2006) MGED ontology: a resource for semantics-based description of microarray experiments. *Bioinformatics* **22**: 866–873.

Willats WG, Rasmussen SE, Kristensen T, Mikkelson JD (2002) Sugar-coated microarrays: a novel slide surface for the high-throughput analysis of glycans. *Proteomics* **2**: 1666–1671.

Wong CW, Albert TJ, Vega VB, Norton JE, Cutler DJ, Richmond TA, Stanton LW, Liu ET, Miller LD (2004) Tracking the evolution of the SARS coronavirus using high-throughput, high-density resequencing arrays. *Genome Res* **14**: 398–405.

Yang YH, Speed T (2002) Design issues for cDNA microarray experiments. *Nat Rev Genet* **3**: 579–588.

Yang YH, Buckley MJ, Dudoit S, Speed T (2000) *Comparison of Methods for Image Analysis on cDNA Microarray Data.* Technical report #584. University of California, Berkeley.

Yang YH, Buckley MJ, Dudoit S, Speed T (2002) Comparison of methods for image analysis on cDNA microarray data. *J Comput Graph Stat* **11**: 108–136.

Young BD, Anderson MLM (1985) Quantitative analysis of solution hybridization. In: *Nucleic Acid Hybridization – A Practical Approach* (eds B.D. Hames and S.J. Higgins). IRL Press, Oxford, pp. 47–71.

Zhang A, Wassarman KM, Rosenow C, Tjaden BC, Storz G, Gottesman S (2003) Global analysis of small RNA and mRNA targets of Hfq. *Mol Microbiol* **50**: 1111–1124.

Zheng D, Constantinidou C, Hobman JL, Minchin SD (2004) Identification of the CRP regulon using *in vitro* and *in vivo* transcriptional profiling. *Nucleic Acids Res* **32**: 5874–5893.

Zheng T, Peelen D, Smith LM (2005) Lectin arrays for profiling cell surface carbohydrate expression. *J Am Chem Soc* **127**: 9982–9983.

Zhu H, Snyder M (2001) Protein arrays and microarrays. *Curr Opin Chem Biol* **5**: 40–45.

Zhu H, Bilgin M, Bangham R *et al.* (2001) Global analysis of protein activities using proteome chips. *Science* **293**: 2101–2105.

Zwick ME, McAffee F, Cutler DJ, Read TD, Ravel J, Bowman GR, Galloway DR, Mateczun A (2004) Microarray-based resequencing of multiple *Bacillus anthracis* isolates. *Genome Biol* **6**: R10.

Immunoprecipitation with microarrays to determine the genome-wide binding profile of a DNA-associated protein

2

Joseph T. Wade

2.1 Introduction

Protein–DNA interactions play a crucial role in many major cellular processes, e.g. transcription, DNA replication, DNA repair, DNA recombination, and genome packaging. It is therefore vitally important to know the genomic location of these interactions in order to fully understand the mechanisms involved. Many *in vitro* and *in vivo* assays have been developed to determine the level and position of protein association with DNA. One such method is chromatin immunoprecipitation (ChIP; reviewed by Orlando, 2000). ChIP is a versatile technique that can be used to determine the position and strength of protein–DNA interactions *in vivo*. ChIP can also be coupled with microarray analysis: ChIP-chip (also known as ChIP-on-chip or ChIP[2]). ChIP-chip allows identification of all protein–DNA interactions for a given protein on a genome-wide level. In principle, ChIP-chip can be used to map the genome-wide location of any protein that associates with DNA. This rapidly emerging technique has greatly advanced our understanding of a number of important cellular processes, including transcription (Lee *et al.*, 2002; Harbison *et al.*, 2004), DNA replication (Wyrick *et al.*, 2001; Katou *et al.*, 2003) and genome packaging (Bernstein *et al.*, 2004; Glynn *et al.*, 2004; Lee *et al.*, 2004; Sekinger *et al.*, 2005). Related techniques have also been used to study the association of proteins with RNA *in vivo* (Takizawa *et al.*, 2000; Hieronymus and Silver, 2003; Shepard *et al.*, 2003; Zhang *et al.*, 2003) and of proteins with DNA *in vitro* (Mukherjee *et al.*, 2004; Liu *et al.*, 2005).

2.2 Background

Many proteins associate directly or indirectly with DNA *in vivo*. In order to understand the mechanism of action of these proteins it is essential to know which genomic regions they associate with. Many techniques have been developed to determine the position and/or strength of protein–DNA interactions both *in vitro* and *in vivo*. However, only ChIP-chip can be used to determine the genome-wide association of a protein *in vivo*. ChIP is the most powerful technique for studying protein–DNA interactions *in vivo* and ChIP is the only technique that can be coupled with microarray technology to allow protein–DNA interactions to be studied on a genome-wide scale. ChIP-chip is a rapidly emerging technique: the first published ChIP-chip experiments were performed less than 7 years ago and to date over 100 proteins have been studied using ChIP-chip in many different organisms. ChIP-chip is rapidly becoming a standard technique for studying DNA–protein interactions.

Work by Lieb *et al.* (2001) represents one of the earliest uses of ChIP-chip, and provides an excellent example of the power of the technique. Lieb *et al.* (2001) aimed to identify all the DNA sites for the Rap1, Sir2, Sir3, and Sir4 proteins across the *Saccharomyces cerevisiae* genome. Rap1 is a sequence-specific DNA-binding protein that was already known to associate with both transcriptionally silenced regions at the telomeres and ribosomal DNA loci (rDNA), and with the promoters of many genes, including the majority of the 137 ribosomal protein genes. Sir2, Sir3, and Sir4, on the other hand, interact indirectly with DNA through contacts with Rap1 bound at transcriptionally silenced regions and are important in establishing this silencing. By identifying the genomic locations of these proteins Lieb *et al.* (2001) were able to make a number of important findings relating to the mechanism of action of these proteins and transcription factors in general (see Section 2.7).

2.2.1 Description of ChIP-chip

A visual description of ChIP-chip is shown in *Figure 2.1* (reviewed by Orlando, 2000). The ChIP procedure begins by crosslinking growing cells using formaldehyde. Formaldehyde is a crosslinking agent that reversibly crosslinks proteins to DNA and proteins to proteins. The precise crosslinks that occur upon formaldehyde addition are not known. However, formaldehyde crosslinking causes DNA-bound proteins to be covalently attached to their DNA site. Additionally, proteins bound indirectly to DNA (i.e. through protein–protein interactions) will be crosslinked indirectly to their DNA site. Crosslinked cells are then lysed and the DNA is sheared by sonication to an average size of ~400 bp. The crosslinked cell extracts are then immunoprecipitated with antibodies against the protein of interest. Since DNA-associated proteins are covalently crosslinked to their DNA sites the immunoprecipitation step will also enrich for DNA regions bound by the protein of interest. After the immunoprecipitation step the crosslinks are reversed by heat and the DNA is purified. The relative amounts of different genomic regions can then be determined using quantitative PCR (standard ChIP). As a control for PCR efficiency, 'input' DNA is used. This is generated by reversing the crosslinks of a fraction of the crosslinked cell extract before immunoprecipitation. Input DNA is simply sheared genomic DNA with no enriched regions. Therefore any enrichment of regions in the immunoprecipitated DNA relative to the input DNA must be due to a protein–DNA interaction. An alternative control DNA sample can be generated by a mock immunoprecipitation (see Section 2.3 for more details on control samples).

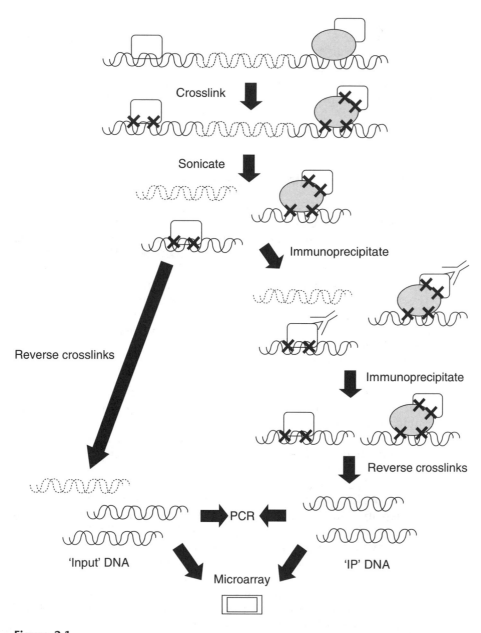

Figure 2.1

Schematic of ChIP and ChIP-chip. Protein–DNA complexes are crosslinked with formaldehyde and DNA is fragmented by sonication. A fraction of the crosslinked cell extract is taken and the crosslinks are reversed to generate the 'Input' DNA. The remainder of the crosslinked cell extract is immunoprecipitated with antibodies raised against the protein of interest. Crosslinks are reversed in the immunoprecipitated sample to generate the 'IP' DNA. Enrichment of specific DNA regions in IP DNA relative to Input DNA can then be determined using PCR. Alternatively, Input and IP DNA can be hybridized to a microarray (ChIP-chip).

ChIP-chip combines ChIP and microarray technology. DNA generated by ChIP can be amplified, labeled, and hybridized to a microarray. The control DNA sample described for standard ChIP is also amplified, labeled, and hybridized to a microarray (preferably the same microarray as the sample of interest), and serves as a control for microarray hybridization. Regions bound *in vivo* by the protein of interest will be represented by probes on the microarray that have a relatively high ratio of sample/control binding. In this way it is possible to determine the association of a protein across the whole genome.

2.2.2 Advantages of ChIP-chip over related techniques

ChIP-chip has many advantages over other techniques that are used to study protein–DNA interactions. The major advantage is that ChIP-chip allows the study of protein–DNA interactions across the whole genome. Other techniques only allow the study of proteins with DNA sites located over relatively short regions of the genome. The development of microarrays has allowed huge amounts of information to be generated in a single experiment and this can now be applied to DNA–protein interactions. As microarray technology improves it will be possible to visualize DNA–protein interactions using ChIP-chip at higher density and over larger regions, such as large eukaryotic genomes.

A second important advantage of ChIP-chip over many other techniques is that it is preformed *in vivo*. Most other techniques that are used to study protein–DNA interactions are performed *in vitro*, e.g. electromobility shift assay, *in vitro* footprinting techniques. These techniques can be used to determine the precise specificity of a protein for different DNA sequences *in vitro*, but they do not necessarily reflect the situation *in vivo*. It has been shown that protein–DNA associations can differ considerably *in vitro* and *in vivo* due to considerations other than DNA sequence such as DNA accessibility (Lieb *et al.*, 2001). Furthermore, *in vitro* techniques are not well suited to the study of indirect protein–DNA interactions whereas ChIP-chip can be used to identify both direct and indirect DNA–protein interactions. Some footprinting techniques can also be used to study protein–DNA interactions *in vivo*. However, data from these experiments is often difficult to interpret and it is hard to determine which proteins are responsible for changes in footprint patterns. ChIP, on the other hand, can be used to determine both the position and strength of protein–DNA interactions *in vivo*. Since ChIP involves an immunoprecipitation step that is specific to the protein of interest, it provides protein-specific information.

In addition to being an *in vivo* technique, ChIP provides a 'snapshot' of the cell at any given time. Since formaldehyde can diffuse rapidly into cells and causes covalent crosslinking of proteins and DNA, addition of formaldehyde to cells immediately crosslinks proteins and DNA. Therefore all interactions are preserved throughout the rest of the ChIP procedure.

ChIP-chip can be used not only to study the association of proteins with DNA through indirect interactions but also to study the association of specifically modified proteins. Antibodies can be raised against proteins that have been modified in specific ways, e.g. by phosphorylation, acetylation or methylation. Therefore it is possible to use ChIP-chip to determine the genome-wide association of a specifically modified protein. An example of such a modified protein is human histone H3. Histone H3 can be mono-, di- or tri-methylated at lysine 4. Bernstein *et al.* (2005) used ChIP-chip to determine the association of di- and tri-methlyated histone H3 across chromosomes 21 and 22 using antibodies specific to each of these modifications. In this way it is possible to compare the genome-wide association of different isoforms of the same protein using ChIP-chip.

Microarray technology has been used extensively to measure global changes in mRNA levels and hence transcription. This information can be used to determine which genes are transcriptionally regulated by a given transcription factor. However, ChIP-chip can be used to the same effect and has two significant advantages. Firstly, microarray expression analysis identifies changes in transcription that are due to indirect effects, e.g. the transcription factor of interest regulates a second transcription factor that regulates a distinct subset of genes. In this case a mutation in the transcription factor of interest would result in both direct and indirect transcriptional changes. ChIP-chip, on the other hand, only identifies regions bound by the protein of interest and hence will not be subject to indirect effects. Secondly, ChIP-chip can be performed with wild type cells whereas microarray expression analyses often involve the use of mutant proteins. These mutations may themselves cause indirect effects and also may be difficult or even impossible to generate. Despite the drawbacks of microarray expression analysis it remains a powerful technique and, when coupled with ChIP-chip, can provide an extraordinary amount of information about the mechanisms of transcriptional regulation as well as other cellular processes (Gao *et al.*, 2004; Tachibana *et al.*, 2005).

One disadvantage of ChIP-chip is that it does not allow sites of protein–DNA interaction to be mapped at the same resolution as other techniques such as *in vitro* footprinting. However, with advances in microarray technology more probes can be placed on a single microarray, thereby increasing the resolution of ChIP-chip to comparable levels (see Section 2.4).

2.3 Experimental design

There are a number of factors to consider when designing a ChIP-chip experiment.

2.3.1 Immunoprecipitation considerations

In the case of the ChIP-chip experiments carried out by Lieb *et al.* (2001), antibodies raised against native proteins were used during the ChIP procedure. These antibodies have been used extensively for ChIP experiments (Reid *et al.*, 2000). There are commercially available antibodies raised against many different DNA-associated proteins in many different organisms. Antibodies can also be raised against purified proteins or peptides in just a few months. However, antibodies raised against native proteins are of variable quality and in some cases are not suitable for ChIP. There is no general rule to determine which antibodies work well and which work poorly or not at all. Trial and error is often the only way to determine the utility of an antibody for ChIP. Additionally, it is important to determine that an antibody is specific for the protein of interest, e.g. by western blot. There are alternatives to using antibodies raised against native proteins, namely epitope-tagging proteins or fusing proteins to domains that can be affinity purified.

There are numerous examples of epitope-tagged proteins being used for ChIP (Lee *et al.*, 2002; Harbison *et al.*, 2004). Commonly used epitope tags are HA, myc and FLAG. Epitope-tagged proteins can be immunoprecipitated using well-characterized monoclonal antibodies that are known to work well for ChIP. Also, epitope-tagged proteins can be eluted during the ChIP procedure using a peptide rather than a standard heat elution, thereby reducing the experimental background. However, epitope-tagging can affect protein function due to structural changes resulting from the additional residues or may alter expression levels due to changes in mRNA stability. For nonessential proteins, loss of function may be difficult to determine. Additionally, epitope tagging may be difficult or impossible in some organisms. In this case, an antibody raised against the native protein is essential.

Proteins can also be fused to domains that can be affinity-purified, e.g. glutathione-*S*-transferase (GST), maltose-binding protein (MBP) or tandem affinity purification (TAP) tag. ChIP can then be performed using an affinity purification step rather than an immunoprecipitation (Wade *et al.*, 2004). These fusion domains are generally larger than epitope tags and therefore are more likely to affect protein function. For some organisms, e.g. *Saccharomyces cerevisae*, there are commercially available collections of strains that contain proteins tagged with epitope tags or with domains that can be affinity-purified (Ghaemmaghami *et al.*, 2003). These collections provide an excellent resource for ChIP-chip experiments.

2.3.2 Microarray considerations

The choice of microarray is a very important step in the ChIP-chip procedure. This decision is often determined by the availability and/or cost of different microarrays. There are essentially two kinds of microarrays suitable for ChIP-chip: PCR product microarrays that represent open reading frames (ORFs), intergenic regions, or both, and tiled oligonucleotide microarrays that cover both ORFs and intergenic regions. Most ChIP-chip studies have used PCR product microarrays, primarily due to the high cost of tiled oligonucleotide microarrays. The choice of microarray depends largely on the protein to be studied. If nothing is known about the pattern of association of the protein with the genome then tiled oligonucleotide microarrays provide the best option since these give an unbiased coverage of the entire genome. However, in most cases spotted PCR product microarrays are sufficient. If the protein of interest is known to bind primarily to ORFs or intergenic regions (which is likely for most transcription factors) then spotted PCR product microarrays representing either ORFs or intergenic regions should be used. If nothing is known about the association of the protein with DNA or if it is known to associate with both ORFs and intergenic regions then a microarray containing both should be used, or two microarrays, one containing ORF PCR products and the other, intergenic regions. If only ORF or intergenic region spotted PCR product microarrays are available then it is possible to use these to study the binding of any protein. This is due to the fact that ChIP generates DNA fragments of ~300–500 bp in length. Therefore, if a protein binds to a DNA site located in an intergenic region, it may be detectable by ChIP-chip using a spotted PCR product microarray containing only ORFs if the DNA site is located sufficiently close to an ORF.

For ChIP-chip studies of higher eukaryotes the genome size limits the choice of microarray. Spotted microarrays contain a limited number of features, usually less than 20 000. For most higher eukaryotes this is insufficient to cover the entire genome. This leaves two options: use a microarray that corresponds to a specific fraction of the genome, e.g. CpG islands (reviewed by Blais and Dynlacht, 2005), or use a high-density tiled microarray. Recent advances in microarray technology mean that it is now possible to cover large regions of the genomes of higher eukaryotes on a single high-density tiled microarray (Cawley *et al.*, 2004; ENCODE Project Consortium, 2004; Bernstein *et al.*, 2005).

Lieb *et al.* (2001) chose to use spotted PCR product microarrays containing both ORFs and intergenic regions. They were studying the binding of Rap1, Sir2, Sir3, and Sir4 in *S. cerevisiae*. Due to the small genome size of *S. cerevisiae*, spotted PCR product microarrays are sufficient to cover the entire genome. Although the microarrays used in this study contained both ORF and intergenic region PCR products, the vast majority of targets for these proteins were found to be in intergenic regions. For this reason, ChIP-chip studies of transcription factors in most lower eukaryotes and prokaryotes can be performed using spotted PCR product microarrays containing only intergenic regions.

2.3.3 Amplification methods

The vast majority of published ChIP-chip studies have used amplified DNA generated from ChIP material. There are three established methods of amplification (*Figure 2.2*). The primary method of amplification is random priming followed by PCR (*Figure 2.2A*). This was the method used by Lieb *et al.* (2001). An alternative method that has also been used extensively is ligation-mediated PCR in which fixed-sequence linkers are ligated to the ends of ChIP DNA fragments, followed by amplification using PCR (*Figure 2.2B*) (Lee *et al.*, 2002; Harbison *et al.*, 2004). Both of these methods are equally as effective at amplifying ChIP DNA. However, both involve a PCR step. Therefore, any bias of this process will be exponentially amplified. An alternative method has been proposed that does not involve PCR (*Figure 2.2C*) (Liu *et al.*, 2003; Bernstein *et al.*, 2004, 2005). T7 promoter sequences are attached to each DNA fragment. Addition of T7 RNA polymerase results in RNA synthesis from each tagged DNA fragment and this RNA is then reverse transcribed. This method has been shown to have greater reproducibility and fidelity than the random priming/PCR method (Liu *et al.*, 2003). Finally, methods are being developed that allow ChIP-chip to be performed without an amplification step (Carter and Vetrie, 2004).

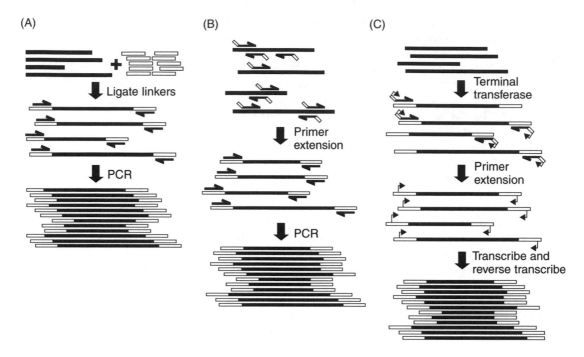

Figure 2.2

Methods of DNA amplification for ChIP-chip. (A) Ligation-mediated PCR. Fixed-sequence linkers are ligated to the ends of ChIP DNA. The DNA is then amplified by PCR using a primer specific to the linkers. (B) Random priming followed by PCR. ChIP DNA is copied by primer extension using random primers with a fixed-sequence tail. DNA is then amplified by PCR using a primer specific to the fixed-sequence tail. (C) T7 transcription method. Poly-dT is annealed to the ends of ChIP DNA using terminal transferase. The DNA is copied by primer extension using oligo-dA primers with a fixed-sequence tail that contains a T7 promoter. Many copies of complementary RNA are made using T7 RNA polymerase and the RNA is reverse transcribed to make cDNA.

2.3.4 Choice of control

All ChIP-chip experiments require the hybridization of two samples to a microarray (or microarrays): the ChIP material for the protein of interest and a control sample. This control sample can be generated in a number of different ways each of which have their advantages. Both the ChIP sample and the control sample are amplified in the same way.

The simplest control is the 'total' or 'input' DNA used in the ChIP procedure. This involves reversing the crosslinks in a sample of the crosslinked cell extract used for the immunoprecipitation step (see Section 1.2). This control has been widely used for ChIP-chip (Lee *et al.*, 2002; Harbison *et al.*, 2004). The advantages of this control are that it requires relatively little crosslinked cell extract (1/100th of that used for the immunoprecipitation is sufficient) and it does not require an additional immunoprecipitation step. However, this control does not account for any artefacts due to immunoprecipitation, e.g. affinity of specific DNA regions for the immunoprecipitation matrix or antiserum. To control for these potential artefacts a control immunoprecipitation can be performed either using no antibody (Lieb *et al.*, 2001), an irrelevant antibody, e.g. an antibody raised against a protein from a different organism (Lieb *et al.*, 2001; Grainger *et al.*, 2004), or pre-immune serum from the organism in which the antiserum was raised (Horak *et al.*, 2002). If the ChIP-chip experiment involves a tagged fusion protein then an alternative control is to perform an identical ChIP experiment using an untagged variant (Hall *et al.*, 2004).

In practice there is little difference between these different controls. Indeed, published ChIP-chip experiments have been performed with each of these controls. Lieb *et al.* (2001) chose to use both a no antibody and an irrelevant antibody control. No significant difference was observed with the two different controls.

2.4 Data acquisition

ChIP-chip data is generated using standard methods by scanning a microarray that has been hybridized with a Cy3- and a Cy5-labeled ChIP sample (or two microarrays hybridized with single labeled ChIP samples). For each ChIP-chip experiment two datasets are required: one for the protein of interest and one for the control. No specialized technology is required.

2.5 Theory of data analysis

ChIP-chip data can be analyzed in a number of different ways. The main goal of ChIP-chip experiments is usually to identify the regions of the genome that have the highest association with the protein of interest. All methods rely on the fact that the target regions for a given DNA-binding protein will be represented on the microarray by probes that have a high ratio of sample/input signal. The specific method of analysis used depends to some extent on the aim of the experiment and also the type of array used.

2.5.1 Analysis of data generated using spotted PCR product microarrays

To study the binding of Rap1, Lieb *et al.* (2001) performed six independent ChIP-chip experiments using a variety of controls. In order to generate a list of Rap1 targets from the microarray data they first ranked the probes from each microarray by sample/control ratio. They then calculated the median percentile rank from all experiments for each probe and then ranked these values. They selected the top 8% of probes from this list as 'targets' of Rap1.

A related method is to normalize the ratios for each probe from a single microarray such that the median value for each experiment is the same. The normalized ratios for each probe are then averaged from all experiments. This list of averaged ratios is rank ordered and a cut-off is defined such that ratios above the cut-off value are listed as 'targets'. The main advantage of both of these methods is their simplicity. For this reason they are the most widely used methods for analyzing ChIP-chip data. All calculations can be easily performed without the need for statistical or computational expertise.

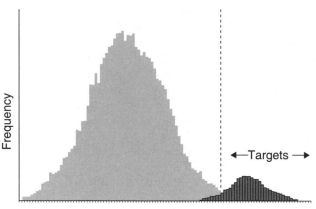

Figure 2.3

Frequency distribution of median percentile rank or normalized ratio for each microarray probe from a ChIP-chip experiment. The scale on the x-axis is an indication of how well different regions are bound by the protein of interest. Probes representing non-targets and targets are shown in light and dark gray respectively. The cut-off used to determine target regions is shown as a vertical dashed line.

Despite their simplicity they are also both effective methods for generating lists of targets for a given DNA-binding protein. Using a median percentile rank rather than a normalized ratio controls for variability in the efficiency of the ChIP procedure. ChIP experiments with varying efficiencies are weighted equivalently. Hence a single good or bad experiment will have less effect on the final list of targets. A disadvantage of the median percentile rank method as opposed to the normalized ratio method is that it requires multiple (≥ 3) experiments. However, most ChIP-chip experiments require at least three replicates to be reliable.

For both of these techniques the data can be plotted as a histogram representing the frequencies of median percentile rank or mean normalized ratio (*Figure 2.3*). ChIP-chip data should form a bimodal distribution. Probes representing the nontarget regions of the genome should form a separate distribution to those representing the target regions. The target regions should be visible as a separate distribution with high ranks/ratios. From this histogram it is easy to determine a suitable cut-off for targets and nontargets (*Figure 2.3*). In the case of the study by Lieb *et al.* (2001) this cut-off was the top 8% of the data.

These methods are simple and effective but they do not account for error resulting from the microarray itself. In particular, microarray probes that have low signals are prone to higher ratio variability (Roberts *et al.*, 2000). A number of methods have been developed to incorporate this error into the generation of a list of targets.

2.5.2 Error model

A number of studies have used a single array error model to determine the variability of data from individual probes on the microarray (Lee *et al.*,

2002; Harbison *et al.*, 2004). The method uses the data from a single microarray and the individual probe ratios are weighted according to the absolute intensity of the signal. The higher the signal, the greater the weight given to the ratio. The weighted data are then ranked as described above. The single array error model works well in situations where only a small fraction of the probes represent bound DNA since it relies on the assumption that the distribution of the sample/control ratios is symmetric. An alternative way to generate a single array error model is to use only the probes that give the bottom 50% of the ratios (Li *et al.*, 2003; Gibbons *et al.*, 2005). These ratios can be used to model the distribution of the probes representing unbound DNA in the top 50% of the ratios.

2.5.3 Analysis of data generated using tiled oligonucleotide microarrays

Analysis methods for ChIP-chip experiments using spotted PCR product microarrays are not suitable for those using tiled oligonucleotide microarrays. Since ChIP generates DNA fragments of ~400 bp in size, the microarray features for spotted PCR product arrays are, on average, larger than the DNA fragments being hybridized. However, the opposite is true for tiled oligonucleotide microarrays. Thus, using the methods described above, multiple adjacent probes would be identified as separate targets rather than as a single target encompassing multiple probes. Therefore, alternative methods have been developed for analysis of ChIP-chip data using these arrays.

At present there have been very few ChIP-chip studies using tiled oligonucleotide microarrays and hence the methods for analyzing this data are limited. Individual probes on tiled oligonucleotide microarrays generally give less reliable data than individual probes on spotted PCR product arrays. However, information from multiple adjacent oligonucleotide probes provides significantly more information than a single PCR product probe. Therefore, ChIP-chip studies using tiled oligonucleotide microarrays have used a sliding window approach to analyze the data (Cawley *et al.*, 2004; Bernstein *et al.*, 2005; Buck *et al.*, 2005). A window of fixed length is moved across the genome and ratios for probes within the window are averaged. Targets are defined as regions where the average ratio for a defined number of adjacent windows reaches a defined cut-off value. The width and amplitude of the target indicates the level of association with any given region of the genome. Use of a sliding window is a powerful method to analyze ChIP-chip data because it integrates data from multiple adjacent probes rather than relying on a single probe. This also reduces error caused by isolated 'rogue' probes that are not representative of the surrounding probes.

Very few methods have been developed to analyze ChIP-chip data from studies using tiled oligonucleotide microarrays by taking into account the probe-specific variance. However, a recent study has used 'estimated probe behavior' to normalize for differences in probe-specific variance (Li *et al.*, 2005). This method compares the behavior of each individual probe over several control experiments and uses this to estimate the probe-specific variance. This variance is then used to weight the data from each ChIP-chip experiment. In this way, unreliable probes are given less weight and vice versa.

2.5.4 Accounting for dye bias

In any microarray-based experiment there is often a nonlinear relationship between the probe intensity and the ratio of the two dye channels. This is due to biases in the efficiency of dye incorporation and detection. It is possible to counter this problem by first normalizing the data. A standard method is Lowess normalization (Quackenbush, 2002; Wade *et al.*, 2005). Lowess controls for intensity-dependent dye biases by carrying out a locally weighted linear regression. This is then used to normalize the data over different intensity ranges. A simpler method for removing dye bias from ChIP-chip experiments is to perform dye-swap experiments, i.e. in different replicate experiments swap the dyes used for the experimental and control samples. This method was used by Lieb *et al.* (2001).

2.5.5 Alternative analyses

Whilst almost all ChIP-chip experiments aim to identify the regions of the genome most strongly associated with the protein of interest, other features of the ChIP-chip data can be analyzed.

2.5.6 Comparing subsets of the genome

For some proteins that associate with DNA, especially those that are not transcription factors, it is interesting to determine if there is a specific association of the protein with a subset of the genome. For example, some factors associate specifically with promoter regions as opposed to coding sequence (Lee *et al.*, 2004; Sekinger *et al.*, 2005), whereas other factors may bind across the genome in a periodic fashion (Glynn *et al.*, 2004). Clearly, there are many different analyses that can be performed to compare the binding of a protein with different subsets of the genome.

2.5.7 Identifying specific binding sites

Most ChIP-chip studies have focused on sequence-specific DNA binding proteins. For these proteins it is useful to know the precise location of the DNA site for each bound region identified by the ChIP-chip experiment. For ChIP-chip experiments using spotted PCR product arrays this can only be done using alignment programs (reviewed by Liu *et al.*, 2004). The sequences of regions identified by the ChIP-chip experiment as having the highest association with the protein of interest are fed directly into these programs which identify overrepresented motifs. Hence predictions can be made as to the precise location of the DNA sites. For ChIP-chip experiments using tiled oligonucleotide microarrays it is theoretically possible to identify the precise position of the DNA sites in an unbiased way without using sequence information. The ChIP procedure generates a population of DNA fragments of ~400 bp centered on the DNA site for the protein of interest. Therefore, when hybridized to a tiled oligonucleotide microarray, these

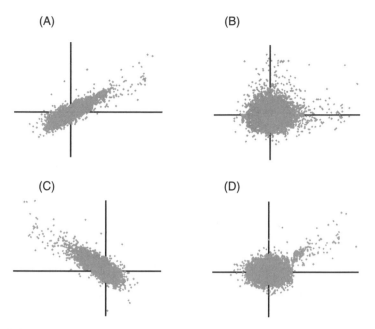

Figure 2.4

Inferring relationships between proteins by comparing ChIP-chip data. Normalized log ratios for every probe from ChIP-chip experiments for one protein plotted against those for another protein. Each data point represents the log ratio for a single probe from two ChIP-chip experiments, each with a different protein. (A) Pattern expected for proteins that bind DNA predominantly as a complex. There is a high correlation between the two datasets. (B) Pattern expected for proteins that bind to DNA independently. There is no correlation between the two datasets. (C) Pattern expected for proteins that bind to DNA mutually exclusively. There is a negative correlation between the two datasets. (D) Pattern expected for proteins that bind to DNA independently but have shared targets. There is a high correlation between the two datasets for probes with the highest log ratios. For the remainder of the data there is no correlation.

fragments should result in a peak of binding centered over the DNA site (Li *et al.*, 2005). In practice, the accuracy of mapping binding sites in this way depends largely on the density of the probes represented on the microarray. A ChIP-chip study of LexA in *Escherichia coli* using Affymetrix microarrays allowed determination of LexA binding sites to an average of ~150 bp (Wade *et al.*, 2005).

2.5.8 Comparing ChIP-chip data for different proteins

ChIP-chip can provide a wealth of information on the DNA-association of a protein. By comparing ChIP-chip data for different proteins it is also possible to draw conclusions about the interactions of different proteins with each other. Two proteins that form a complex will bind with very similar profiles across the genome. This can be visualized by plotting the log ratios for each microarray probe against each other (*Figure 2.4*). The two datasets should also be positively correlated. An example of this is different members of the RSC nucleosome remodeling complex in *S. cerevisiae* (Ng *et al.*, 2002) (*Figure 2.4A*). Alternatively, if the two proteins bind completely independently of one another there should be no correlation between the datasets (*Figure 2.4B*) (Ng *et al.*, 2002). Two other potential scenarios are that the two proteins bind mutually exclusively (*Figure 2.4C*; negative correlation) or that the two proteins bind independently but share several common sites, e.g. two transcription factors that co-regulate a subset of promoters (*Figure 2.4D*). Thus a very simple analysis of ChIP-chip data can yield important information on the interactions of different proteins.

2.6 Data analysis

2.6.1 Spotted PCR product microarrays

There are a number of different methods that have been employed to analyze data from spotted PCR product microarrays. The advantages of these different methods are described in Section 2.5. The simplest method is to rank order the data and select the top targets above a defined threshold.

(i)   Assign genes or intergenic regions to each probe represented on the microarray. Note that intergenic regions may be associated with one or two genes.
(ii) For each probe on the microarray calculate the ratio for IP/input samples. Remove flagged data.
(iii) Calculate the median ratio for the whole microarray. Normalize the data such that the median value is 1.
(iv) Arrange the data in rank order.
(v)   Define a cut-off value for the ratio and select only the data higher than this value. This represents the targets for your protein of interest. The cut-off chosen will affect the number of false positives and false negatives. For most experiments a suitable cut-off is 3 standard deviations above the median value.
(vi) To incorporate data from multiple experiments, average the normalized ratios and rank order the data. The more datasets that make up the final ordered list, the lower the cut-off that can be used.

2.6.2 Tiled oligonucleotide microarrays

There are very few examples of ChIP-chip studies using tiled oligonucleotide microarrays. Therefore a step-by-step protocol for analysis of data from these experiments does not exist. Potential methods for analysis are described in Section 2.5.

2.7 Summary of the results, conclusions, and related applications

Lieb *et al.* (2001) identified the binding sites for Rap1, Sir2, Sir3, and Sir4 in *S. cerevisiae* on a genome-wide scale. It was already known that telomeres contained transcriptionally silenced regions bound by Sir2, Sir3, and Sir4. However, it was not known how far these silenced regions extended from the telomere. Therefore, by mapping the localization of Sir2, Sir3, and Sir4 using ChIP-chip, Lieb et al. (2001) were able to identify the boundaries of the telomeric silenced loci. Additionally, a number of novel targets for Sir2, Sir3, and Sir4 were identified at various positions across the genome.

Rap1 (and not Sir2, Sir2 or Sir4) binds to many sites at promoters. A number of these sites had previously been identified using other techniques and many more had been predicted using computational analysis (Lascaris et al., 1999). Rap1 is essential for cell viability making studies of Rap1-regulated genes difficult. By using ChIP-chip, Lieb et al. (2001) were able to identify all of the DNA sites for Rap1, including novel binding sites located upstream of 185 ORFs. These novel sites included the promoters of a number of genes involved in ribosomal RNA synthesis, a major cellular process not previously known to be regulated by Rap1. Additionally, the ChIP-chip data for Rap1 were used to generate an unbiased consensus DNA site for Rap1. For sequence-specific DNA-binding proteins, DNA targets identified by ChIP-chip can be aligned using a number of different programs to generate a position weight matrix and hence a consensus motif (reviewed by Liu et al., 2004). There are several examples of proteins whose DNA-binding specificity has been first determined using ChIP-chip (Wade et al., 2004; Ben-Yehuda et al., 2005). In the case of Rap1, Lieb et al. (2001) used the position weight matrix to predict binding sites for Rap1 across the genome. Lieb et al. (2001) showed that whilst potential DNA sites for Rap1 were present in both promoter regions and coding sequences, Rap1 specifically associated with DNA sites in promoter regions. This strongly suggests the existence of a higher order chromatin structure that occludes the binding of Rap1, and presumably other DNA-binding proteins, to nonpromoter regions (Lee et al., 2004; Sekinger et al., 2005)

The work by Lieb et al. (2001) provides an excellent example of using ChIP-chip to study the mechanism of action of transcription factors. Indeed, the majority of ChIP-chip studies have focused on transcription factors since these represent the largest group of DNA-binding proteins. Two studies in particular have used ChIP-chip to identify the genome-wide location of over 100 transcription factors in S. cerevisiae (Lee et al., 2002; Harbison et al., 2004), some under a variety of different growth conditions. The data has provided an invaluable resource to the yeast community and to the transcription community as a whole. A number of subsequent studies have drawn directly from the results of these large-scale ChIP-chip studies (Wade et al., 2004; Tachibana et al., 2005).

Most transcription factors bind primarily to sites located in promoter regions. However, nontranscription factor DNA-associated proteins do not necessarily have a preference for promoter regions. ChIP-chip can be used to determine patterns of binding of any DNA-associated protein across the genome. For example, Glynn et al. (2004) studied the DNA association of the cohesin complex in S. cerevisiae using ChIP-chip. Their results showed that the cohesin complex binds to regions across the entire genome with a semi-periodic spacing of 11 kb. Ben-Yehuda et al. (2005) studied the DNA association of the chromosome-anchoring protein, RacA, in Bacillus subtilis. Their results showed that RacA binds to sites spread across a ~600 kb region centered on the origin. In this way, ChIP-chip has been used to determine the pattern of genome-wide association of many nontranscription factor proteins.

Much has been made in this chapter of the power of ChIP-chip in studies of DNA-binding proteins. However, ChIP-chip is most useful when combined with other techniques. Whilst it is of great importance to know the specific DNA targets of a given protein this is not sufficient to determine its function at these locations. ChIP-chip is therefore often used in conjunction with other methods, in particular microarray expression analysis. In fact, due to the large number of published microarray expression analyses, there is a significant amount of information to be gained from comparing ChIP-chip data with existing microarray expression analysis datasets. An excellent example of this is a study by Kurdistani *et al.* (2004) in which the authors use ChIP-chip to determine the association of eleven specifically modified histones across the *S. cerevisiae* genome. By comparing their data with existing microarray expression analyses they were able to draw important conclusions about the association of different histone modifications with specific subsets of co-regulated genes. Thus combining ChIP-chip analysis with other microarray-based analyses can provide extraordinary insights into the functional mechanisms of DNA-binding proteins. In fact, there is already sufficient data from ChIP-chip experiments to perform these kinds of analyses without doing any benchwork. A recent bioinformatics study (Gao *et al.*, 2004) has combined published microarray expression analysis and ChIP-chip data to elucidate global regulatory mechanisms in *S. cerevisiae*.

2.7.1 Alternatives to ChIP-chip

Two groups have recently developed a method which they have called genome-wide mapping technique (GMAT) or sequence tag analysis of genomic enrichment (STAGE) that, like ChIP-chip, allows the mapping of DNA-associated proteins across the genome (Roh *et al.*, 2004, 2005; Kim *et al.*, 2005). They couple ChIP with a method based on serial analysis of gene expression (SAGE) that allows DNA generated by ChIP to be concatenated, amplified, and sequenced. In principle these techniques can provide as much information as ChIP-chip. However, the number of targets identified is limited by the number of sequencing reactions performed. Without access to an affordable high-throughput sequencing facility these methods are impractical. The main advantage of these techniques over ChIP-chip is that they can be used to study protein–DNA association in any organism, regardless of genome size and regardless of whether the complete sequence is known. However, it does not provide comprehensive coverage of the whole genome in the same way that ChIP-chip does. Also, as microarray technology continues to improve, manufacturing microarrays that cover large genomes will become easier and cheaper, and the number of sequenced genomes increases at a rapid rate.

It is also possible to clone and sequence DNA generated by ChIP (Weinmann and Farnham, 2002). This is impractical given the number of sequencing reactions required to generate an equivalent amount of information to a ChIP-chip reaction. However, with advances in sequencing technology it may soon be possible to simultaneously sequence millions of short regions of DNA fragments generated by ChIP (Brenner *et al.*, 2000; Shendure *et al.*, 2004). In principle, only 11–16 bases of sequence information per DNA molecule represented in the ChIP sample would be enough to map protein–DNA associations across an entire genome.

2.7.2 Techniques related to ChIP-chip

RNA immunoprecipitation-chip

A technique based on ChIP-chip has been developed to determine the association of proteins with RNA on a transcriptome-wide level. (Takizawa *et al.*, 2000; Hieronymus and Silver, 2003; Shepard *et al.*, 2003; Zhang *et al.*, 2003). The technique, RNA immunoprecipitation-chip (RIP-chip), is very similar to ChIP-chip: RNA–protein complexes are immunoprecipitated from cell extracts and amplified labeled cDNA is hybridized to a microarray. Published studies using RIP-chip have not included a crosslinking step. However, RIP can be performed using formaldehyde crosslinking which allows detection of weak protein–RNA interactions (Gilbert *et al.*, 2004).

DNA-immunoprecipitation-chip and protein-binding microarrays

ChIP-chip allows the identification of protein–DNA interactions *in vivo*. A technique, DNA immunoprecipitation-chip (DIP-chip), has been developed to identity protein–DNA interactions *in vitro*. Purified, fragmented genomic DNA is incubated with purified protein of interest. The protein–DNA complexes are then immunoprecipitated and the amplified, labeled DNA is hybridized to a microarray. DIP-chip can be used to determine the DNA-binding specificity of a given protein. Since this is an *in vitro* technique, the protein and genomic DNA need not be from the same organism. The DNA-binding specificity of a given protein can be determined by identifying association of the protein with genomic DNA from any organism. *In vitro* DNA-binding specificity can also be determined using protein binding microarrays (PBMs) (Mukherjee *et al.*, 2004). PBMs are DNA microarrays that are hybridized with purified protein of interest and fluorescent antibodies raised against the protein of interest. Protein–DNA interactions are identified by visualizing protein binding directly to the microarray. In combination with ChIP-chip, DIP-chip or PBMs could be used to compare the *in vitro* and *in vivo* DNA-binding profiles of a given protein (Liu *et al.*, 2005).

ROMA

ROMA (run-off microarray analysis) is a technique that determines the transcriptional changes brought about by specific transcription factors in a defined *in vitro* system, on a genomic scale (for details see Chapter 5). It is analogous to more traditional transcriptomic analyses but eliminates the possibility of indirect effects on transcription. ROMA can provide similar information to ChIP-chip since transcriptionally induced or repressed genes must be regulated by transcription factors bound at their promoters. As with DIP-ChIP, ROMA could be used in combination with ChIP-chip to compare the *in vitro* and *in vivo* binding specificities of a given transcription factor. ROMA does not identify protein–DNA interactions that do not affect transcription *in vitro*, so the information provided by ROMA is subtly different to that provided by DIP-ChIP.

References

Ben-Yehuda S, Fujita M, Liu XS *et al.* (2005) Defining a centromere-like element in *Bacillus subtilis* by identifying the binding sites for the chromosome-anchoring protein RacA. *Mol Cell* **17**: 773–782.

Bernstein BE, Kamal M, Lindblad-Toh K *et al.* (2005) Genomic maps and comparative analysis of histone modifications in human and mouse. *Cell* **120**: 169–181.

Bernstein BE, Liu CL, Humphrey EL, Perlstein EO, Schreiber SL (2004) Global nucleosome occupancy in yeast. *Genome Biol* **5**: R62.

Blais A, Dynlacht BD (2005) Devising transcriptional regulatory networks operating during the cell cycle and differentiation using ChIP-on-chip. *Chromosome Res* **13**: 275–288.

Brenner S, Johnson M, Bridgham J *et al.* (2000) Gene expression analysis by massively parallel signature sequencing (MPSS) on microbead arrays. *Nat Biotechnol* **18**: 630–634.

Buck MJ, Nobel AB, Lieb JD (2005) ChIPOTle: a user-friendly tool for the analysis of ChIP-chip data. *Genome Biol* **6**: R97.

Carter NP, Vetrie D (2004) Applications of genomic microarrays to explore human chromosome structure and function. *Hum Mol Genet* **2**: R297–R302.

Cawley S, Bekiranov S, Ng HH *et al.* (2004) Unbiased mapping of transcription factor binding sites along human chromosomes 21 and 22 points to widespread regulation of non-coding RNAs. *Cell* **116**: 499–509.

ENCODE Project Consortium (2004) The ENCODE (ENCyclopedia Of DNA Elements) Project. *Science* **306**: 636–640.

Gao F, Foat BC, Bussemaker HJ (2004) Defining transcriptional networks through integrative modeling of mRNA expression and transcription factor binding data. *BMC Bioinformatics* **5**: 31.

Ghaemmaghami S, Huh WK, Bower K, Howson RW, Belle A, Dephoure N, O'Shea EK, Weissman JS (2003) Global analysis of protein expression in yeast. *Nature* **425**: 737–741.

Gibbons FD, Proft M, Struhl K, Roth FP (2005) Chipper: discovering transcription-factor targets from chromatin immunoprecipitation microarrays using variance stabilization. *Genome Biol* **6**: R96.

Gilbert C, Kristjuhan A, Winkler GS, Svejstrup JQ (2004) Elongator interactions with nascent mRNA revealed by RNA immunoprecipitation. *Mol Cell* **14**: 457–464.

Glynn EF, Megee PC, Yu HG, Mistrot C, Unal E, Koshland DE, DeRisi JL, Gerton JL (2004) Genome-wide mapping of the cohesin complex in the yeast *Saccharomyces cerevisiae*. *PLoS Biol* **2**: E259.

Grainger DC, Overton TW, Reppas N *et al.* (2004) Genomic studies with *Escherichia coli* MelR protein: applications of chromatin immunoprecipitation and microarrays. *J Bacteriol* **186**: 6938–6943.

Hall DA, Zhu H, Zhu X, Royce T, Gerstein M, Snyder M (2004) Regulation of gene expression by a metabolic enzyme. *Science* **306**: 482–484.

Harbison CT, Gordon DB, Lee TI *et al.* (2004) Transcriptional regulatory code of a eukaryotic genome. *Nature* **431**: 99–104.

Hieronymus H, Silver PA (2003) Genome-wide analysis of RNA–protein interactions illustrates specificity of the mRNA export machinery. *Nat Genet* **33**: 155–161.

Horak CE, Mahajan MC, Luscombe NM, Gerstein M, Weissman SM, Snyder M (2002) GATA-1 binding sites mapped in the beta-globin locus by using mammalian chIp-chip analysis. *Proc Natl Acad Sci USA* **99**: 2924–2929.

Katou Y, Kanoh Y, Bando M, Noguchi H, Tanaka H, Ashikari T, Sugimoto K, Shirahige K (2003) S-phase checkpoint proteins Tof1 and Mrc1 form a stable replication-pausing complex. *Nature* **424**: 1078–1083.

Kim J, Bhinge AA, Morgan XC, Iyer VR (2005) Mapping DNA–protein interactions in large genomes by sequence tag analysis of genomic enrichment. *Nat Methods* **2**: 47–53.

Kurdistani SK, Tavazoie S, Grunstein M (2004) Mapping global histone acetylation patterns to gene expression. *Cell* **117**: 721–733.

Lascaris RF, Mager WH, Planta RJ (1999) DNA-binding requirements of the yeast protein Rap1p as selected *in silico* from ribosomal protein gene promoter sequences. *Bioinformatics* **15**: 267–277.

Lee CK, Shibata Y, Rao B, Strahl BD, Lieb JD (2004) Evidence for nucleosome depletion at active regulatory regions genome-wide. *Nat Genet* **36**: 900–905.

Lee TI, Rinaldi NJ, Robert F *et al.* (2002) Transcriptional regulatory networks in *Saccharomyces cerevisiae*. *Science* **298**: 799–804.

Li W, Meyer CA, Liu XS (2005) A hidden Markov model for analyzing ChIP-chip experiments on genome tiling arrays and its application to p53 binding sequences. *Bioinformatics* **1**: i274–i282.

Li Z, Van Calcar S, Qu C, Cavenee WK, Zhang MQ, Ren B (2003) A global transcriptional regulatory role for c-Myc in Burkitt's lymphoma cells. *Proc Natl Acad Sci USA* **100**: 8164–8169.

Lieb JD, Liu X, Botstein D, Brown PO (2001) Promoter-specific binding of Rap1 revealed by genome-wide maps of protein–DNA association. *Nat Genet* **28**: 327–334.

Liu CL, Schreiber SL, Bernstein BE (2003) Development and validation of a T7 based linear amplification for genomic DNA. *BMC Genomics* **4**: 19.

Liu X, Noll DM, Lieb JD, Clarke ND (2005) DIP-chip: rapid and accurate determination of DNA-binding specificity. *Genome Res* **15**: 421–427.

Liu Y, Wei L, Batzoglou S, Brutlag DL, Liu JS, Liu XS (2004) A suite of web-based programs to search for transcriptional regulatory motifs. *Nucleic Acids Res* **32**: W204–W207.

Mukherjee S, Berger MF, Jona G, Wang XS, Muzzey D, Snyder M, Young RA, Bulyk ML (2004) Rapid analysis of the DNA-binding specificities of transcription factors with DNA microarrays. *Nat Genet* **36**: 1331–1339.

Ng HH, Robert F, Young RA, Struhl K (2002) Genome-wide location and regulated recruitment of the RSC nucleosome-remodeling complex. *Genes Dev* **16**: 806–819.

Orlando V (2000) Mapping chromosomal proteins in vivo by formaldehyde-crosslinked-chromatin immunoprecipitation. *Trends Biochem Sci* **25**: 99–104.

Quackenbush J (2002) Microarray data normalization and transformation. *Nat Genet* **32**: 496–501.

Reid JL, Iyer VR, Brown PO, Struhl K (2000) Coordinate regulation of yeast ribosomal protein genes is associated with targeted recruitment of Esa1 histone acetylase. *Mol Cell* **6**: 1297–1307.

Roberts CJ, Nelson B, Marton MJ *et al.* (2000) Signaling and circuitry of multiple MAPK pathways revealed by a matrix of global gene expression profiles. *Science* **287**: 873–880.

Roh TY, Ngau WC, Cui K, Landsman D, Zhao K (2004) High-resolution genome-wide mapping of histone modifications. *Nat Biotechnol* **22**: 1013–1016.

Roh TY, Cuddapah S, Zhao K (2005) Active chromatin domains are defined by acetylation islands revealed by genome-wide mapping. *Genes Dev* **19**: 542–552.

Sekinger EA, Moqtaderi Z, Struhl K (2005) Intrinsic histone–DNA interactions and low nucleosome density are important for preferential accessibility of promoter regions in yeast. *Mol Cell* **18**: 735–748.

Shendure J, Mitra RD, Varma C, Church GM (2004) Advanced sequencing technologies: methods and goals. *Nat Rev Genet* **5**: 335–344.

Shepard KA, Gerber AP, Jambhekar A, Takizawa PA, Brown PO, Herschlag D, DeRisi JL, Vale RD (2003) Widespread cytoplasmic mRNA transport in yeast: identification of 22 bud-localized transcripts using DNA microarray analysis. *Proc Natl Acad Sci USA* **100**: 11429–11434.

Tachibana C, Yoo JY, Tagne JB, Kacherovsky N, Lee TI, Young ET (2005) Combined global localization analysis and transcriptome data identify genes that are directly coregulated by Adr1 and Cat8. *Mol Cell Biol* **25**: 2138–2146.

Takizawa PA, DeRisi JL, Wilhelm JE, Vale RD (2000) Plasma membrane compartmentalization in yeast by messenger RNA transport and a septin diffusion barrier. *Science* **290**: 341–344.

Wade JT, Hall DB, Struhl K (2004) The transcription factor Ifh1 is a key regulator of yeast ribosomal protein genes. *Nature* **432**: 1054–1058.

Wade JT, Reppas NB, Church GM, Struhl K (2005) Genomic analysis of LexA binding reveals the permissive nature of the *Escherichia coli* genome and identifies unconventional target sites. *Genes Dev* **19**: 2619–2630.

Weinmann AS, Farnham PJ (2002) Identification of unknown target genes of human transcription factors using chromatin immunoprecipitation. *Methods* **26**: 37–47.

Wyrick JJ, Aparicio JG, Chen T, Barnett JD, Jennings EG, Young RA, Bell SP, Aparicio OM (2001) Genome-wide distribution of ORC and MCM proteins in *S. cerevisiae*: high-resolution mapping of replication origins. *Science* **294**: 2357–2360.

Zhang A, Wassarman KM, Rosenow C, Tjaden BC, Storz G, Gottesman S (2003) Global analysis of small RNA and mRNA targets of Hfq. *Mol Microbiol* **50**: 1111–1124.

Array-based comparative genomic hybridization as a tool for solving practical biological and medical questions

3

David Blesa, Sandra Rodríguez-Perales, Sara Alvarez,
Cristina Largo, and Juan C. Cigudosa

3.1 Introduction

Genetic aberrations, as losses of genetic material (deletions) or localized gains that affect certain regions of the genome, have been shown to be the basis of many diseases or human pathologies. Rare diseases, such as developmental abnormalities or mental retardation, or much more prevalent pathologies, such as cancer, are characterized by the occurrence of one or more of such genetic alterations in the genome that lead to changes in DNA sequence copy number.

Comparative genomic hybridization (CGH) has been one of the methods for the identification and further characterization of these genomic copy number changes. CGH is a molecular cytogenetic technique that allows the analysis of DNA gains and losses in the entire genome in a single hybridization experiment. It is based on the co-hybridization of two differentially fluorescence labeled DNAs to normal human metaphase chromosomes. Equal amounts of the labeled test and reference DNAs compete to hybridize proportionally to the copy numbers of the sequences present in the target chromosomes. In this way, the relative fluorescence intensity of the test to reference is determined along the length of the target chromosomes and differences between the abundance of complementary sequences in the hybridized DNAs are localized and quantified. This technique was developed by Kallioniemi *et al.* (1992) and, since then, it has contributed to the knowledge of the chromosomal aberrations present in many constitutional diseases and tumors. The sensitivity of the CGH technique depends on the degree of condensation of the chromosomes and on the size of the chromosomal aberration, something that limits

CGH's power of resolution to approximately five to ten megabasepairs (Mb) of DNA sequence. Changes (deletions, gains or amplifications) that affect genomic regions smaller than this size are not readable or efficiently detected by chromosome CGH.

Through the introduction, use and management of genome-based tools, research into genetic alterations that give rise to these diseases, as common as cancer, has undergone a technical revolution comparable to the advance of microscopy in the laboratory. Now, the study of the gene–disease relationship can be achieved by analyzing the behavior of thousands of genes, the complete genome if possible, in a simultaneous form. These systems, generically called arrays are changing the way we pose problems and draw conclusions from experiments, since they offer us a complex picture of the genome as a whole.

This change to genome-based approaches has had an immediate effect also in chromosome CGH. Metaphase spreads are being replaced as targets for hybridization by genomic microarrays. The limitations in the power of resolution of the chromosome CGH have been easily overcome by substituting the chromosome by small fragments of DNA arrayed onto a solid support: large-insert clones (BAC/PAC clones, i.e. bacterial artificial chromosomes, ~150 kb in length), complementary DNA (cDNA) clones, or oligonucleotides (Solinas-Toldo *et al.*, 1997; Pinkel *et al.*, 1998; Pollack *et al.*, 1999; Lindblad-Toh *et al.*, 2000; Lucito *et al.*, 2000; Mei *et al.*, 2000). Resolution is now limited by the type, amount, and distribution through the genome of the clones that are included in the array. Typically, most studies utilize whole genome microarrays comprising large-insert BAC or PAC clones spaced at approximately one clone per megabasepair (Fiegler *et al.*, 2003a), but higher resolution arrays comprising overlapping clone sets from specific regions (Buckley *et al.*, 2002) are also being employed. There are published experiments with array platforms that include from a few hundred clones to over 30 000 clones covering the complete human genome (Carter and Vetrie, 2004; Ishkanian *et al.*, 2004). The jump from chromosomes to clones as targets for CGH has also changed the name of the technique that can now be frequently mentioned as array based CGH, matrix-CGH, or, simply, array CGH.

The nature of the genetic aberrations that take place in diseases such as cancer and genetically determined mental retardation, that is amplifications, genomic gains and/or deletions, makes array CGH the most adequate approach to investigate them. An important advantage is that array CGH requires only DNA, which can be isolated from routine paraffin embedded pathology samples. This allows for studies on samples that have been stored for many years.

In this chapter, we present data about the use of different array CGH platforms to unveil the genomic abnormalities that can take place in two pathological conditions. First, we describe the use of array CGH to search for possible DNA copy number changes involving the subtelomeric regions of acute myeloid leukemia (AML) cells with apparently normal karyotype. Second, we describe the use of array CGH to speed up the process of cloning a familial chromosome translocation, t(3;8)(p14;q24), which is associated with the onset of renal cancer in those members of the family that are carriers of the chromosome aberration.

3.2 Scientific background

3.2.1 Leukemic cells with normal karyotype may show cytogenetically undetectable DNA copy number changes

The first case study deals with AML and normal karyotype. AML is a type of hematological tumor characterized by the proliferation of undifferentiated myeloid precursor stem cells (Huntly and Gilliland, 2005). The proliferating clone replaces normal stem cells production in the bone marrow resulting in a defective hematopoietic homeostasis with severe clinical consequences. This proliferation and blocked differentiation is commonly sustained or caused by a genetic molecular or chromosomal mutation that can be detected for proper diagnosis and monitored for therapeutic purposes. In fact, there is much information regarding the chromosomal changes that take place in AML: specific chromosome translocations, deletions, trisomies, for example (Heim and Mitelman, 1995). Most of these genetic aberrations are well characterized, even at the gene rearrangement level. They can be detected in a routine cytogenetic analysis (the karyotype) or, when the involved gene is known, by a fluorescence *in situ* hybridization (FISH) assay. These chromosome rearrangements may be used as prognostic factors and, in fact, they are currently used in clinical practice to classify patients in risk groups. Successful therapies for AML greatly rely on the correct classification of a patient in a determined group risk (Lowenberg, 2001). However, it is accepted that around 50% of the *de novo* cases of AML do not show chromosome rearrangements that are reliably detected by conventional cytogenetics or FISH assays. With the large amount of information that came from the Human Genome project and the availability of genomic analysis systems, we should explore the presence of other genetic aberrations that may take place in AML and which are not disclosed by the cytogenetic analysis due to its low resolution power. A CGH array works with genomic DNA and is then a very adequate tool to investigate gains (amplifications) and losses (deletions) that may be present in the target samples.

3.2.2 The cloning of a familial translocation associated with renal cell carcinoma

Renal cell carcinoma (RCC) comprises a heterogeneous group of tumors that have been divided into different subtypes based on histological features. Clear-cell RCC (CC-RCC, also known as nonpapillary RCC) is the most common type (75% of all RCCs). Although CC-RCCs mostly occur in a sporadic form, several familial cases have been reported. The most common form of familial CC-RCC is in association with the dominantly inherited von Hippel–Lindau (VHL) cancer syndrome. The other form is composed of families that show segregation of CC-RCC with constitutive balanced translocations involving chromosome #3. Previously, we described a Spanish family carrying a constitutional t(3;8)(p14.1;q24.32) translocation (Melendez *et al.*, 2003) (*Figure 3.1*). All CC-RCC patients in this family were carriers of this rearrangement. We speculated that deregulation of a gene(s) located at or near the translocation breakpoints may play

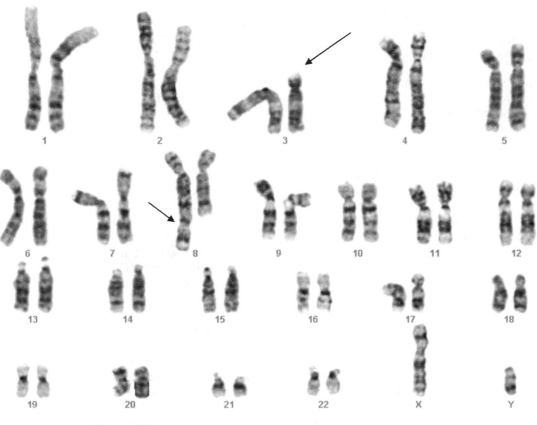

Figure 3.1

G-banded karyotype of a member of the family carrying the
t(3;8)(p14.1;q24.32) translocation. Arrows identify the derivative chromosomes
#3 and #8.

a role in the development of RCC in this family. A scientific project was
then conducted to clone the chromosomal breakpoints that were involved
in the translocation and to identify or detect the existence of any genes that
may have been affected as a result of this rearrangement. This hypothesis
was supported by the existence of breakpoint spanning genes with biological significance that are disrupted in some previously reported familial
translocations involving chromosome #3 and associated with RCC: *FHIT*
(located at 3p14), *TRC8* (8q24.1), *DIRC1* (2q33), *DIRC2* (3q21), *DIRC3*
(2q35), *LSAMP* (3q13.3), and *NORE* (1q32.1) (Rodriguez-Perales *et al.*, 2004).

3.3 Design of the experiments

The alternative possibilities to study the presence of genomic DNA gains
and/or losses in a given sample come from the different available platforms,
and their specific features, that may be used in array CGH experiments.

The panel of platforms can be initially divided into three groups regarding the type and size of the clones arrayed: large BAC/PAC clones of a mean

size of 150 kb of genomic sequence, cDNA clones with sizes around hundreds of basepairs (bp) covering the coding sequence of genes, and oligonucleotides with sizes around 40–60 bp. Each type of platform has its own advantages and limitations. In recent years we have observed in the literature and in scientific forums a competition among providers and users of different platforms in order to demonstrate which one was the most powerful or efficient, or both, to detect DNA gains and losses. Regardless of economic or other reasons, there is a substantial amount of published work on this issue that allow us to draw some simple recommendations.

BAC arrays, also called genomic arrays, were the first introduced platform (Solinas-Toldo *et al.*, 1997). They are the most commonly used for their robustness and good signal in the hybridization experiments. Because of this good hybridization yield, they are especially useful in the detection of genomic losses (deletions) and of alterations affecting a single element of the array. Their two main disadvantages are that the DNA production for spotting onto the array is expensive and that each element (i.e. a BAC or PAC clone) usually contains several genes. If the research involves the localization of altered genes, each gene within the altered BAC/PAC clone has to be tested (Albertson and Pinkel, 2003; Fiegler *et al.*, 2003a, 2003b; Carter and Vetrie, 2004).

cDNA arrays used as CGH arrays are a second option. In this platform the clones are expressed sequences that have been obtained from a previously defined library. cDNA arrays, originally developed and extensively used for expression profiling, are not a first choice for studying copy number changes. They are widely available but were not intended for CGH analysis and yield poor signal to noise ratios for many clones and require a large amount of DNA (~10 µg) for hybridization. They are indicated for certain experiments for localization of genes that are simultaneously overexpressed and amplified. However, the cDNA approach is clearly not recommended if the objective is the detection of small altered regions or single copy gains or losses (Monni *et al.*, 2001; Hyman *et al.*, 2002; Pollack *et al.*, 2002; Clark *et al.*, 2003; Hedenfalk *et al.*, 2003).

Finally, oligonucleotide-based CGH arrays are the most recent approach. It implies the use of short sequences of new synthesized fragments of DNA (oligonucleotides) of 40 or 60 bp in length (40-mer or 60-mer oligos) as targets for hybridization in the slides. The main advantages of this type of array are the high coverage density that can be easily achieved (it is rather normal that they contain 40 000 clones), each oligonucleotide can be designed to yield the best possible hybridization result, can cover poor gene regions, and all genes can be represented in the array at one time. The main disadvantages are the high cost (which eventually will decrease) and, more importantly, the variable and not so robust yield of the hybridization signal that will require some improvements of the protocols and analysis software. They also require large amounts of DNA for hybridization and are not reliable for single element alterations (although their high density design overcomes, in part, this problem) (Barrett *et al.*, 2004; Bignell *et al.*, 2004; Huang *et al.*, 2004; Rauch *et al.*, 2004; Wong *et al.*, 2004; Herr *et al.*, 2005).

The old controversy about the low density of the BAC genomic arrays, mostly based on clone collections offering a theoretical density of 1 clone per Mb, versus the high density oligo arrays, that began offering clones covering thousands of genes, has been overcome by the design and production of genomic arrays displaying the complete tiling-path of the genome in over 30 000 BAC/PAC clones (Barrett *et al.*, 2004; Ishkanian *et al.*, 2004).

With all this in mind, the choice of a specific platform should be made according to the question that we are trying to answer, biological and practical issues. Are we interested in the characterization of global genomic alteration profiling or are we looking for the detection of small altered regions? Are we interested in the genes within the alteration or in its boundaries? What are the quality and nature of the sample to be analyzed? These generic questions will lead to one or another platform.

3.3.1 Acute myeloid leukemia with normal karyotype

Around 50% of AML cases show myeloid blasts, which are the proliferating cells in this disease, with a normal karyotype. Routine cytogenetic analysis may detect very efficiently genetic aberrations such as chromosome translocations and gains or losses of complete chromosomes, even deletions or duplications can also be detected provided than their size is larger than one chromosome band (medium size, 5 Mb). However, DNA copy number changes encompassing smaller fragments are beyond the scope of microscopic analysis and their identification should be approached by other methods.

In the search for genetic aberrations that may be associated with some of the AML cases with normal karyotype, one reasonable approach is to analyze the presence of DNA copy number changes by array CGH. There is just a single report in the literature where AML with normal karyotype has been studied with array CGH (Raghavan *et al.*, 2005). In this study, in which the authors used oligo arrays for detecting loss of heterozygosity, 20% of normal karyotype AMLs were found to have uniparental disomy, a genomic condition not detectable by conventional cytogenetics.

The first question to address for this study was the type of platform to be used. AML is a malignant disease that is frequently characterized by deletions (Cigudosa *et al.*, 2003). Of the three types of array CGH approaches BAC arrays seem to be the most reliable to detect losses and low copy number changes, so we decided to use them in our study. In the search for genetic markers in AML, we concentrated on some particular segments of the genome, the subtelomeric regions, which comprise the sequences that are placed immediately after the telomeres towards the centromere. Whereas chromosome telomeres consist of hundreds of repeats of the same sequence (TTAGGG) and their role seems to be to protect chromosome ends through the cell cycle (Blasco, 2005), the genomic regions that immediately follow the repetitive sequences, the subtelomeric regions, contain a high density of genes and segmental duplications (Bailey *et al.*, 2002). Within this context, we have been collaborating with other laboratories in preparing a BAC clone collection that completely covered, with overlapped clones, the first megabasepair of the subtelomeric regions, plus a partial coverage of the five following megabases, at a density of one clone per megabase, of all human chromosomes (*Figure 3.2*). We reasoned that studying the subtelomeric regions of AML cases with normal karyotype could provide some useful information regarding the actual genomic status of these otherwise phenotypic abnormal proliferating clones.

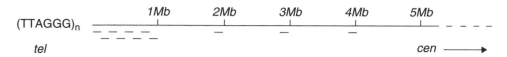

Figure 3.2

Schematic coverage of subtelomeric regions by the BAC clones in the subtelomeric array. The $(TTAGGG)_n$ sequence represents the telomere. Small bars below the line represent BAC clones. The first sequence megabase is completely covered by clones and the following four megabases at 1 clone/Mb. *tel*: telomere; *cen*: centromere.

3.3.2 Cloning of the translocation t(3;8)(p14.1;q24.32)

To clone a translocation point the first approach is to construct a physical map of the breakpoints by placing in order the genomic clones (yeast artificial chromosomes (YACs), BAC, PAC or cosmids) that cover the rearranged chromosomal regions in a comprehensive manner so the region is covered with overlapping clones (contigs). Clones can be easily selected from public databases and obtained for their use as probes for FISH analysis of the aberrant chromosomes. Breakpoint spanning clones can then be identified in a systematic way. However, such FISH investigations typically require several rounds of hybridization starting with clones relatively widely spaced followed by clones at increasingly higher densities until the aberration is defined. This process can be labor intensive and time consuming as many clones have to be hybridized to the patient's chromosomes. The process can be accelerated if we can take advantage of array CGH derived experiments (Fiegler *et al.*, 2003a, 2003b). The complete approach combined array CGH, FISH, polymerase chain reaction (PCR) on flow sorted derivative chromosomes, long-range PCR and sequencing.

Again for this specific experimental design, our choice was a BAC array with a 1 Mb resolution. The choice was due to the availability, robustness, and previously accumulated experience for this kind of experiment.

3.4 Data acquisition

3.4.1 Acute myeloid leukemia with normal karyotype

Material

We selected a series of 16 cases of AML samples with normal karyotype and collected at diagnosis. Array CGH reference DNAs were two separate pools of DNAs (10 female and 10 male) from healthy donors. An AML sample with known chromosome abnormalities as detected by conventional cytogenetic analysis is used to illustrate the resolution power of the technique.

Array CGH platform: subtelomeric array

A specific subtelomeric array CGH platform was constructed in collaboration with Dr Klaas Kok, from the Department of Human Genetics from the

University Hospital of Groningen. This project has been fostered within the European COST B19 Action 'Molecular Cytogenetics of Solid Tumors'. The array included a total of 494 BAC clones that cover 41 subtelomeric regions with a mean of 12 clones/subtelomere. It also contains another 25 control clones from #1 and X chromosomes plus several other clones for quality evaluation of the hybridization. The subtelomeric clone collection was designed to cover the first sequence megabase of each human subtelomeric region in a continuous manner and the following four megabases at a density of one clone per megabase. The array production, that included clone DNA extraction, purification, degenerate oligo priming PCR amplification, as well as printing the DNA onto the slides, was performed essentially as previously described (Westra *et al.*, 2005).

DNA labeling and hybridization

We labeled 1 μg of the sample and reference DNA with Cy3 and Cy5 fluorochromes respectively (Array CGH Protocol 6). Both labeled DNAs were co-hybridized onto the arrays during 40 h. Slides were then washed (Protocol 6, p. 251) and scanned (DNA Microarray Scanner BA, Agilent, Palo Alto, CA, USA) to obtain an image of the signal on each spot of the array. Images were analyzed with GenePix Pro 5.0 (Axon Instruments, Union City, CA, USA), which calculates the signal intensities for each fluorochrome.

3.4.2 Cloning of the translocation t(3;8)(p14.1;q24.32)

In this experimental approach, the array CGH is used as a tool for cloning a translocation breakpoint and the material to be hybridized on the array is DNA obtained from selected chromosome material, specifically those chromosomes that have been rearranged by the translocation event. These experimental procedures have been called *array painting* and they are essentially described by Fiegler *et al.* (2003b).

Material

A lymphoblastoid cell line transformed by Epstein–Barr virus was established from one of the members affected by CC-RCC of the family that carries the translocation. This cell line was the source of chromosomes and DNA for the cloning procedure. To perform the array painting we need to have individualized chromosomes that may be obtained by flow sorting (Carter, 1994). In our experiment, we flow-sorted approximately 500 copies of the rearranged chromosomes #3 and #8, der(3) and der(8), from the patient's cell line and used them as templates for degenerated oligonucleotide priming (DOP)-PCR as described (Telenius *et al.*, 1992). DOP-PCR products from the two derivative chromosomes were differentially labeled with biotin- and digoxigenin-dUTPs by a second round of PCR cycles and hybridized to normal metaphase spreads. Only chromosomal regions comprising the derivative chromosomes showed hybridization signals (*Figure 3.3*).

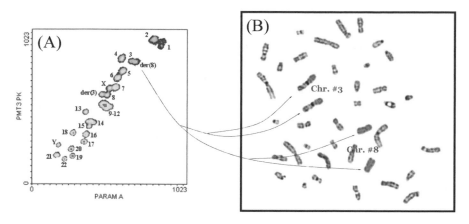

Figure 3.3

(A) Karyotype of the lymphoblastoid cell line from a CC-RCC patient generated by flow sorting. Derivative chromosomes #3 and #8 are indicated. (B) Chromosome painting of the purified der(8) hybridized against a normal metaphase showing that the der(8) is composed of chr. #8 and part of chr. #3. (A color version of this figure is available at the book's website, www.garlandscience.com/9780415378536)

Array CGH platform: genomic array

We used the whole genome array developed at the Sanger Institute (Fiegler *et al.*, 2003a) comprising clones selected to be spaced at approximately 1 Mb intervals across the human genome. Only clones corresponding to the sequences present in the sorted chromosomes showed fluorescence and the fluorescence ratio can be either high or low depending on which derivative chromosome the sequence of the clone corresponds to. If a clone on the array spans the breakpoint, sequences from both the derivatives hybridize generating intermediate ratio values.

DNA labeling and hybridization

The two derivative chromosomes were differentially labeled, and hybridized onto the genomic BAC array. Details are given in Protocol 6, p. 249.

3.5 Theory of data analysis

Chromosome aberrations are the basis of developmental abnormalities and cancer because they lead to gains and losses of part of the genome and they include interstitial deletions and duplications, nonreciprocal translocations and gene amplifications. Data output of CGH analysis are ratios between the fluorescence intensity values of test and reference DNAs. The ratio should be 1 when two chromosomal copies of the test and reference DNAs are present in the hybridization reaction. When one of the two copies of a given segment of DNA is lost (heterozygous deletion or mosonomy) the ratio test/reference decreases to 0.5 and it will eventually go to zero where there is complete loss of both copies (homozygous deletion or nulisomy). One copy gain of a segment (duplication) will produce a three to two ratio (1.5) and to increase the number of copies will raise the ratios following the same scale: four copies, ratio 2; five copies, ratio 2.5; six copies, ratio 3, and so on. Actually, repetitive sequences within the BAC clones and probes, sequence homologies within the genome and normal DNA from stroma or polyclonal nature of the tumor sample reduce the resolution power of the technique, the usual values for one copy gain or loss being 1.4 and 0.6, respectively. Moreover, small variations introduced during the whole labeling and hybridization process generates some dispersion of the data that accounts for ± 0.15 of the expected value. So, normal values can range from 0.85 to 1.15 and clones belonging to altered regions from 1.25 to 1.55 and from 0.45 to 0.75. These ratios are usually expressed as $\log_2$ values.

3.6 Data analysis

The data analyses of both experiments are conducted in the same way. Data ratios between test and reference signals are normalized by print-tip loss with DNMAD (http://bioinfo.cnio.es/, see reference manual for details); this process generates $\log_2$ values. Each clone is spotted in triplicate onto the array and average ratios for these replicas are calculated; if their variation coefficient is higher than 0.2 this average is not calculated and the clone is discarded for the analysis.

Normalized, averaged, clone data ratios are imported into an MS Excel data sheet and related to their precise chromosomal position in the genome. The data is then ordered by this position and plotted (ratio vs. chromosome position). There exist some noncommercial software programs that normalize and/or represent the data in a more graphical view with some added analysis capabilities. Some of these programs are already available on the web, some examples are: CGH-Explorer (at http://www.ifi.uio.no/bioinf/Papers/CGH/); SeeGH (at http://www.bccrc.ca/ArrayCGH); arrayCGHbase (at http://medgen.ugent.be/arrayCGHbase); CAP (on request at bioinfo-cgh@curie.fr) or CGH-Plotter (at http://sigwww.cs.tut.fi/TICSP/CGH-Plotter). In arrays that include well-characterized clones, the clones whose sequence is located on regions or chromosomes with gain/amplification or loss in the test DNA will display ratios that clearly diverge from the normal range ($\log_2$ values: 0.0 ± 0.1). In the AML experiments, the threshold for considering a clone as altered was established for each hybridization as two standard deviations (± 2 SD) of the mean of all clone ratio values. Additionally, there had to be at least two consecutive clones with abnormal ratios to consider a region as altered in its copy number. In the translocation cloning experiment, the sharp transition in ratio values pinpoints the localization of the breakpoint and no threshold values are needed for the analysis. The position of each clone in the genome is known precisely and it allows delimiting of the boundaries of the copy number alteration and also the disclosure of the structure of complex amplicons in which some DNA fragments can be much more amplified than others.

It has to be taken into consideration that DNAs obtained from tumors of polyclonal origin (with different alterations in each cellular clone), or contaminated with normal DNA from surrounding stroma cells, will show lower resolution in array CGH analysis. It is very important to have this consideration in mind when designing array CGH experiments.

3.7 Summary of the results

3.7.1 Acute myeloid leukemia with normal karyotype

An AML case with a complex karyotype displaying several known chromosome abnormalities was analyzed with the subtelomeric array. The subtelomere hybridization of this control sample confirmed some of the abnormalities expected in these regions as detected by the cytogenetic analysis, allowed the description of new abnormalities that cannot be detected by this method and showed some discrepancies that can be attributed to the inherent technical differences (*Figure 3.4*).

We have looked for DNA copy number changes in the subtelomeric regions of 16 samples of AML patients with normal karyotype at diagnosis. As a previous step, four normal donor DNAs were hybridized against the

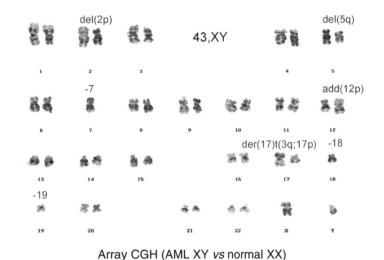

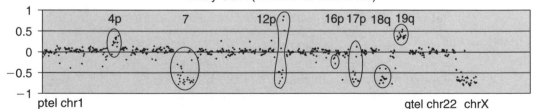

Figure 3.4

G-banded karyotype and subtelomeric array CGH of an AML case with complex karyotype. Major changes are indicated on the G-karyotype and clones localized on duplicated or lost regions are highlighted in the array. Each spot represents a clone in the array, ratio vs. position is plotted. Clones are ordered from the telomere of the *p* arm of chr. #1 to the telomere of the *q* arm of chr. X along the abscissa axis. Array CGH clearly shows complex rearrangements (as in 12p), cryptic gains (4p) and deletions (16p).

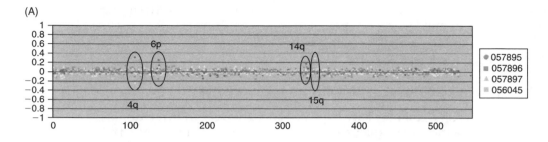

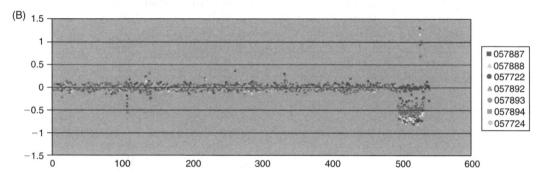

Figure 3.5

Array CGH results for normal and AML cases. Clones are plotted as in *Figure 3.4*.
(A) Hybridization results for four normal vs. pooled normal samples. Clones within
possible polymorphic regions are indicated. (B) Hybridization results for seven AML
cases in which several regions seem to be recurrently altered (including the
possible polymorphic ones). In this example, clones belonging to the X and Y
chromosomes show ratios clearly different from 0 in samples XY vs. pool XX
references (all of them except case #057894). (A color version of this figure is
available at the book's website, www.garlandscience.com/9780415378536)

normal pooled reference DNAs. In these control hybridizations four genomic
regions, comprising a few BAC clones, showed copy number polymorphisms. These regions were located at the subtelomeres 4q, 6p, 14q and 15q
and the polymorphisms were detected as clones that showed abnormal
ratios in some of the normal samples but not in others. In the AML cases,
these polymorphic regions were also found to be altered in many samples.
Thirteen other clones, in 10 different subtelomeric regions, were recurrently
detected as abnormal in two to six different cases. These 10 regions were
located within the subtelomeres from chromosomes: 4p, 6p, 7p, 15q, 18p,
18q, 19p and 21q (*Figure 3.5*). To investigate if the detected recurrently
altered clones are a common feature of AML or a normal polymorphic
characteristic of the human population, FISH analysis with probes covering
altered regions can be the first option and also the technique of choice for
analyzing detected copy number changes in large sets of AML cases.

3.7.2 Cloning of the translocation t(3;8)(p14.1;q24.32)

In our experiment, a sharp transition in ratio values define the translocation breakpoint. The profiles for chromosomes #3 and #8 are shown in

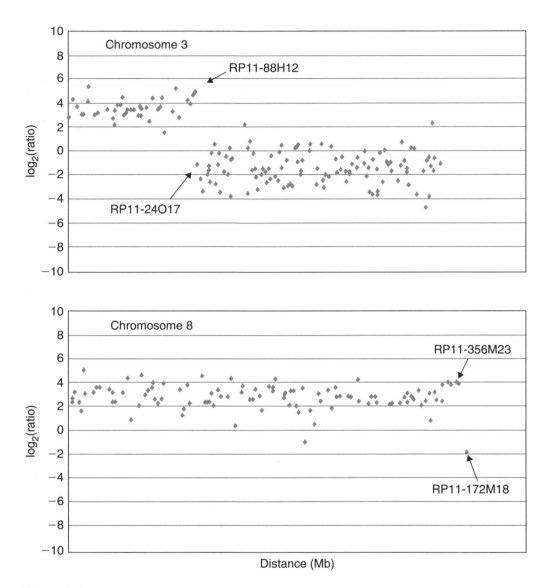

Figure 3.6

Array painting results for chromosomes #3 and #8. Each spot represents a clone in the array. Ratio vs. position is plotted. Clones are ordered from the telomere of the *p* arm to the telomere of the *q* arm along abscissa axes. Sharp ratio transitions between clones localize the translocation breakpoints. Flanking clones are indicated, their exact position in the genome is known.

Figure 3.6. Thus, the chromosome #3 breakpoint lies between clones RP11-88H12 and RP11-24O17; and the clones that defined the chromosome #8 breakpoint are RP11-356H23 and RP11-172M18. After the identification of the BAC clones that span or flank the breakpoints, the cloning process was followed by other molecular approaches.

The complete molecular description of the rearrangement has been essentially reported (Rodriguez-Perales *et al.*, 2004). The analysis of the sequence of the junction of both derivative chromosomes revealed a 5 kb micro-deletion at the chromosome #3 breakpoint together with a high density of repetitive motifs and an AT-rich region. No gene had been described at any of the breakpoints and both chromosome #3 and #8 regions flanking the breakpoints were very poor in gene content. Expression analysis of these genes was carried out by RT-PCR but no change was detected in cell lines and patients carrying the translocation.

3.8 Conclusions and suggestions for the general implementation of the case study

In conclusion, the AML study shows how array CGH can uncover genetic/genomic changes that cannot be detected by conventional cytogenetics techniques. For the translocation t(3;8) analysis, the use of the array CGH technique exemplifies how the time consuming FISH mapping procedure to delimit the 1 Mb sequence flanking the translocation site can be reduce to just one assay.

Any biological or medical question that may be caused by genomic copy changes or suspected unbalanced chromosomal rearrangements are suitable to be characterized by array CGH. Translocations, which many times occur with small duplications or deletions in their boundaries, and oncogenes or tumor suppressor genes that can be overexpressed or underexpressed by copy number alterations, are a common feature in tumoral processes. Array CGH is a global genomic technique and allows for global screening and profiling of copy number alterations. This, and the precise localization of the detected changes are the strengths of this kind of array technology. The main effort should be taken in the design of the experiment in relation to the type of array and the resolution needed. Some studies will require specific arrays that cover small chromosomic regions at high clone densities; others, will need whole genome low-density clone configurations. In any case, it will depend on the biological material available and the questions to answer.

Acknowledgments

This work has been partially supported by grants PI040555 and GR/SAL/0219/2004 from Fondo de Investigaciones Santinarias (Instituto de Salud Carlos III) and Comunidad de Madrid, respectively. D.B., S.R.P. and J.C.C. have also been financed by the European COST B19 Action 'Molecular Cytogenetics of Solid Tumours'. Within this Action, we are very grateful to Dr Klaas Kok (Human Genetics Department, University Hospital, Groningen, Netherlands) who has been our main collaborator and helped to develop the subtelomeric array. We have to acknowledge the technical and intellectual support from N.P. Carter from the Wellcome Trust Sanger Institute for the t(3;8) cloning project. We are grateful to Drs M. Urioste and J. Benitez (from the Department of Human Genetics, CNIO) for the t(3;8) translocation patient material and family studies, and to Drs M.J. Calasanz, M.D. Odero, and J. Cervera for providing the AML samples and clinical information.

References

Albertson DG, Pinkel D (2003) Genomic microarrays in human genetic disease and cancer. *Hum Mol Genet* **12(Spec No 2)**: R145–R152.

Bailey JA, Gu Z, Clark RA *et al.* (2002) Recent segmental duplications in the human genome. *Science* **297**: 1003–1007.

Barrett MT, Scheffer A, Ben-Dor A *et al.* (2004) Comparative genomic hybridization using oligonucleotide microarrays and total genomic DNA. *Proc Natl Acad Sci USA* **101**: 17765–17770.

Bignell GR, Huang J, Greshock J *et al.* (2004) High-resolution analysis of DNA copy number using oligonucleotide microarrays. *Genome Res* **14**: 287–295.

Blasco MA (2005) Mice with bad ends: mouse models for the study of telomeres and telomerase in cancer and aging. *EMBO J* **24**: 1095–1103.

Buckley PG, Mantripragada KK, Benetkiewicz M *et al.* (2002) A full-coverage, high-resolution human chromosome 22 genomic microarray for clinical and research applications. *Hum Mol Genet* **11**: 3221–3229.

Carter NP (1994) Bivariate chromosome analysis using a commercial flow cytometer. *Methods Mol Biol* **29**: 187–204.

Carter NP, Vetrie D (2004) Applications of genomic microarrays to explore human chromosome structure and function. *Hum Mol Genet* **13(Spec No 2)**: R297–R302.

Cigudosa JC, Odero MD, Calasanz MJ *et al.* (2003) De novo erythroleukemia chromosome features include multiple rearrangements, with special involvement of chromosomes 11 and 19. *Genes Chromosomes Cancer* **36**: 406–412.

Clark J, Edwards S, Feber A *et al.* (2003) Genome-wide screening for complete genetic loss in prostate cancer by comparative hybridization onto cDNA microarrays. *Oncogene* **22**: 1247–1252.

Fiegler H, Carr P, Douglas EJ *et al.* (2003a) DNA microarrays for comparative genomic hybridization based on DOP-PCR amplification of BAC and PAC clones. *Genes Chromosomes Cancer* **36**: 361–374.

Fiegler H, Gribble SM, Burford DC *et al.* (2003b) Array painting: a method for the rapid analysis of aberrant chromosomes using DNA microarrays. *J Med Genet* **40**: 664–670.

Hedenfalk I, Ringner M, Ben-Dor A *et al.* (2003) Molecular classification of familial non-BRCA1/BRCA2 breast cancer. *Proc Natl Acad Sci USA* **100**: 2532–2537.

Heim S, Mitelman F (1995) Acute myeloid leukemia. In: *Cancer Cytogenetics* (eds S. Heim and F. Mitelman). Wiley-Liss, New York, pp. 69–140.

Herr A, Grutzmann R, Matthaei A, Artelt J, Schrock E, Rump A, Pilarsky C (2005) High-resolution analysis of chromosomal imbalances using the Affymetrix 10K SNP genotyping chip. *Genomics* **85**: 392–400.

Huang J, Wei W, Zhang J *et al.* (2004) Whole genome DNA copy number changes identified by high density oligonucleotide arrays. *Hum Genomics* **1**: 287–299.

Huntly B, Gilliland DG (2005) Pathobiology of acute myeloid leukemia. In: *Hematology: Basic Principles and Practice* (eds R. Hoffman, E.J.J. Benz, S.J. Shattil *et al.*). Elsevier Inc., Philadelphia, pp. 1057–1069.

Hyman E, Kauraniemi P, Hautaniemi S *et al.* (2002) Impact of DNA amplification on gene expression patterns in breast cancer. *Cancer Res* **62**: 6240–6245.

Ishkanian AS, Malloff CA, Watson SK *et al.* (2004) A tiling resolution DNA microarray with complete coverage of the human genome. *Nat Genet* **36**: 299–303.

Kallioniemi A, Kallioniemi OP, Sudar D, Rutovitz D, Gray JW, Waldman F, Pinkel D (1992) Comparative genomic hybridization for molecular cytogenetic analysis of solid tumors. *Science* **258**: 818–821.

Lindblad-Toh K, Tanenbaum DM, Daly MJ *et al.* (2000) Loss-of-heterozygosity analysis of small-cell lung carcinomas using single-nucleotide polymorphism arrays. *Nat Biotechnol* **18**: 1001–1005.

Lowenberg B (2001) Prognostic factors in acute myeloid leukaemia. *Best Pract Res Clin Haematol* **14**: 65–75.

Lucito R, West J, Reiner A, Alexander J, Esposito D, Mishra B, Powers S, Norton L, Wigler M (2000) Detecting gene copy number fluctuations in tumor cells by microarray analysis of genomic representations. *Genome Res* **10**: 1726–1736.

Mei R, Galipeau PC, Prass C *et al.* (2000) Genome-wide detection of allelic imbalance using human SNPs and high-density DNA arrays. *Genome Res* **10**: 1126–1137.

Melendez B, Rodriguez-Perales S, Martinez-Delgado B *et al.* (2003) Molecular study of a new family with hereditary renal cell carcinoma and a translocation t(3;8)(p13;q24.1). *Hum Genet* **112**: 178–185.

Monni O, Barlund M, Mousses S *et al.* (2001) Comprehensive copy number and gene expression profiling of the 17q23 amplicon in human breast cancer. *Proc Natl Acad Sci USA* **98**: 5711–5716.

Pinkel D, Segraves R, Sudar D *et al.* (1998) High resolution analysis of DNA copy number variation using comparative genomic hybridization to microarrays. *Nat Genet* **20**: 207–211.

Pollack JR, Perou CM, Alizadeh AA, Eisen MB, Pergamenschikov A, Williams CF, Jeffrey SS, Botstein D, Brown PO (1999) Genome-wide analysis of DNA copy-number changes using cDNA microarrays. *Nat Genet* **23**: 41–46.

Pollack JR, Sorlie T, Perou CM *et al.* (2002) Microarray analysis reveals a major direct role of DNA copy number alteration in the transcriptional program of human breast tumors. *Proc Natl Acad Sci USA* **99**: 12963–12968.

Raghavan M, Lillington DM, Skoulakis S, Debernardi S, Chaplin T, Foot NJ, Lister TA, Young BD (2005) Genome-wide single nucleotide polymorphism analysis reveals frequent partial uniparental disomy due to somatic recombination in acute myeloid leukemias. *Cancer Res* **65**: 375–378.

Rauch A, Ruschendorf F, Huang J, Trautmann U, Becker C, Thiel C, Jones KW, Reis A, Nurnberg P (2004) Molecular karyotyping using an SNP array for genomewide genotyping. *J Med Genet* **41**: 916–922.

Rodriguez-Perales S, Melendez B, Gribble SM *et al.* (2004) Cloning of a new familial t(3;8) translocation associated with conventional renal cell carcinoma reveals a 5 kb microdeletion and no gene involved in the rearrangement. *Hum Mol Genet* **13**: 983–990.

Solinas-Toldo S, Lampel S, Stilgenbauer S, Nickolenko J, Benner A, Dohner H, Cremer T, Lichter P (1997) Matrix-based comparative genomic hybridization: biochips to screen for genomic imbalances. *Genes Chromosomes Cancer* **20**: 399–407.

Telenius H, Pelmear AH, Tunnacliffe A, Carter NP, Behmel A, Ferguson-Smith MA, Nordenskjold M, Pfragner R, Ponder BA (1992) Cytogenetic analysis by chromosome painting using DOP-PCR amplified flow-sorted chromosomes. *Genes Chromosomes Cancer* **4**: 257–263.

Westra JL, Boven LG, van der Vlies P *et al.* (2005) A substantial proportion of microsatellite-unstable colon tumors carry TP53 mutations while not showing chromosomal instability. *Genes Chromosomes Cancer* **43**: 194–201.

Wong KK, Tsang YT, Shen J, Cheng RS, Chang YM, Man TK, Lau CC (2004) Allelic imbalance analysis by high-density single-nucleotide polymorphic allele (SNP) array with whole genome amplified DNA. *Nucleic Acids Res* **32**: e69.

Use of single nucleotide polymorphism arrays: Design, tools, and applications

4

Mercedes Robledo, Anna González-Neira, and Joaquín Dopazo

4.1 Introduction

The genetic diversity among individuals is a field of intense investigation, mainly because, in addition to harboring a record of human migrations, it explains the basis of inherited variation in disease susceptibility (Wang *et al.*, 1998). About 90% of human variation has been ascribed to single nucleotide polymorphisms (SNPs) (Collins *et al.*, 1998). SNPs are single base pair positions in genomic DNA at which different sequence (single nucleotide) alternatives (alleles) exist in normal individuals in some populations. In most cases SNPs are di-allelic, with two alternative bases occurring at an appreciable frequency (at least 1%). Tri- and tetra-allelic SNPs are rare in humans. Single base insertion and deletion variants, referred as 'indels', are not formally be considered to be SNPs (Brookes, 1999).

SNPs and point mutations are structurally identical, differing only in their frequency. In general, variations that occur in 1% or less of a population are considered point mutations related to a phenotype, and those occurring in more than 1% are considered to be polymorphisms. The effect of SNPs will be discussed in this chapter. An SNP located in a coding region of the genome (cSNPs) is referred to as a nonsynonymous variant if it generates a change of the corresponding amino acid, and as synonymous otherwise. SNPs may also be located in regulatory regions that govern gene expression (rSNPs), but are most commonly found in intronic regions, in which case they are referred to as anonymous SNPs or intronic SNPs.

One of the applications where SNPs receive most attention is their use as genetic markers for the study of complex human diseases and pharmacogenomics (Hirschhorn and Daly, 2005). Over the last two decades, research on inherited cancer susceptibility has focused on the identification of mutations segregating with disease in large family pedigrees. Genetic linkage analysis has led to localization of high penetrance mutations in genes for several common cancers such as breast and ovarian cancer (Hall *et al.*, 1990; Wooster *et al.*, 1994), colon cancer with adenomatous polyposis coli (Bodmer, 1987), HNPCC (Lindbolm *et al.*, 1993) and melanoma (Cannon-Albright *et al.*, 1992), and also for rare syndromes such as multiple endocrine neoplasia type 2 (Mathew *et al.*, 1987; Simpson *et al.*, 1987) and Von Hippel–Lindau disease (Hosoe *et al.*, 1990; Seizinger *et al.*, 1991). However, these rare high-penetrance mutations are not responsible for all the familial cases of the corresponding diseases, nor do they explain the overall prevalences of these diseases. The explanation is found in more common low penetrance variants in genes which most likely account for these common diseases, so their identification is of great practical importance.

Association studies are particularly efficient for identification of genes with relatively common variants that confer a modest or small effect on disease risk, which are precisely the types of genes expected to be responsible for most complex disorders such as cancer (Risch and Merikangas, 1996; Collins *et al.*, 1997). SNPs have several characteristics which make them particularly valuable as genetic markers in the search for low-penetrance variants in genes using association studies. While microsatellite markers, typically used in linkage analysis (Kenealy *et al.*, 2004; Sawcer *et al.*, 2004), are more informative, they are less abundant than SNPs for which almost 10 million have been mapped in the human genome. Furthermore, SNPs exhibit a widespread distribution across the genome, and even more importantly they are easy to work with in genotyping, and have greater potential for automation (Kennedy *et al.*, 2003).

Association studies do not typically involve analysis of large family pedigrees. The simplest and more common association studies are case–control studies, based on unrelated individuals, in which the prevalence of a particular genetic marker, or set of markers, in affected versus unaffected individuals is compared (Healey *et al.*, 2000; Auranen *et al.*, 2005; Hong *et al.*, 2005). However, to avoid spurious association due to population stratification (or confounding by ethnicity), Falk and Rubinstein (1987) data recommended also genotyping the parents of each case subject and using the nontransmitted parental alleles as a control sample. In doing this, the cases and controls are matched in genetic ancestry and the analysis is therefore robust to population stratification. Using this idea, Spielman *et al.* (1993) constructed a joint test of linkage and association, called the 'transmission/disequilibrium test' (TDT), which attempts to identify preferential transmission of alleles from parent to affected child within different triads (each comprising an affected child and their two parents). While the major advantage of family-based analyses is that they are not prone to false-positive findings due to population stratification, they are very difficult to carry out because both parents may not be available for genotyping, particularly for late-onset diseases. Recently, Epstein *et al.* (2005) demonstrated that combining both triad and case–control studies in association analyses can substantially increase the power to identify disease-influencing variants. Also, samples of triads can be used to confirm previous association results that were found using unrelated cases and controls, since replication using triads ensures that associations observed in the case–control study are due to real associations and not stratification.

One of the unanswered questions regarding association studies is the dilemma of choosing a direct versus an indirect approach. In this chapter,

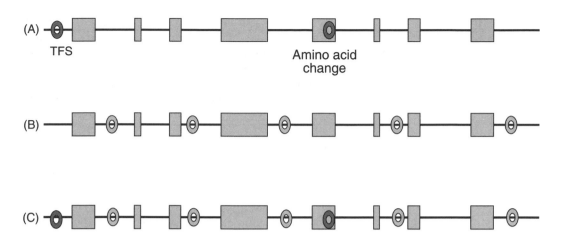

Figure 4.1

Approaches to the association study. (A) Direct approach, where candidate SNPs are directly tested for association with a phenotype. SNPs are selected on the basis of their potential effect on the protein functionality. (B) Indirect approach, where SNPs are selected on the basis of linkage disequilibrium (LD) patterns in a specific gene, or region of the genome and independently of functionality. (C) The combined direct–indirect approach. Darker grey corresponds to potential functional SNPs.

in addition to describing the philosophy of both approaches, we argue that both strategies are complementary and do not exclude each other.

4.1.1 Direct association approach

Under the direct approach, potentially functional SNPs are studied, based on the hypothesis that when a SNP occurs in coding DNA, the implications are readily apparent as a change in the amino acid make-up of the translated protein (Ramensky *et al.*, 2002) (*Figure 4.1*). Even where the SNP remains silent at the protein level, as in the case of synonymous variants, the effect of the polymorphism can be assayed at the level of mRNA (Knight, 2003) (see *Figure 4.1A*).

The use of functional SNPs could be an important factor in significantly increasing the sensitivity of association tests. In fact, several complex genetic disorders such as Alzheimer's disease (Strittmatter *et al.*, 1993) and Crohn's disease (Hugot *et al.*, 2001) have been associated with functional SNPs, lending credence to strategies that give priority to candidate markers based on predictable function.

The increasing volume of available information, together with the development of more sophisticated methods of protein structure prediction has led to different attempts at relating the effect of amino acid changes to structural distortions and, consequently, possible phenotypic effects. Two different approaches taken have included on the one hand, the study of the conservation of residues in homologous proteins (Ng and Henikoff, 2001), and, on the other, the study of changes in the stability (Chasman and Adams, 2001; Guerois, *et al.*, 2002) and other properties of the protein due to changes in amino acids (Ferrer-Costa, *et al.*, 2002; Sunyaev, *et al.*, 2000a). Nevertheless, there are different ways by which the functionality of a gene product can be affected without requiring a punctual amino acid change in the protein. There is increasing evidence that many human disease genes harbor exonic or noncoding mutations that affect pre-mRNA splicing (Cartegni *et al.*, 2002). Alternative splicing produced by mutations in intron/exon junctions, or in distinct binding motifs such as exonic splicing enhancers (ESEs) (Colapietro *et al.*, 2003), to which different proteins involved in splicing bind, may also be involved in the etiology of different diseases (Cartegni *et al.*, 2002). In fact, it has been estimated that 15% of point mutations that result in human genetic diseases cause RNA splicing defects (Krawczak *et al.*, 1992).

Alterations in the level of expression of the gene products can also cause diseases. There are different SNPs associated with alterations in gene expression (Yan *et al.*, 2002) and in some cases they are known to alter a regulatory sequence motif. A recent large-scale screening over a set of 16 chromosomes found SNPs in the promoter regions of 35% of the genes and experimental evidence suggested that around a third of promoter variants may alter gene expression to a functionally relevant extent (Hoogendoorn *et al.*, 2003). Therefore, the inclusion of other possible causes of loss of functionality in gene products, beyond the simple estimation of the possible phenotypic effect of an amino acid change, considerably increases the number of SNPs with potential phenotypic effect to be considered for the design of direct association studies.

Direct association mapping based on testing only known functional variants has the advantage of biological plausibility although it is dependent on further *in vivo* studies to demonstrate their biological relevance. This approach is also limited by incomplete knowledge about functional variation. The development of efficient bioinformatic tools to identify these critical polymorphisms *in silico* is therefore essential.

4.1.2 Indirect association approach

The indirect approach is based on the idea that the DNA variation found in a population is the result of the past transmission of that variation through generations of that population, and that this process produces a structure to the SNP variation that can be of considerable value in addressing the primary goal of finding variants associated with disease risk (Clark, 2004). In this way the indirect method relies on linkage disequilibrium (LD) between a disease locus and either a marker allele or multilocus haplotype (combination of alleles that are inherited together on the same chromosome) (see the section 'Linkage disequilibrium' on p. 94 for detailed definitions). Two or more alleles are in LD when they are inherited together with a higher frequency than would be expected based on chance alone (Pritchard and Przeworski, 2001; Ardlie *et al.*, 2002). LD reduces the number of polymorphic markers that need to be studied in order to detect an association between a gene and a trait when random markers are used. Following this strategy, SNPs are selected throughout the candidate gene or region independently of their potential function. Associations could be observed with causal variants if they are measured, but also with measured variants that are in LD with the causal variant (see *Figure 4.1B*). In other words, an SNP found to be associated with disease could be functionally important or serve merely as a genetic marker, with the functional locus co-inherited with the polymorphic allele (Knight, 2003). In the absence of LD, only a causative polymorphism will show any appreciable association with disease. However, in the presence of LD, polymorphisms that are physically near to a causal polymorphism will also show an association.

4.1.3 Combined approach

Although it has been suggested that analyses using a haplotype-based or indirect strategy rather than considering candidate SNPs individually (direct strategy) could increase the possibility of detecting associations for complex traits (Morris and Kaplan, 2002), we consider that the combination of both strategies is optimal. That is, for a given candidate gene, by considering both LD and all functional SNPs available, the advantages of both strategies can be utilized: the need to genotype fewer markers without missing information and the inclusion of SNPs with biologically plausible links to the disease in question (see *Figure 4.1C*).

The first association studies were based on the analysis of very few markers at a time, most of them considered as the putative causal variant. This was mainly a consequence of limitations in genotyping technologies. The recent development of new high-throughput technologies for SNP genotyping has opened up the possibility of taking a genome-wide approach to the study of polymorphisms. For the first time the investigation of low-penetrance variants contributing to complex diseases or multigenic disorders can be approached in a realistic manner.

4.2 Scientific background: Essential steps to be considered in the design of the experiment

We now focus on the process of marker selection and on the available tools for data analysis, applied to the identification of candidate genes by using a technology based on SNP arrays.

4.2.1 Candidate gene selection

The selection of candidate genes to perform an association study could be based on: (1) an in-depth knowledge of the specific disease biology of interest, selecting genes which are involved in the same biological network; (2) previous genetic studies in a model organism; or (3) genes selected from a candidate region identified by a linkage study. For example, in medullary thyroid cancer susceptibility (MTC) it is known that the *RET* proto-oncogene is related to both familial and sporadic forms of the disease (Mulligan *et al.*, 1993; Zedenius *et al.*, 1994). Taking this knowledge into account, an appropriate strategy would be to select genes belonging to the *RET* pathway itself (Airaksinen *et al.*, 1999; Manie *et al.*, 2001). In addition, any gene with a function relevant to any aspect of carcinogenesis (for example, genes related to cell cycle control) could be considered as a strong candidate.

4.2.2 Selection criteria for single nucleotide polymorphisms

There are four important properties of an SNP for it to be considered as an optimal candidate for large-scale genotyping purposes: functional effect, LD with other SNPs in the same gene, minor allele frequency, and the probability of successful genotyping. Finding ideal SNPs is not always possible, but computer-aided methods can facilitate the selection process in order to achieve a final collection of SNPs bearing the optimal information. We will consider the difficulties often encountered in retrieving such information as well as solutions that can be implemented in available software.

Functional impact of the polymorphism

Much attention has been focused on a range of methods to model the possible phenotypic effect of SNPs that cause amino acid changes (Sunyaev *et al.*, 2000a and b; Chasman and Adams, 2001; Miller and Kumar, 2001; Ng and Henikoff, 2001; Ferrer-Costa *et al.*, 2002, 2004; Guerois *et al.*, 2002), and only recently the interest has broadened to include functional SNPs affecting regulatory regions or the splicing process (Conde *et al.*, 2004, 2005). It has recently been postulated that additional putative functional sequences in the genome could be contained in nonexonic regions conserved across species and therefore may represent functionally important regulatory sites to be studied (Pennacchio and Rubin, 2001; Cheremushkin and Kel, 2003) and there is increasing evidence that many human disease genes are due to noncoding mutations affecting regulatory regions (Prokunina *et al.*, 2002; Yan *et al.*, 2002; Hudson, 2003). It is therefore important that both coding and noncoding regions are considered in searching for potentially functional SNPs.

Linkage disequilibrium

Linkage disequilibrium (LD) is the degree of correlation between two neighboring genetic variants in a specific population. The most common statistics used for LD measurement are the disequilibrium parameter (D') and the disequilibrium coefficient (r^2) (Hill and Robertson, 1968). A haplotype is the combination of alleles at different markers along the same chromosome that are transmitted together. Each individual has two haplotypes that represent the paternal and maternal chromosomes transmitted from his or her parents.

Since the power of an association test declines rapidly as LD diminishes (Houlston and Peto, 2004), the previous knowledge of the extent of LD between markers across a region is essential for association studies based on haplotypes. The extent of LD varies significantly between regions of the genome (Reich *et al.*, 2001, 2002), in part as a result of heterogeneity in recombination rates (Wall and Pritchard, 2003), but also due to other factors such as population structure and population size (Pritchard and Przeworski, 2001). This local variation in LD patterns consists of hotspots of recombination separating regions with low levels of haplotype diversity called 'haplotype blocks' (Gabriel *et al.*, 2002). The rationale behind haplotype-based analyses is that within haplotype blocks (with high LD), the analysis of only a reduced number of SNPs (called 'tagSNPs') (see *Figure 4.2*) is sufficient to define the total haplotype variability (Ahmadi *et al.*, 2005; Barrett *et al.*,

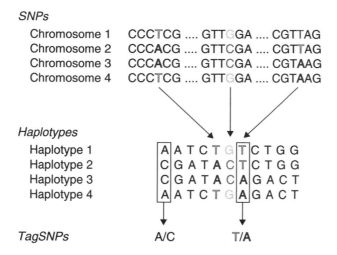

Figure 4.2

Example of taqSNP selection. The selected taqSNPs allow the inference of genotypes of neighboring SNPs in LD with them.

2005). This idea has formed the basis of one of the primary objectives of the HapMap project (www.hapmap.org), which has determined a detailed map of fine-scale variation in LD and recombination across the genome for each of four reference populations. Marker density is currently one SNP every 5 kb except for some parts of the genome with much higher recombination rates, where higher SNP density has been implemented. This information is already freely available to the scientific community and so a more comprehensive ('tagging') approach to testing candidate genes or even the entire genome may be, and is being, considered.

Frequency information and validation status

There are more than 10 million SNPs stored in the most recent version of dbSNP (build 124), and more than half of them have been validated by different means (http://www.ncbi.nlm.nih.gov/SNP/snp_summary.cgi). In addition to being validated, an SNP must exist in the population at a frequency that makes it a suitable marker. Very rare SNPs (minor allele lower than 5%) are not suitable for association or linkage studies. It has been postulated under 'the common variant, common disease (CV/CD) hypothesis', that susceptibility to a given disease may be explained by modest effects of a number of relatively common variants (Pritchard and Cox, 2002).

Probability of successful genotyping in an adequate genotyping platform

Adequate oligonucleotide design is one factor that significantly contributes to the success of the genotyping assay. Crucial in the design of probe assays is the use of software that evaluates sequences flanking the targeted SNP, including repeated sequences, palindromic sequences, GC and AT content and neighboring polymorphisms (other SNPs and small indels). In the case of the Illumina genotyping platform, for example, each individual assay design is linked to scoring information that quantifies the likelihood of success for that design.

In summary, the selection of SNPs by combining LD information (selecting 'tagSNPs'), the prediction of functionality and phylogenetic conservation, SNP allele frequency and validation status, and also the probability of obtaining a quality genotyping result, makes this process not one of selection 'at random', but rather selection using sound criteria to obtain the maximum yield from the generated data.

4.3 Use of microarray technology: The Illumina platform as an example

Until relatively recently, popular genotyping methods were designed to analyze only one or a few SNPs per assay. These include Taqman technology, mass spectrometry analysis, Invader assay, and others. Even the most conservative estimates propose that several hundred SNP markers genotyped in many thousands of individuals would be required for association studies in candidate genes to have adequate power to identify low-penetrance genes related to complex traits (Glazier *et al.*, 2002).

Advances in array technology now allow us to efficiently genotype hundreds of thousand of SNPs simultaneously. Basically, SNP-array technology is based on solid-surface-mediated detection, where hundreds of thousands of predefined sequences can be arrayed in a very small area. These predefined sequences sort the products of allele-specific reactions via DNA hybridization. When allele-specific reactions are coupled with fluorescent labeling, either on the nucleotides or on the primers, measurement of fluorescence intensities at distinct points on a microarray identifies the form of the product of allele-specific reactions and therefore the genotype (Cutler *et al.*, 2001; Shen *et al.*, 2005). These new array technologies provide several advantages: (1) fast genotyping; (2) cost effectiveness; (3) highly accurate SNP genotyping; and (4) robustness.

Currently, one of the systems most used for SNP genotyping is the Illumina microarray system. This platform combines a highly multiplexed assay with an accurate technology based on random arrays of DNA-coated beads. Well-characterized and unique DNA sequences are attached to micro-beads with unique fluorescence signatures. Illumina's assay protocol, the Golden Gate assay, is based on a highly multiplexed allele-discriminating and extension reaction, allowing the genotyping of up to 1536 SNPs simultaneously and minimizing the time, reagent volumes and material requirements of the process. This technology was also designed in order to minimize sources of variability and human error to ensure optimal data quality and reproducibility. For more details, see Protocol 8 on p. 259.

4.4 Tools for data acquisition and data analysis

We will now describe the different tools used for the optimal SNP selection, and SNP analysis, according to the criteria described above.

4.4.1 Computational tools for selecting optimal SNPs: two-step protocol

We will first describe the use of two programs that allow us to carry out a two-step approach to the selection of an optimal set of SNPs for the genotyping. PupaSNP (http://www.pupasnp.org) (Conde *et al.*, 2004) has been designed so that users can input lists of genes to produce lists of all SNPs contained therein, including information on putative functional properties, validation status, and corresponding population allele frequencies. In most cases this information is sufficient for selecting the SNPs. If too many SNPs are reported for a gene it is possible to refine this selection according to a range of filters, using the program PupasView (http://pupasview.bioinfo.cipf.es) (Conde *et al.*, 2005). Filters can be interactively applied according to functional properties, cross-species conservation, population frequency, and available LD. This refinement permits a final selection of a minimum number of SNPs with optimal properties in terms of population frequencies and potential phenotypic effect.

PupaSNP and PupasView use a precompiled database which contains a collection of entries of dbSNP mapped to the Golden Path genome assembly, as implemented in the human section of Ensembl (http://www.ensembl.org). SNPs have been labeled according to their potential effects on the phenotype. The programs analyze both transcriptional and gene product levels. Promoter regions (10 000 bp upstream) of genes are scanned for the presence of possible different regulatory motifs. These include alterations in the following.

* *Transcription factor binding sites*. The program Match (Kel *et al.*, 2003) which uses only high quality matrices from the Transfac database (Wingender *et al.*, 2000) and cut-offs to minimize false positives is used for this purpose.
* *Intron/exon border consensus sequences*. The Ensembl database is used to extract the intron/exon organization of the genes and the corresponding sequences. The two conserved nucleotides on each side of the splicing point that constitute the splicing signal (Krawczak *et al.*, 1992) are then located and all the SNPs altering these signals are recorded.
* *Exonic splicing enhancers*. Mutations that inactivate or activate an ESE sequence may result in exon skipping, errors in alternative splicing patterns or malformation. Different classes of ESE consensus motifs have been described, but they are not always easily identified. Exon sequences are therefore scanned to identify putative ESE sites responsive to the human SR proteins SF2/ASF, SC35, SRp40, and SRp55 by using the available weight matrices (Cartegni *et al.*, 2002). This gives scores related to the likelihood that the site in question is a real ESE. Only ESE sites with scores over a predetermined threshold (see Cartegni *et al.*, 2002 for details) are selected for further analyses.
* *Triplex-forming oligonucleotide target sequences*. It has been found that the population of triplex-forming oligonucleotide target sequences (TTSs) is much more abundant than that expected from simple random models (Goñi *et al.*, 2004). Furthermore, it has recently been suggested that such motifs might play a role in the control of gene expression (Goñi *et al.*, 2004). Although the role of TTS in regulation is still a matter of

speculation, the program also reports SNPs that disrupt these structures.

- *SNPs in exons that cause an amino acid change with putative pathological effect.* The putative pathologic effect of an amino acid change can be predicted as described in Ferrer-Costa *et al.* (2002, 2004).
- *Human–mouse conserved regions.* Untranslated whole genome comparisons by BLASTz are performed for species pairs which are thought to be similar enough to be able to detect homology directly at the DNA level (Schwartz *et al.*, 2003). Of particular interest, because of their phylogenetic position with respect to humans, is mouse (or rat): distant enough to interpret conservation as important but not so distant for similarity to be lost. The phenotypic effect of a change in such regions is quite speculative but cross-species conservation can be useful in cases in which no other information is available. This information is useful for reinforcing the likelihood of other predictions (e.g. an ESE in a conserved region is more likely to be real than one in a nonconserved region).

Step one: The web interface of PupaSNP

Input data

PupaSNP has been designed for high throughput screening of functional SNPs. Thus, the input consists of a list of genes. The list can be directly provided as a collection of gene identifiers (Ensembl IDs, or external IDs, which include GenBank, Swissprot/TrEMBL and other gene IDs supported by Ensembl) or can be specified by chromosomal location (cytobands or chromosomal coordinates). In the latter case, PupaSNP extracts all genes contained in the specified location. Ensembl coordinates are used to extract the genes. Only Ensembl annotated genes, but not predictions, are extracted.

The web interface

PupaSNP inputs lists of genes, which can be defined by chromosome position specified by means of cytoband units or in absolute chromosomal position (as mapped in the corresponding Ensembl assembly). Lists of genes can be typed, uploaded or pasted into the input box. PupaSNP finds all the SNPs mapping in locations that might cause a loss of functionality in the genes. Functional information for the genes is obtained from OMIM and Gene Ontology and reported. Information on homologous genes can also be retrieved. Finally, SNPs do not need to be annotated in the genome to be included in the query tool. Results include SNPs in a the promoter region of the genes, SNPs located at intron boundaries, SNPs located at exonic splicing enhancers and coding SNPs located at Interpro domains. *Figure 4.3* shows part of the results provided by the program for the SNPs with possible phenotypic effects on genes in the p36.33 cytoband of the chromosome 1. *Figure 4.3C* is especially interesting because it shows how the scores obtained by the motif scanning method can be used to assess the possible impact of the polymorphism in the recognition of the ESE motif by the cellular machinery. Both the SNPs and the genes found are linked to the Ensembl Genome Browser. Genes are also linked to PupasView.

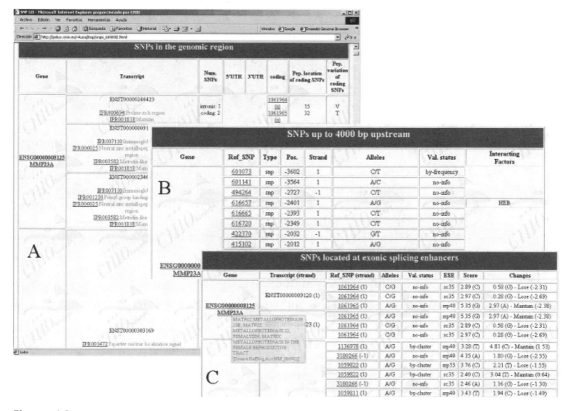

Figure 4.3

Example of results from a search using PupaSNP. (A) List of genes and the corresponding transcripts with the SNPs mapped in the different regions, including coding, 5'-UTR and 3'-UTR regions. For coding SNPs, the position within the transcript and change produced (if any) is reported. (B) SNPs located in the promoter regions, with a limit of 4000 bp. Disruptions of predicted TFBS are listed. The validation status of the SNP, (no-information, by-submitter, by-frequency, by-cluster – see DBSNP web page) is also provided. (C) SNPs located at exonic splice enhancers. The scores refer to the closeness of the site to the motif. If the polymorphism has a site with a lower score, that would generally imply poorer recognition of the site by the cellular machinery and consequently a putative alteration in the normal splicing process. When the cursor is placed over the gene name, additional information is displayed. (A color version of this figure is available on the book's website, www.garlandscience.com/9780415378536)

Step two: The web interface of PupasView. Refining the selection of SNPs

The main purpose of PupasView is to provide the user with an optimal set of SNPs for genotyping experiments by filtering the annotated SNPs with a series of filters related to their impact in protein functionality and pathology, their population frequency and LD.

The input data is a list of gene identifiers (as for PupaSNP). PupasView can also be invoked directly from PupaSNP. The program then presents a list of filter options that can be selected and applied as many times as desired. The options include:

(i)   validation status obtained from dbSNP,
(ii) type of SNP according to its position in the gene (coding, intron, utr, local),
(iii) minor allele frequency,
(iv) population,
(v)   Functional properties as follows:
  * Non-synonymous SNPs. All or only those predicted as pathologic by the pmut algorithm (Ferrer-Costa *et al.*, 2002, 2004).
  * SNPs disrupting predicted transcription factor binding sites (all or only those that are in regions conserved in the mouse genome).
  * SNPs disrupting predicted ESEs (all or only those that are in regions conserved in the mouse genome).
  * SNPs disrupting potential triplex forming regions (all or only those that are in regions conserved in the mouse genome).
  * SNPs disrupting intron/exon boundaries.
  * Regions conserved in mouse.

There are also options for the way in which haplotype blocks are constructed:

* confidence intervals (Gabriel *et al.*, 2002);
* four gamete rule (Wang *et al.*, 2002);
* solid spine of LD (The International HapMap Consortium, 2003).

Figure 4.4 shows an example of results from PupasView. *Figure 4.4A* corresponds to running PupasView on the gene TP53 without applying any filter. All the SNPs in the gene and the neighborhood are displayed. If the cursor is maintained over an SNP, information on it is displayed by means of pop-up text. *Figure 4.4B* shows a sub-group of these SNPs after selecting only SNPs for which population frequency was available. Finally, *Figure 4.4C* shows the selection obtained if only SNPs with putative functional effects are chosen, constituting the final, reduced subset of optimal SNPs. The upper horizontal bar below the figure represents LD parameters (which can be obtained individually by placing the cursor over them), while the lower horizontal bar represents haplotype blocks identified using the selected algorithm. The blocks are displayed graphically with brown rectangles going from the first to the last SNP within the block. When the cursor is placed over the rectangles a *tooltip* text pops up, showing the SNPs and the haplotypes formed in the block, with their estimated HapMap frequencies in parentheses. Tag SNPs are signaled with an exclamation mark (!).

4.4.2 A computational tool for SNP genotyping analysis: Gencall software

Gencall is an automated genotyping software designed specifically for the Illumina platform which assigns a call score to each genotype produced. This score indicates the degree of separation between homozygote and heterozygote clusters for each SNP and the position of the particular genotype call within a cluster, thereby quantifying the quality and robustness of each SNP

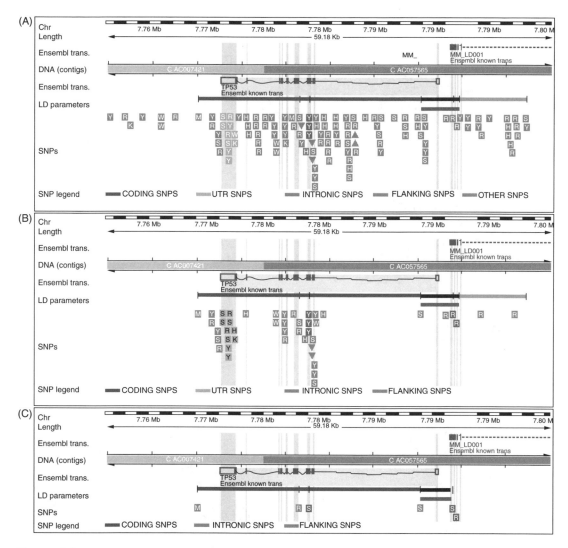

Figure 4.4

Sequential application of filters using PupasView. (A) All available SNPs in the gene *TP53*. (B) Those SNPs with population frequencies available. (C) Those SNPs with any functional characteristic. The appearance of the PupasView results varies depending on the versions of Ensembl and dbSNP used. (A color version of this figure is available on the book's website, www.garlandscience.com/9780415378536)

assay and the genotypes determined across samples. In *Figure 4.5A* we give a visual example of the genotype clusters when the variation within a cluster is very small and clusters are well separated. It is well known that the quality of DNA can influence genotyping results, both in the quality and the quantity of successful genotypes. Poorer quality DNA can manifest itself as more disperse clusters (see *Figure 4.5B*). The main consequence of this cluster dispersion is the generation of genotyping errors. In fact, several studies have

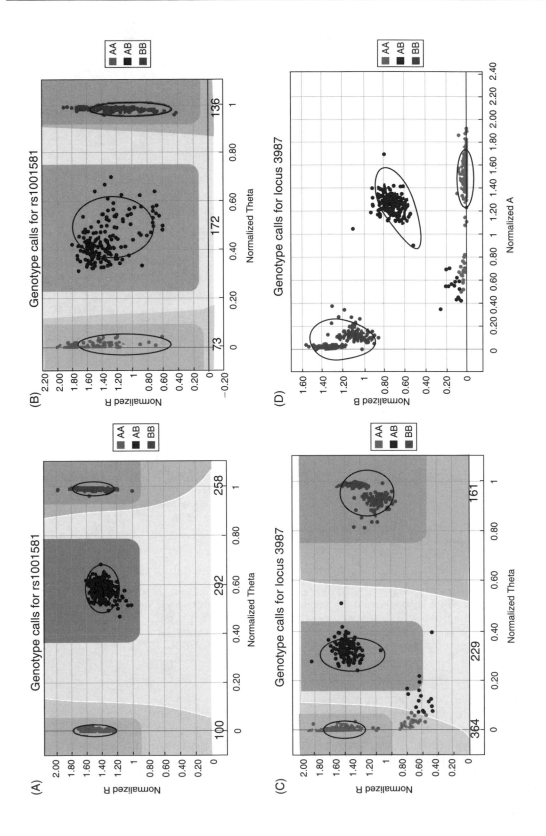

demonstrated that the genotyping error rate has an important role in influencing the power of association studies. Akey *et al.* (2001) demonstrated that genotyping errors rates as small as 3% can have serious affects on LD measures such as D' or r^2, and therefore influence the ability to identify associations of marker SNPs in LD with nongenotyped causal SNPs. It is therefore crucial to minimize, if not eliminate, the extent of genotyping error, particularly where an indirect approach is taken. For this reason, it is essential to include duplicate samples and related individuals (trios or nuclear families) in the assay, in order to check and quantify the degree of genotyping error. The success of an assay can also vary between SNPs genotyped. *Figure 4.5C* and *D* depict an extreme example, referred to as the 'extra' cluster phenomenon, where the assay fails. This phenomenon can be easily explained by the existence of a linked polymorphism under one of the allele-specific oligos (ASOs) decreasing the intensities and thereby creating this multiclustering pattern (see Protocol 8).

4.5 Summary and conclusions

The ability to analyze hundreds of thousands of SNPs that high-throughput genotyping technologies offer opens up the possibility of performing massive studies, but at the same time poses new challenges in terms of data management and data analysis. As a consequence of this, there is now a growing interest in searching for low-penetrance genes that are responsible for complex traits.

It is evident that the explosion of association studies and their huge promise have been possible because of the improved detail and resolution of genetic maps (HapMap project), the development of efficient bioinformatic tools and the fast development of high-throughput genotyping technologies (microarray technologies). With the use of high-throughput methods based on arrays, the bottleneck in the genotyping process has moved to a different stage in the process. A non-negligible problem in this new scenario now lies in the step of selection among large lists of SNPs to be used in the genotyping experiment. To address this problem, we propose a simple two-step protocol to select optimal SNPs (see Section 4.4.2). SNPs are chosen according to their minor allele frequency, validation status, probability of successful genotyping, potential functionality, and also the LD pattern with other SNPs to avoid redundant information.

Figure 4.5

GenCall software-produced plots of samples in polar coordinate representation. Each representation is a graph for a single locus with each 'dot' representing a DNA sample. The GenCall software has also automatically grouped the samples for each locus into the two homozygote clusters (light and medium gray) and the heterozygote cluster (black). (A) GenCall image obtained for a single SNP genotyped using high quality DNA samples, (B) GenCall image obtained for the same SNP using poor quality DNA samples. (C) GenCall image when multiclustering occurs, causing the assay to fail. (D) Cartesian coordinate representation of the same multiclustering pattern. (A color version of this figure is available on the book's website, www.garlandscience.com/9780415378536)

Under this protocol, which combines aspects of direct and indirect approaches, associations may be observed directly with functional variants, but also with those genotyped SNPs that are in LD with a causal variant. Furthermore, haplotype combinations are commonly used to localize a disease-conferring gene or locus because, as mentioned previously, it has been suggested that this strategy is more powerful in localizing susceptibility loci that confer moderate risk or protection for common disease than strategies using individual SNPs.

With regard to high-throughput genotyping methodology, the Illumina platform exhibits the following characteristics: highly efficient, accurate and robust SNP genotyping that is fast and cost effective. For any chosen platform, robustness and efficiency are the most important attributes. In the case of Illumina, both of these are near to 100%. Thanks to genotyping quality controls such as PCR contamination detection controls and internal assay controls to assess the assay and array hybridization at various experimental steps, including gDNA/oligo annealing, PCR, array hybridization, and imaging, the data is highly accurate.

In conclusion, while there may be several techniques that are appropriate for genotyping SNPs, the microarray approach using multiplexed assays seems to be the most successful one in terms of timeliness and cost effectiveness for genotyping a large number of SNPs in association studies.

Acknowledgments

This work is supported by grants from the Fundació La Caixa and Fundación Ramón Areces and grants PI020919 and PI041313 from the FIS (Fondo de Investigaciones Sanitarias). The Functional Genomics node of the INB (Instituto Nacional de Bioinformática) and CeGen (Centro Nacional de Genotipado) are funded by the Fundación Genoma España.

References

Ahmadi KR, Weale ME, Xue ZY et al. (2005) A single-nucleotide polymorphism tagging set for human drug metabolism and transport. Nat Genet 37: 84–89.

Airaksinen MS, Titievsky A, Saarma M (1999) GDNF family neurotrophic factor signaling: four masters, one servant? Mol Cell Neurosci 13: 313–325.

Akey JM, Zhang K, Xiong M, Doris P, Jin L (2001) The effect that genotyping errors have on the robustness of common linkage-disequilibrium measures. Am J Hum Genet 68: 1447–1456.

Ardlie KG, Kruglyak L, Seielstad M (2002) Patterns of linkage disequilibrium in the human genome. Nat Rev Genet 3: 299–309.

Auranen A, Song H, Waterfall C et al. (2005) Polymorphisms in DNA repair genes and epithelial ovarian cancer risk. Int J Cancer (published online 27 May 2005).

Barrett JC, Fry B, Maller J, Daly MJ (2005) Haploview: analysis and visualization of LD and haplotype maps. Bioinformatics 21: 263–265.

Bodmer WF (1997) Genetic diversity and disease susceptibility. Philos Trans R Soc Lond B Biol Sci 29; 352: 1045–1050.

Brookes A (1999) The essence of SNPs. Gene 234: 177–186.

Cannon-Albright LA, Goldgar DE, Meyer LJ et al. (1992) Assignment of a locus for familial melanoma MLM, to chromosome 9p13-p22. Science 258: 1148–1152.

Cartegni L, Chew SL, Krainer AR (2002) Listening to silence and understanding nonsense: exonic mutations that affect splicing. *Nat Rev Genet* 3:285–298.

Chasman D, Adams RM (2001) Predicting functional consequences of non-synonymous single nucleotide polymorphisms: structure-based assessment of amino acid variation. *J Mol Biol* **307**: 683–706.

Cheremushkin E, Kel A (2003) Whole genome human/mouse phylogenetic footprinting of potential transcription regulatory signals. *Pac Symp Biocomput* 8: 291–302.

Clark AG (2004) The role of haplotypes in candidate gene studies. *Genet Epidemiol* **27**: 321–333.

Colapietro P, Gervasini C, Natacci F, Rossi L, Riva P, Larizza L (2003) NF1 exon 7 skipping and sequence alterations in exonic splice enhancers (ESEs) in a neurofibromatosis 1 patient. *Hum Genet* **113**: 551–554.

Collins FS, Guyer MS, Chakravarti A (1997) Variations of a theme: Cataloguing human DNA sequence variation. *Science* **278**: 1580–1581.

Collins FS, Brooks LD, Chakravarti A (1998) A DNA polymorphism discovery resource for research on human genetic variation. *Genome Res* **8**: 1229–1231.

Conde L, Vaquerizas JM, Santoyo J, Al-Shahrour F, Ruiz-Llorente S, Robledo M, Dopazo J (2004) PupaSNP Finder: a web tool for finding SNPs with putative effect at transcriptional level. *Nucleic Acids Res* **32**: W242–W248 (web server issue).

Conde L, Vaquerizas J, Ferrer-Costa C, Orozco M, Dopazo J (2005) PupasView: a visual tool for selecting suitable SNPs, with putative pathologic effect in genes, for genotyping purposes. *Nucleic Acids Res* **33**: W501–W505 (web server issue).

Cutler DJ, Zwick ME, Carrasquillo MM *et al.* (2001) High-throughput variation detection and genotyping using microarrays. *Genome Res* **11**: 1913–1925.

Epstein MP, Veal CD, Trembath RC, Barker JNWN, Li C, Satten GA (2005) Genetic association analysis using data from triads and unrelated subjects. *Am J Hum Genet* **76**: 592–608.

Falk CT, Rubinstein P (1987) Haplotype relative risks: an easy reliable way to construct a proper control sample for risk calculations. *Ann Hum Genet* **51**: 227–233.

Ferrer-Costa C, Orozco M, de la Cruz X (2002) Characterization of disease-associated single amino acid polymorphisms in terms of sequence and structure properties. *J Mol Biol* **315**: 771–786.

Ferrer-Costa C, Orozco M, de la Cruz X (2004). Sequence-based prediction of pathological mutations. *Proteins* **57**: 811–819.

Gabriel SB, Schaffner SF, Nguyen H *et al.* (2002) The structure of haplotype blocks in the human genome. *Science* **296**: 2225–2259.

Glazier AM, Nadeau JH, Aitman TJ (2002) Finding genes that underlie complex traits. *Science* **298**: 2345–2349.

Goñi JR, de la Cruz X, Orozco M (2004) Triplex-forming oligonucleotide target sequences in the human genome. *Nucleic Acids Res* **32**: 354–360.

Guerois R, Nielsen JE, Serrano L (2002) Predicting changes in the stability of proteins and protein complexes: a study of more than 1000 mutations. *J Mol Biol* **320**: 369–387.

Hall JM, Lee MK, Newman B, Morrow JE, Anderson LA, Huey B, King MC (1990) Linkage of early-onset familial breast cancer to chromosome 17q21. *Science* **250**: 1684–1689.

Healey CS, Dunning AM, Teare MD *et al.* (2000) A common variant in BRCA2 is associated with both breast cancer risk and prenatal viability. *Nat Genet* **26**: 362–364.

Hill WG, Robertson A (1968). Linkage disequilibrium in finite populations. *Theor Appl Genet* **38**: 226–231.

Hirschhorn JN, Daly MJ (2005) Genome-wide association studies for common diseases and complex traits. *Nat Rev Genet* **6**: 95–108.

Hong YC, Lee KH, Kim WC, Choi SK, Woo ZH, Shin SK, Kim H (2005) Polymorphisms of XRCC1 gene, alcohol consumption and colorectal cancer. *Int J Cancer* **116**: 428–432.

Hoogendoorn B, Coleman SL, Guy CA, Smith K, Bowen T, Buckland PR, O'Donovan MC (2003) Functional analysis of human promoter polymorphisms. *Hum Mol Genet* **12**: 2249–2254.

Hosoe S, Brauch H, Latif F *et al.* (1990) Localization of the von Hippel–Lindau disease gene to a small region of chromosome 3. *Genomics* **8**: 634–640.

Houlston RS, Peto J (2004) The search for low-penetrance cancer susceptibility alleles. *Oncogene* **23**: 6471–6476.

Hudson TJ (2003) Wanted: regulatory SNPs. *Nat Genet* **33**: 439–440.

Hugot JP, Chamaillard M, Zouali H *et al.* (2001). Association of NOD2 leucine-rich repeat variants with susceptibility to Crohn's disease. *Nature* **411**: 599–603.

Kel AE, Gößling E, Reuter I, Cheremushkin E, Kel-Margoulis OV, Wingender E (2003) MATCHTM: a tool for searching transcription factor binding sites in DNA sequences. *Nucl Acids Res* **31**: 3576–3579.

Kenealy SJ, Babron MC, Bradford Y *et al.* (2004) A second-generation genomic screen for multiple sclerosis. *Am J Hum Genet* **75**: 1070–1078.

Kennedy GC, Matsuzaki H, Dong S *et al.* (2003) Large-scale genotyping of complex DNA. *Nat Biotechnol* **21**: 1233–1237.

Knight JC (2003) Functional implications of genetic variation in non-coding DNA for disease susceptibility and gene regulation. *Clin Sci* **104**: 493–501.

Krawczak M, Reiss J, Cooper DN (1992) The mutational spectrum of single base-pair substitutions in mRNA splice junctions of human genes: causes and consequences. *Hum Genet* **90**: 41–54.

Lindblom A, Tannergard P, Werelius B, Nordenskjold M (1993) Genetic mapping of a second locus predisposing to hereditary non-polyposis colon cancer. *Nat Genet* **5**: 279–282.

Manie S, Santoro M, Fusco A, Billaud M (2001) The RET receptor: function in development and dysfunction in congenital malformation. *Trends Genet* **17**: 580–589.

Mathew CG, Chin KS, Easton DF *et al.* (1987) A linked genetic marker for multiple endocrine neoplasia type 2A on chromosome 10. *Nature* **328**: 527–528.

Miller MP, Kumar S (2001) Understanding human disease mutations through the use of interspecific genetic variation. *Hum Mol Genet* **10**: 2319–2328.

Morris RW, Kaplan NL (2002) On the advantage of haplotype analysis in the presence of multiple disease susceptibility alleles. *Genet Epidemiol* **23**: 221–233.

Mulligan LM, Kwok JB, Healey CS *et al.* (1993) Germ-line mutations of the RET proto-oncogene in multiple endocrine neoplasia type 2A. *Nature* **363**: 458–460.

Ng PC, Henikoff S (2001) Predicting deleterious amino acid substitutions. *Genome Res* **11**: 863–874.

Pennacchio LA, Rubin EM (2001) Genomic strategies to identify mammalian regulatory sequences. *Nat Rev Genet* **2**: 100–109.

Pritchard JK, Cox NJ (2002) The allelic architecture of human disease genes: common disease-common variant ... or not? *Hum Mol Genet* **11**: 2417–2423.

Pritchard JK, Przeworski M (2001) Linkage disequilibrium in humans: models and data. *Am J Hum Genet* **69**: 1–14.

Prokunina L, Castillejo-Lopez C, Oberg F *et al.* (2002) A regulatory polymorphism in PDCD1 is associated with susceptibility to systemic lupus erythematosus in humans. *Nat Genet* **32**: 666–669.

Ramensky V, Bork P, Sunyaev S (2002) Human non-synonymous SNPs: server and survey. *Nucleic Acid Res* **30**: 3894–3900.

Reich DE, Cargill M, Bolk S *et al.* (2001) Linkage disequilibrium in the human genome. *Nature* **411**: 199–204.

Reich DE, Schaffner SF, Daly MJ, McVean G, Mullikin JC, Higgins JM, Richter DJ, Lander ES, Altshuler D (2002) Human genome sequence variation and the influence of gene history, mutation and recombination. *Nat Genet* **32**: 135–142.

Risch N, Merikangas K (1996) The future of genetic studies of complex human diseases. *Science* **13**: 1516–1517.

Sawcer SJ, Maranian M, Singlehurst S *et al.* (2004) Enhancing linkage analysis of complex disorders: an evaluation of high-density genotyping. *Hum Mol Genet* **13**: 1943–1949.

Schwartz S, Kent WJ, Smit A, Zhang Z, Baertsch R, Hardison RC, Haussler D, Miller W (2003) Human–mouse alignments with BLASTZ. *Genome Res* **13**: 103–107.

Seizinger BR, Smith DI, Filling-Katz MR *et al.* (1991) Genetic flanking markers refine diagnostic criteria and provide insights into the genetics of Von Hippel–Lindau disease. *Proc Natl Acad Sci USA* **88**: 2864–2868.

Shen R, Fan JB, Campbell D *et al.* (2005) High-throughput SNP genotyping on universal bead arrays. *Mutation Res* **573**: 70–82.

Simpson NE, Kidd KK, Goodfellow PJ *et al.* (1987) Assignment of multiple endocrine neoplasia type 2A to chromosome 10 by linkage. *Nature* **328**: 528–530.

Spielman RS, McGinnis RE, Ewens WJ (1993) Transmission test for linkage disequilibrum: the insulin gene region and insulin-dependent diabetes mellitus (IDDM). *Am J Hum Genet* **52**: 506–516.

Strittmatter WJ, Saunders AM, Schmechel D, Pericak-Vance M, Enghild J, Salvesen GS, Roses AD (1993). Apolipoprotein E: high-avidity binding to beta-amyloid and increased frequency of type 4 allele in late-onset familial Alzheimer disease. *Proc Natl Acad Sci USA* **90**: 1977–1981.

Sunyaev S, Ramensky V, Koch I, Lathe W, Kondrashov AS, Bork P (2000a) Prediction of deleterious human alleles. *Hum Mol Genet* **10**: 591–597.

Sunyaev S, Ramensky V, Bork P (2000b) Towards a structural basis of human non-synonymous single nucleotide polymorphisms. *TIG* **16**: 198–200.

The International HapMap Consortium (2003) The International HapMap Project. *Nature* **426**: 789–796.

Wall JD, Pritchard JK (2003) Haplotype blocks and linkage disequilibrium in the human genome. *Nat Rev Genet* **4**: 587–597.

Wang DG, Fan JB, Siao CJ *et al.* (1998) Large-scale identification, mapping, genotyping of single-nucleotide polymorphisms in the human genome. *Science* **280**: 1077–1082.

Wang N, Akey JM, Zhang K, Chakraborty R, Jin L (2002) Distribution of recombination crossovers and the origin of haplotype blocks: The interplay of population history recombination, mutation. *Am J Hum Gen* **71**: 1227–1234.

Wingender E, Chen X, Hehl R *et al.* (2000) TRANSFAC: an integrated system for gene expression regulation. *Nucleic Acids Res* **28**: 316–319.

Wooster R, Neuhausen SL, Mangion J *et al.* (1994) Localization of a breast cancer susceptibility gene BRCA2, to chromosome 13q12-13. *Science* **265**: 2088–2090.

Yan YW, Velculescu VE, Vogelstein B, Kinzler KW (2002) Allelic variation in human gene expression. *Science* **297**: 1143.

Zedenius J, Wallin G, Hamberger B, Nordenskjold M, Weber G, Larsson C (1994) Somatic and MEN 2A de novo mutations identified in the RET proto-oncogene by screening of sporadic MTCs. *Hum Mol Genet* **3**: 1259–1262.

In vitro analysis of gene expression

5

Donling Zheng, Chrystala Constantinidou, Jon L. Hobman, and Steve D. Minchin

5.1 Introduction

The ability to profile global changes in the levels of gene expression under different conditions makes microarrays the method of choice in many areas of modern biology. They have been used to study differential gene expression in different growth conditions, for example *Escherichia coli* grown in minimal versus rich media (Tao *et al.*, 1999) and glucose versus acetate as the sole carbon source (Oh *et al.*, 2002). In addition, changes in response to chemical exposure or environmental stress have also been investigated; for example, the response of *E. coli* to hydrogen peroxide (Zheng *et al.*, 2001) and to UV light (Courcelle *et al.*, 2001). An alternative approach is to compare expression profiles between wild type and mutant stains. Gene expression profiling of *E. coli* mutants defective in IHF (Arfin *et al.*, 2000) and H-NS (Hommais *et al.*, 2001), have identified genes that are either positively or negatively regulated in the absence of these transcription factors.

A novel technique, run-off transcription microarray analysis (ROMA), has been developed by John Helmann and colleagues which combines the conventional *in vitro* run-off transcription assay with macroarray analysis (Cao *et al.*, 2002). An *in vitro* run-off transcription reaction containing genomic DNA template and RNA polymerase in the presence and absence of a particular transcription factor is used to generate ^{33}P-labeled RNA. The labeled RNA is then analyzed by hybridization to macroarray membranes. Direct visual comparison of RNA polymerase with and without the specific transcription factor allows the identification of genes subject to regulation by that factor. The ROMA procedure has been further developed by Zheng *et al.* (2004) to exploit glass-based oligonucleotide microarrays. It is important to note that if you wish to get the maximum information regarding a particular regulon or transcription factor the use of several approaches is advisable. The three main approaches would be *in vivo* transcription profiling, ROMA, and genome sequence analysis. In both the study by Cao *et al.* (2002) and the one by Zheng *et al.* (2004) ROMA was used to complement conventional *in vivo* transcription profiling and genome sequence analysis.

The three possible approaches, *in vivo* transcription profiling, ROMA, and genome sequence analysis, have different advantages and disadvantages. Genome sequence analysis has been successful for identifying promoters regulated by alternative sigma factors as well as conventional transcription factors (Petersohn *et al.*, 1999; Paget *et al.*, 2001; Cao *et al.*, 2002). However, a significant disadvantage is that the DNA sequence that is bound by the transcription factor needs to be characterized; it is therefore hard to apply this method to a novel regulator. In addition, binding sites for any protein factor are not all identical and some will match the consensus better than others, so a decision must be made as to what is a significant match to the consensus sequence. Therefore, some poorly conserved sites may be missed or pseudo sites that have a significant match to consensus but have no biological role may be identified. *In vivo* microarray experiments involving a direct comparison of the transcriptome of wild-type and mutant strains provide experimental evidence for a regulon. However, great care must be taken when designing the experiments to minimize the interference of other factors. Genes under complex regulation could be missed. In ROMA, promoters could be missed because they require additional positive activators, negative supercoiling, or other factors that are missing from the defined *in vitro* system. However, compared with the other two approaches, ROMA not only provides experimental evidence for a particular regulon but by its very nature will only identify direct transcriptional effects by a specific factor. Finally, these approaches are complementary to each other, thus a combination of them is helpful to obtain a more precise picture of the function of a regulator. We have developed and exploited ROMA as a tool to study the CRP regulon of *E. coli* (Zheng *et al.*, 2004).

5.2 Scientific background

The *E. coli* cyclic AMP receptor protein (CRP) regulates transcription initiation for more than 100 genes mainly involved in catabolism of carbon sources other than glucose (Kolb *et al.*, 1993). CRP regulates transcription by binding to CRP binding sites within target promoters. The CRP binding site is well characterized and has the consensus sequence 5'-AAA**TGTGA**TCTAGA**TCACA**TTT-3'). During glucose starvation the level of the messenger molecule cyclicAMP (cAMP) increases and CRP binds cAMP to form a CRP–cAMP complex; it is the CRP–cAMP complex that is active in gene regulation. At most promoters, CRP works by activating transcription, but it is known to repress transcription from a small subset of promoters. At CRP-dependent promoters CRP activates transcription by making direct protein–protein contacts with RNA polymerase (RNAP). Class I CRP-dependent promoters, e.g. *lac*, contain a single CRP binding site upstream of the DNA binding sites for RNAP. At these promoters, CRP activates transcription by interacting with the C-terminal domain of the RNAP α subunit (αCTD) via a surface-exposed patch, known as activating region 1 (AR1). Class II CRP-dependent promoters, e.g. *gal*P1, contain a CRP binding site overlapping the –35 hexamer for RNAP. At these promoters, CRP makes multiple interactions with RNAP, including between AR1 and αCTD, the activating region 2 (AR2) and the N-terminal domain of the α subunit (αNTD) and between activating region 3 (AR3) and the RNAP σ^{70} subunit region 4. Class III promoters contain tandem CRP sites; CRP activation at these promoters involves a combination of class I and class II mechanisms. Additionally, many CRP-dependent promoters are co-dependent on a second activator (Busby and Ebright, 1999).

The CRP regulon is one of the best characterized regulons in *E. coli*; it has been subject to extensive biochemical and genetic analysis. Microarray technology has been used to help define and fully characterize this complex regulon.

5.3 Design of the experiment

In vivo transcription profiling would appear to be a good approach to fully characterize the CRP regulon, but because of the level of complexity of this regulon this approach has only limited value. Many CRP-regulated promoters require additional activator proteins, whereas others are subject to repression; therefore individual operons often require multiple signals for activation (Busby and Ebright, 1999). To turn on all genes that have the potential to be activated by CRP would require a complex mix of signals and this would not be practical. If multiple signals, such as complex sugar mixtures, were added to the system it would almost certainly lead to changes in gene expression independent of CRP. It is therefore almost impossible to design an *in vivo* microarray experiment to distinguish direct effects from indirect effects. ROMA is better suited for identifying the regulon of a factor that is part of a complex regulatory network. However, combining several approaches such as *in vivo* profiling, sequence analysis and ROMA increases the likelihood of obtaining a more accurate definition of a complex regulon.

Cao *et al.* initially developed ROMA to exploit nylon-based macroarrays (Cao *et al.*, 2002). We have modified the original ROMA protocol in order to use oligonucleotide-based microarrays (glass slides with oligonucleotide probes). This type of array has several advantages over the membrane-based macroarrays. The glass-based arrays allow the utilization of fluorescence labeling, enabling test and control samples to be differentially labeled and then hybridized to the same array to allow direct comparison of the two samples on the same slide. A second benefit of microarrays is that they can accommodate a larger number of probes and therefore allow analysis of a larger number of genes. Finally, the use of fluorescence rather than radioisotopes produces higher resolution and lower background as well as being safer and easier to handle. Oligonucleotide-based microarrays do have some disadvantages, the first is that the transcription process is not as robust *in vitro* compared to *in vivo* and since the oligonucleotides represent only a small section of the transcript there may be some signal loss due to premature termination of transcription before the section of gene represented by the oligonucleotide is transcribed. In addition, unlike PCR products that contain both DNA strands (sense and antisense), the oligunucleotide array has been developed for 'standard' *in vivo* transcription profiling where a cDNA copy of the transcript is generated and hybridized to the array. The oligonucleotides are therefore sense, i.e. the same sequence as the RNA; it is therefore not possible to label the RNA during transcription and hybridize it directly to the array. Instead, several purification steps are required to isolated the RNA and generate labeled cDNA suitable for hybridization to the array.

Our analysis using ROMA has been complemented by an *in vivo* transcription profiling approach. Both a wild-type *E. coli* strain and a mutant strain in which the *crp* gene had been deleted (Δcrp) were grown in minimal media containing fructose and then pulsed with glucose. CRP should be active when *E. coli* is growing on fructose, whereas in the presence of glucose cAMP levels will be too low for formation of the active cAMP–CRP complex. There may be some CRP-independent effects of glucose and therefore in addition to analyzing the transcription profile in the presence and absence of glucose we have also analyzed glucose regulation in a Δcrp strain. We have therefore defined the CRP-regulated genes as those genes regulated by glucose in the wild-type strain, but not regulated in the Δcrp strain.

5.4 Data acquisition

The basic experimental procedure requires purified genomic DNA template, purified RNA polymerase, and transcription factors. In the case study described below we have used purified wild-type and mutant derivatives of CRP. RNA is then purified from *in vitro* transcription reactions. The RNA is reverse transcribed and fluorescently labeled. The labeled cDNA is then hybridized to the array and the array scanned to give a readout of the amount of labeled cDNA for each gene which, in turn, presents the level of transcription of that gene in the *in vitro* reaction.

MG1655 genomic DNA, digested with *Eco*RI, was used as template for *in vitro* transcription by RNAP containing σ^{70} in the presence of NTPs (ATP, CTP, GTP, and TTP). Transcription factors were applied to the reactions as appropriate. Two reaction conditions were set up for each experiment, with one a control reaction and the other a test reaction. After the transcription reaction, the RNA was purified from the reaction mixture and the genomic DNA templates removed by DNase I digestion. The purified RNA transcripts with and without DNase treatment were checked on a 1% agarose gel a typical gel is shown in *Figure 5.1*. The genomic DNA templates were shown as fragments more than 500 bp in length (lane 2) and generated RNA transcripts that were 500–2000 bases long according to DNA markers (lanes 3 and 4). After RNA purification and DNase digestion, the bands corresponding to template DNA disappeared (lanes 5 and 6), which indicated the complete removal of the genomic templates.

The RNA transcripts were then reverse transcribed into aminoallyl-labeled cDNA by reverse transcriptase in the presence of a dNTP mixture containing aminoallyl dUTP and the labeled cDNA subsequently coupled with monoreactional CyDyes™ (Cy3 or Cy5). The aminoallyl dUTP is incorporated into DNA by reverse transcriptase with the same efficiency as dTTP and its amine residue can interact with the ester group of the monoreactional CyDye. Products from control reactions were labeled with Cy3 and those from test reactions labeled with Cy5. The fluorescence-labeled cDNA products were purified and co-hybridized onto an *E. coli* oligo array. The resulting signals were digitally visualized, with Cy3 shown in green at 532 nm and Cy5 shown in red at 635 nm on a transformed image. The transcription levels of genes under different conditions can be deduced from the intensities at Cy3 and Cy5 channels.

The *E. coli* oligonucleotide arrays were produced by the Exgen group at the University of Birmingham (UBEC) and contained probes for genes from both *E. coli* K12 and *E. coli* O157. The intensities of different fluorescence at

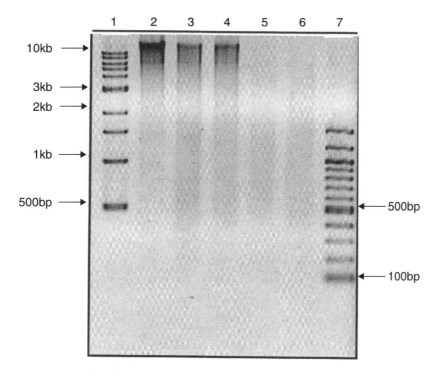

Figure 5.1

Genomic transcription. The gel image shows genomic run-off transcripts before and after purification. Lanes 1 and 7 were DNA markers, the bands corresponding to certain lengths of DNA are indicated; lane 2 EcoRI digested genomic DNA templates; lanes 3 and 4 transcripts from two reactions immediately after transcription; lanes 5 and 6 transcripts after RNA purification and DNA removal.

each spot were analyzed using GenePix Pro 3.0 and further analyzed using GeneSpring 4.2.1.

5.5 Theory of data analysis

5.5.1 Identification of differentially transcribed genes

Differentially transcribed genes were selected using an outlier iteration method (Loos *et al.*, 2001; Britton *et al.*, 2002). The data for each gene were averaged and the geometric mean and standard deviation were calculated for the entire population. Any gene with a log ratio more than 3 standard deviations away from the mean was considered an outlier. Outliers were then removed from the population and retained within the differentially expressed subset and the mean and standard deviations were recalculated for the rest of the data. The step was repeated until few or no outliers were detected. The 99% predictive interval (PI) was set for the final cut-off ratio to define the remaining differentially transcribed genes within the now symmetrical and well-defined distribution.

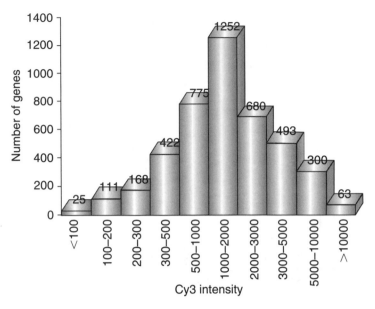

Figure 5.2

Distribution of gene transcription levels by RNAP only. The chart shows the summary of Cy3 intensities determined for 4289 *E. coli* K-12 genes in ROMA experiments. The relative transcription levels driven by RNAP only can be roughly deduced from Cy3 intensities.

The *in vitro* transcription profiles of CRP derivatives were compared with wild-type CRP, and significant changes were identified by outlier iteration. The average of the log ratios for each gene from the CRP derivative experiments was normalized to the wild-type CRP experiment. The geometric mean and standard deviation were calculated for the entire population. The outliers were selected as stated above and the 95% PI was set as a final cut-off. The genes with a log ratio change between two experiments falling in the outlier group or beyond the PI were regarded as differentially regulated by CRP mutants and wild-type CRP.

5.6 Data analysis

Potential random and systematic errors mean that in order to fully exploit the microarray data, rigorous statistical analysis must be completed. The major source of random error is caused by different biological samples that can be influenced by genetic or environmental factors. Random error can also come from replicate experiments on identical samples due to handling errors, but compared with biological variation it is a minor error. The systematic error is dependent on spotting, scanning, and dye labeling, resulting in a constant tendency to over- or underestimate true values (reviewed by Nadon and Shoemaker, 2002). For a simple two-condition comparison, a fixed fold-change (usually 2-fold) was used to identify differential expression in early microarray experiments (Schena *et al.*, 1996;

DeRisi *et al.*, 1997). However, it fails to account for the measurement errors and ignores the fact that a difference less than the cut-off may be biologically significant. In addition, genes with low transcription rates tend to have high ratios. A more sophisticated approach involves calculating the mean and standard deviation (SD) of the distribution of log ratios and defines global or intensity-specific thresholds and confidence. Moreover, the availability of replicate data makes it possible to use statistical methods to identify differentially expressed genes. Here, an outlier method and two-statistic *t*-test methods to identify differentially transcribed genes in the CRP experiments are discussed.

5.6.1 Outlier method

For a standard microarray experiment, the distribution of log ratios is approximately normal. Therefore, a more precise cut-off can be determined by confidence intervals calculated from the mean (should be around zero) and SD of all log ratio values. In addition, intensity-specific thresholds can be further calculated based on the local structure of a data set, which minimizes the effect of variability caused by low intensities (Yang and Speed, 2002). However, these methods are not ideal for a data set that distributes asymmetrically, as the mean deviates from zero. Analysis of gene regulation by CRP is an example of a system when this type of statistical analysis is not ideal because CRP predominantly upregulates transcription (*Figure 5.3A*). Therefore, an alternative method was employed to determine the cut-off.

The outlier method is an adaptation of the normalization procedure which gives an average log ratio centered on zero (Loos *et al.*, 2001; Britton *et al.*, 2002). For a given data set, the extreme high and low values of log ratios outside of defined range (e.g. ± 3 SD) are removed iteratively until the mean stabilizes and approximates to zero. As a result, the distribution of data is approximately normal and a predictive interval (99% in this study) is then set to further select significant values (*Figure 5.3B*). The outliers beyond the thresholds are regarded as deferentially expressed genes.

5.6.2 Standard *t*-test

The *t*-test is a simple, statistically based method for detecting differentially expressed genes from replicate experiments. A standard *t*-test is integrated into the GeneSpring software, which is based on a null hypothesis that there is no expression difference, i.e. the true ratio is equal to 1.0. The resulting *p* value is the probability that the ratio is equal to 1.0. In this study, a *p* value was calculated for each gene, based on three replicate data sets, and a *p* value less than 0.05 was taken as significant. A large number of differentially regulated genes were identified based on $p < 0.05$, some with a ratio close to 1.0. This is because the variances estimated from each gene are not stable. For example, when the estimated variance is small, by chance, the *t* value can be large even though the corresponding fold change is small. A further 1.5-fold change can be applied to restrain the selection. Almost all genes identified by the outlier method had *p* values less than 0.05 (15 genes with $0.05 < p < 0.12$), which indicates that those outliers are statistically significant.

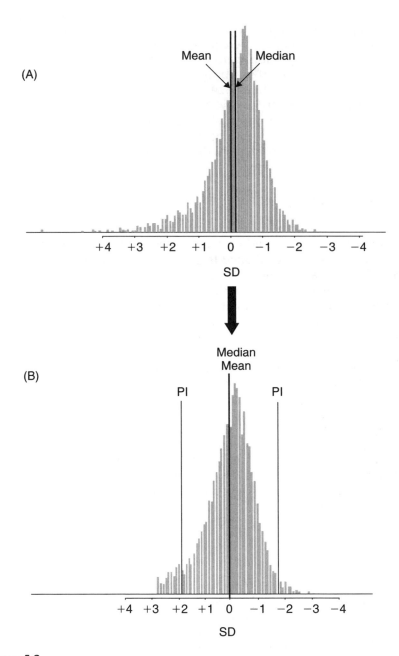

Figure 5.3

CRP data log ratio distribution and outlier iteration result. (A) The asymmetric log ratio distribution of a data set from wild-type CRP experiments. The mean value (dotted line) is deviated from the median value (continuous line) and the standard deviation (SD) is large. (B) The log ratio distribution after outlier iteration. The mean and the median are equal (zero) and the revised SD becomes smaller. A predictive interval (PI, dotted lines) is set for further selection.

Combined with fold-change threshold, the *t*-test also effectively determined the differential transcribed genes but with more false positives. In addition, the *t*-test is vulnerable to instability of variances mainly caused by a small number of replicates. A solution is to compute a global *t*-test, using an estimate of variance across all genes, if it is assumed that the variance is homogeneous between different genes (Arfin *et al.*, 2000; Tanaka *et al.*, 2000). In practice, it may be difficult to apply if the variance is not truly constant for all genes. Moreover, the *p* value is based on a normality assumption, which is also dependent on sample size. Therefore, a standard *t*-test is not satisfactory for analysis of results with few replicates.

5.6.3 SAM statistic analysis

The 'significance analysis of microarrays' (SAM) is a modified *t*-test (*S*-test) which involves the addition of a small positive constant to the denominator of the gene-specific *t*-test (a fudge factor). With this modification, genes with small fold changes will not be selected as significant. It also utilizes permutations of the repeated measurements to estimate the percentage of genes identified by chance (false discovery rate, FDR) for a particular threshold. The user can freely adjust the cut-off threshold to identify larger or smaller sets of genes, and the FDRs are calculated for each set. SAM can be freely downloaded from the internet and runs as a convenient add-in within Excel.

In this study, SAM was also used to analyze the CRP data sets. Since SAM does not have a normalization function, those data sets need to be normalized on GeneSpring before further analysis on SAM. The data from six spots for each gene in a specific spreadsheet format were loaded and analyzed using SAM. A fold change more than 1.5 and a FDR less than 15% were used. SAM resulted in the same hierarchy as the outlier method in terms of the number of upregulated genes observed by wild-type CRP and the two activating region mutants, indicating that SAM does not require a normal distribution assumption.

Although SAM uses a modified *t*-test, it also showed a high false positive rate in the CRP case study. This may be because SAM, like the standard *t*-test, is sensitive to sample variances resulting from a small sample size. Therefore, SAM also requires a large number of replicates.

5.6.4 Operon organization

The operon organization is an important biological factor that should be considered during data analysis. The ratio of the promoter proximal gene determines the actual regulation level at promoter transcription initiation, while the distal genes can show erratic values. However, because of the measurement errors, it is hard to distinguish a low level of regulation from nonregulation. In this case, transcription levels of the following genes might help to make a decision. For example, in the *modABC* operon, the average ratio of *modA* gene was 1.57, which is slightly below the 99% PI threshold, but both downstream genes, *modB* and *modC*, have ratios significantly above the PI (2.15 and 2.33, respectively). It would therefore appear that the *modABC* operon is actually regulated by CRP. In addition, analysis

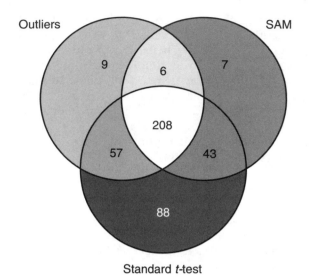

Figure 5.4

CRP-regulated genes identified by each of three methods. This Venn diagram shows differential transcribed genes in response to wild-type CRP using three methods. A total of 281 genes were identified by outlier method (light gray), 264 genes by SAM (fold change >1.5 and FDR $<15\%$; medium gray) and 396 genes by a standard t-test ($p <0.05$ and fold change >1.5; dark gray). There are 208 genes identified by all three approaches, shown in white in the middle.

of all members of an operon can be used to discover false positives, if a promoter proximal gene is not regulated but a promoter distal gene is regulated, the promoter distal gene may represent a false positive.

5.6.5 Comparison of the outlier, standard t-test, and SAM methods

The comparison of the results by the three methods when used to analyze the CRP data set from the case study below are shown in *Figure 5.4* and *Table 5.1*. All methods successfully identified a common set of 208 genes that in general represent the genes that are most significantly regulated. Each method also identified a number of genes that were not identified by one or both of the other methods and this reflects their general ability to identify low levels of regulation as significant. The standard t-test combined with 1.5-fold cut-off identified more genes that were deemed significant, but appeared to result in a higher false positive rate than the outlier method (*Table 5.1*). The outlier method was therefore chosen as the best choice to analyze the CRP data sets in the case study below. This method is easy to perform and independent of data distribution. The disadvantage is that there is no statistical value to indicate the level of confidence, and that it ignores the significance of variations. In contrast, a t-test gives a p value to judge the level of confidence when determining whether a gene is differentially expressed. In addition, SAM utilizes permutations of the repeated

Table 5.1 Summary of results by three methods

Method	Genes (operons)	False positive genes*	False positive rate
Outlier method	280 (168)	29	10.4%
Standard *t*-test and 1.5-fold cutoff	396 (279)	62	15.7%
SAM and 1.5-fold cutoff	264 (191)	44	16.7%

* Genes involved in an operon where the promoter-proximal gene was not regulated in ROMA experiments.

measurements to estimate FDR and combines a cut-off threshold to reject small fold changes that could be selected as significant by a normal *t*-test. SAM is also applicable when the data is distributed with a degree of skewness. However, all *t*-tests require a large number of replicates since they are susceptible to unstable variance and the *p* values are estimated from the normal distribution.

There is no unique method to analyze all array data. Some methods are better for one data set, but may not be good for other data sets. In practice, it is best to use different methods to determine which works best with a particular data set. Here only statistical analysis for simple two-condition comparison is discussed. Multiconditional comparison involves more complicated statistic models, and relies on stable data distribution.

5.7 Summary of results, conclusions, and suggestions for general implementation of the case study

5.7.1 Characterization of the CRP regulon by ROMA

In order to assess the feasibility of using an oligonucleotide-based glass microarray to analyze *in vitro* transcription products we first set up an *in vitro* transcription reaction containing only RNA polymerase. The resulting RNA transcript was purified and Cy3-labeled cDNA prepared. This labeled cDNA was then hybridized to the array. Good hybridization signals were observed at spots containing oligonucleotides representing MG1655 ORFs. No hybridization occurred at blank spots and spots containing negative control genes, which indicated that there was little nonspecific binding. No hybridization was observed to the majority of spots representing ORFs from the O157:H7 strains or their plasmids. However, a few spots containing oligonucleotides representing genes that should be O157:H7 specific did give a hybridization signal. This might be due to cross-hybridization of a MG1655 gene product with an O157:H7 oligonucleotide representing a homologous gene.

Since the DNA templates used in the *in vitro* transcription reactions were purified from the MG1655 strain, only the 4289 *E. coli* K12 MG1655 genes were analyzed. The average transcription levels of these genes are shown in *Figure 5.2* (page 114). More than 4100 genes showed detectable transcripts by RNAP only (background subtracted Cy3 intensity > 100), which indicates that the sensitivity of the system is sufficiently high to detect low level of transcription. There are about 3000 genes with consistently high intensities (>1000), which is nearly 70% of all genes, and more than 300 genes with very high intensities (>5,000). This indicates that 70% of *E. coli* genes can be transcribed relatively efficiently by RNAP containing σ^{70} *in vitro* even in the absence of any other transcription factor.

Having verified that glass-based oligonucleotide arrays could be exploited using the ROMA technique, ROMA was used to characterize the CRP regulon of *E. coli* K12 strain MG1655, this work is reviewed below and was published in *Nucleic Acids Research* (Zheng *et al.*, 2004). Initially the transcription levels of genes with RNA polymerase in the presence and absence of the CRP was determined. Note that cAMP was present in all experiments to ensure that CRP is in the active state. As would be expected levels of transcription of most genes was not effected by the presence of CRP. However, nearly 7% the *E. coli* genes were subject to regulation by CRP in this *in vitro* experiment, in the majority of cases CRP upregulated transcription. A total of 280 genes, present in 188 different operons, had significantly higher transcriptional levels and 20 genes, in 16 operons, had reduced transcriptional levels. Since it is expected that promoter proximal genes should always be upregulated as well as promoter distal genes 12 promoters were excluded from the initial set because promoter proximal genes were not upregulated and so these were expected to be false positives. This left 176 operons identified by ROMA as upregulated by CRP. Although ROMA identified 24 well-characterized CRP regulons, several other known CRP-regulated operons were not identified in our ROMA experiments. There are several reasons that could account for these false negatives. First, some operons require other factors for CRP to activate transcription, e.g. AraC at the *araBAD* promoter (Lobell and Schleif, 1991), which are missing in the ROMA system.

During activation of transcription amino acid residues in two regions of CRP, activating region 1 (AR1) and activating region 2 (AR2), make specific interactions with the transcription machinery. AR1 is required for activation at both class I and class II CRP-dependent promoters, whereas AR2 is only required for activation at class II promoters. Derivatives of CRP with mutations in AR1 will have reduced activity at most CRP-dependent promoters, whereas, AR2 mutants will have reduced activity at class II promoters but give wild-type levels of activity at class I promoters. The effect, on transcription, of two CRP derivatives containing a mutation within either AR1 or AR2 was analyzed. HL159-CRP, which carries a His-to-Leu mutation at position 159 within activation region 1 (AR1), failed to activate transcription of 86% (151/176) of the CRP-dependent genes identified in the initial ROMA experiment. KE101-CRP, which contains a Lys-to-Glu mutation at position 101 within activing region 2 (AR2), failed to activate transcription at 59% (104/176) of CRP-dependent operons. It is therefore possible to predict that 59% of the CRP-dependent promoters identified in the ROMA study contain class II CRP binding sites.

As stated above, 24 of the CRP-activated promoters identified by ROMA have been previously characterized and the transcription start site and position of the CRP binding site mapped. Of these 24 operons, seven contain a single class I CRP binding site, nine contain a single class II CRP binding site, and eight contain a class III promoter with tandem CRP binding sites. These 24 promoters were therefore used to assess the predictive power of the ROMA experiments with mutant derivatives. In the ROMA assay the CRP with a defective AR1 failed to activate 22 out of 24 of these promoters. The two promoters that were still subject to activation were both class III promoters with complex promoter regions which may explain why an AR1 mutant could still function. The CRP with a defective AR2 still activated class I promoters in the ROMA assay but, as would be predicted, failed to activate all class II promoters and half of the class III promoters.

ROMA identified 16 promoters that were downregulated by CRP. Sequence analysis revealed that 11 of these promoters contained an identifiable CRP binding site. At most promoters the CRP binding site overlaps the region between the –35 and –10 hexamers; therefore repression at these operons might involve a simple blocking mechanism. The AR1 and AR2 mutants did not significantly affect the level of repression at most promoters, but four operon repression levels were affected by the AR1 mutation, suggesting some role for AR1 at these operons.

5.7.2 Comparison of ROMA with *in vivo* transcriptional profiling

In addition to using ROMA to characterize the CRP regulon we have used *in vivo* transcription profiling. *E. coli* strains were grown in minimal media containing fructose and then pulsed with glucose. CRP-dependent genes should be upregulated in the absence of glucose. Because glucose may have a CRP-independent effect on transcription the results are compared between a wild-type and *crp* deletion (Δ*crp*) strain. Any CRP-independent glucose effect will be seen in both strains. The CRP-regulated genes were therefore defined as genes regulated by glucose in the wild-type strain, but not regulated in the Δ*crp* strain. Only 17 operons were identified as being activated by CRP and six operons were repressed by CRP. Our *in vivo* transcription profiling failed to identify many operons identified by ROMA and also many known CRP-dependent promoters. In addition to the *in vivo* study completed by Zheng *et al.* (2004), Gosset *et al.* (2004) completed a similar *in vivo* transcriptomic study and identified 39 CRP-activated promoters and 19 CRP-repressed promoters. Of the 58 operons identified by Gosset *et al.*, only eight were identified in the *in vivo* transcriptional profiling experiment carried out by Zheng *et al.* (2004). Gosset *et al.* also failed to identify many known CRP-dependent operons. Direct comparison of the two data sets is difficult because of differences in the strain and the growth conditions used.

In vivo transcription profiling is not an ideal tool for characterization of this regulon because of the level of complexity, in particular, the fact that many operons are subject to regulation by a second transcription factor. Some require an additional activator (e.g. *melAB* requires MelR) (Belyaeva *et al.*, 2000), whereas others are subject to repression (e.g. the *lacZYA* operon is repressed by the lac repressor). It would therefore be difficult, if not impossible, to define a single growth condition that would lead to regulation of all promoters that have the potential to be regulated by CRP *in vivo* without also leading to a number of CRP independent effects.

5.7.3 Comparison of ROMA with genome sequence searching

An alternative approach to identifying CRP-regulated promoters is to search the genome sequence for potential CRP binding sites. Several different algorithms have been used to predict the position of CRP binding sites within the *E. coli* genome (Robison *et al.*, 1998; Tan *et al.*, 2001). Tan *et al.* have predicted 161 strong (including known sites) and 285 weak candidate CRP binding sites. The main issue with sequence searching is that in the cell the DNA is not a uniform structure and therefore there may be local structures that affect CRP binding as well as histone-like proteins. It is therefore difficult to extrapolate sequence composition to strength of CRP binding. Many sites predicted to have a strong CRP binding site may not be positioned correctly within the promoter sequence or may not have the correct spatial arrangement *in vivo*, and may not be functional *in vivo*. Conversely, operons with weak, but functional, sites may not have been identified using sequence searching.

Zheng *et al.* (2001) used sequence searching to help validate the ROMA data. The promoter regions of the 176 operons identified to be regulated by CRP were analyzed in an attempt to identify potential CRP binding sites. A total of 125 operons identified in the ROMA assay contain potential CRP binding sites. However, for 51 operons a potential CRP binding site could not be identified.

We have compared ROMA with both *in vivo* transcription profiling and sequence searching and tried to determine which was the best for characterization of a regulon. It is clear that no one technique is perfect. ROMA is better suited for identifying the regulon of a factor that is part of a complex regulatory network, but combining all three approaches is probably required.

References

Arfin S, Long A, Ito E, Riehle M, Paegle E, Hatfield G (2000) Global gene expression profiling in *Escherichia coli* K12. The effects of integration host factor. *J Biol Chem* **275**: 29672–29684.

Belyaeva T, Wade J, Webster C, Howard V, Thomas M, Hyde E, Busby S (2000) Transcription activation at the *Escherichia coli melAB* promoter: the role of MelR and the cyclic AMP receptor protein. *Mol Microbiol* **36**: 211–222.

Britton RA, Eichenberger P, Gonzalez-Pastor JE, Fawcett P, Monson R, Losick R, Grossman AD (2002) Genome-wide analysis of the stationary-phase sigma factor (sigma-H) regulon of *Bacillus subtilis*. *J Bacteriol* **184**: 4881–4890.

Busby S, Ebright R (1999) Transcription activation by catabolite activator protein (CAP). *J Mol Biol* **293**: 199–213.

Cao M, Kobel PA, Morshedi MM, Wu MF, Paddon C, Helmann JD (2002) Defining the *Bacillus subtilis* σ^W regulon: a comparative analysis of promoter consensus search, run-off transcription/macroarray analysis (ROMA), and transcriptional profiling approaches. *J Mol Biol* **316**: 443–457.

Courcelle J, Khodursky A, Peter B, Brown PO, Hanawalt PC (2001) Comparative gene expression profiles following UV exposure in wild-type and SOS-deficient *Escherichia coli*. *Genetics* **158**: 41–64.

DeRisi JL, Iyer VR, Brown PO (1997) Exploring the metabolic and genetic control of gene expression on a genomic scale. *Science* **278**: 680–686.

Gosset G, Zhang Z, Nayyar S, Cuevas WA, Saier MH (2004) Transcriptome analysis of crp-dependent catabolite control of gene expression in *Escherichia coli*. *J Bacteriol* **186**: 3516–3524.

Hommais F, Krin E, Laurent-Winter C, Soutourina O, Malpertuy A, Le Caer JP, Danchin A, Bertin P (2001) Large-scale monitoring of pleiotropic regulation of gene expression by the prokaryotic nucleoid-associated protein, H-NS. *Mol Microbiol* **40**: 20–36.

Kolb A, Busby S, Buc H, Garges S, Adhya S (1993) Transcriptional regulation by cAMP and its receptor protein. *Annu Rev Biochem* **62**: 749–795.

Lobell R, Schleif R (1991) AraC-DNA looping: orientation of distance-dependent loop breaking by the cyclic AMP receptor protein. *J Mol Biol* **218**: 45–54.

Loos A, Glanemann C, Willis LB, O'Brien XM, Lessard PA, Gerstmeir R, Guillouet S, Sinskey AJ (2001) Development and validation of *Corynebacterium* DNA microarrays. *Appl Environ Microbiol* **67**: 2310–2318.

Nadon R, Shoemaker J (2002) Statistical issues with microarrays: processing and analysis. *Trends Genet* **18**: 265–271.

Oh MK, Rohlin L, Kao KC, Liao JC (2002) Global expression profiling of acetate-grown *Escherichia coli*. *J Biol Chem* **277**: 13175–13183.

Paget MS, Molle V, Cohen G, Aharanowitz Y, Buttner MJ (2001) Defining the disulphide stress response in *Streptomyces coelicolor* A3(2): identification of the SigR regulon. *Mol Microbiol* **42**: 1007–1020.

Petersohn A, Bernhardt J, Gerth U, Hoper D, Koburger T, Volker U, Hecker M (1999) Identification of sigma(B)-dependent genes in *Bacillus subtilis* using a promoter consensus-directed search and oligonucleotide hybridization. *J Bacteriol* **181**: 5718–5724.

Robison K, McGuire AM, Church GM (1998) A comprehensive library of DNA-binding site matrices for 55 proteins applied to the complete *Escherichia coli* K-12 genome. *J Mol Biol* **284**: 241–254.

Schena M, Shalon D, Heller R, Chai A, Brown PO, Davis RW (1996) Parallel human genome analysis: Microarray-based expression monitoring of 1000 genes. *Proc Natl Acad Sci USA* **93**: 10614–10619.

Tan K, Moreno-Hagelsieb G, Collado-Vides J, Stormo GD (2001) A comparative genomics approach to prediction of new members of regulons. *Genome Res* **11**: 566–584.

Tanaka TS, Jaradat SA, Lim MK *et al.* (2000) Genome-wide expression profiling of mid-gestation placenta and embryo using a 15,000 mouse developmental cDNA microarray. *Proc Natl Acad Sci USA* **97**: 9127–9132.

Tao H, Bausch C, Richmond C, Blattner FR, Conway T (1999) Functional genomics: expression analysis of *Escherichia coli* growing on minimal and rich media. *J Bacteriol* **181**: 6425–6440.

Yang YH, Speed T (2002) Design issues for cDNA microarray experiments. *Nat Rev Genet* **3**: 579–588.

Zheng D, Constantinidou C, Hobman JL, Minchin SD (2004) Identification of the CRP regulon using *in vitro* and *in vivo* transcriptional profiling. *Nucleic Acids Res*. **32**: 5874–5893.

Zheng M, Wang X, Templeton LJ, Smulski DR, LaRossa RA, Storz G (2001) DNA microarray-mediated transcriptional profiling of the *Escherichia coli* response to hydrogen peroxide. *J Bacteriol* **183**: 4562–4570.

The analysis of cellular transcriptional response at the genome level: Two case studies with relevance to bacterial pathogenesis

6

Thomas Carzaniga, Donatella Sarti, Victor Trevino,
Christopher Buckley, Mike Salmon, Shabnam Moobed,
David Wild, Chrystala Constantinidou, Jon L. Hobman,
Gianni Dehò, and Francesco Falciani

6.1 Introduction

Expression profiling is by far the most common application of microarray technology. For the majority of model organisms there are microarrays available that represent the entire transcriptional capacity of a cell. On the other hand the relatively low cost of manufacturing a large collection of PCR amplified cDNA clones, or of acquiring an oligonucleotide set designed to the organism of interest makes possible the development of microarrays that have good genome coverage for a large number of non model organisms. As a consequence huge volumes of transcriptomics data have become available in publicly accessible databases such as Gene Expression Omnibus (Barret *et al.*, 2005) or Array Express (http://www.ebi.ac.uk/arrayexpress/). Experimental design and the data analysis strategy are the two single most important factors in the success of an expression profiling experiment. In this chapter we describe two experiments that have been performed in our laboratory to characterize two different aspects of bacterial pathogenesis. These two case studies represent both prokaryotic (the adaptation of *Escherichia coli* cells to temperature shift between 10°C and 37°C) and eukaryotic (the response of host cells to bacterial infection) systems and use two different experimental designs (single- and two-channel microarrays). Although the data analysis techniques used to analyze expression profiling data are usually applicable to both types of microarrays, data processing, especially the normalization procedures, differ substantially.

The aim of this chapter is to give an overview of the main data processing and data analysis techniques that are commonly used in the analysis of microarray data using two different examples of microarray experiments. The examples we have chosen cover: (1) the design of a microarray experiment for describing the dynamic response of a cell to a stimulus and (2) a comparison between two different experimental conditions.

6.2 Scientific background

6.2.1 Case study 1: The response of *E. coli* cells to adaptation to body temperature

Studies on bacterial adaptation to a wide range of stressful conditions have greatly contributed to our understanding of bacterial physiology, regulatory networks, and molecular mechanisms of fundamental relevance to survival in hostile environments.

All living organisms are continuously exposed to temperature variations. Temperature is the environmental factor exerting the most immediate and profound effects on all chemico-physical parameters relevant for cell life. Therefore, temperature variations not only affect biochemical reaction rates, as expected by the Arrhenius equation, but directly influence the structural and functional properties of cellular components such as membrane lipid fluidity and flexibility of secondary and higher order structures of proteins and nucleic acids. As a consequence, adaptation to thermal stress, even within the range compatible with survival of a given species, may involve simultaneous and global adaptations of the cell's most essential machinery, namely that involved in replication, transcription, translation, and membrane biogenesis.

Response to both heat and cold in bacteria follow a similar adaptive pattern: a sudden, transient variation of the expression profile (up- or downregulation) of a number of genes specifically implicated in the acclimatization phase, followed by a new steady-state pattern in adapted cells. Upon heat shock (i.e. a sudden temperature increase) *E. coli* and other bacteria induce the expression of protein chaperones and proteases whose main role is to cope with heat-induced alterations of protein conformation. This is controlled at the transcriptional level by two alternative sigma factors (σ^{32}-*rpoH*, and σ^{24}-*rpoE*) that specifically direct RNA polymerase to transcribe genes implicated, respectively, in cytoplasmic or extracytoplasmic functions from *rpoH* and *rpoE* specific promoters. The heat shock response, therefore, essentially depends on two master regulators; transduction of the 'heat signal' will, in turn, modulate their abundance and activity via complex but fairly well understood mechanisms (reviewed in Narberhaus, 1999; Raivio and Silhavy, 2001).

More complicated appears to be the control of bacterial response to cold shock, where no master regulatory genes have been identified. A temperature downshift causes, in a mesophilic bacterium such as *E. coli*, a transient inhibition of transcription and translation of most genes whereas a few dozen 'cold shock genes' are actively expressed or strongly induced before a new steady state is established a few hours later. At least some of the proteins encoded by these 'cold shock genes' are essential for growth or survival in the cold, bind to nucleic acids and are involved in fundamental functions such as DNA packaging, transcription, RNA degradation, translation, ribosome assembly, etc. Interestingly, post-transcriptional mechanisms (mRNA stabilization and preferential translation), rather than transcriptional activation, play a major role in the increased expression of cold-induced proteins (Gualerzi *et al.*, 2003; Weber and Marahiel, 2003).

Surprisingly, adaptation mechanisms of bacteria shifting from cold to 'warm' (optimum) growth conditions have been much less explored. Nevertheless, physiological rearrangements as drastic as those observed during cold adaptation must occur upon a 'warm shock' and specific regulatory mechanisms are expected to be implicated in 'warm adaptation'. Moreover, transition from cold to 37°C, the mammalian host temperature, may represent an initial signal that triggers invasion and pathogenicity response. Therefore, understanding the 'warm response' may also provide insights into the early stages of bacterial infection of a warm-blooded host.

6.2.2 Case study 2: The response of human intestinal cells to *E. coli* infection

The pathogenic enterohemorrhagic *E. coli* strain (EHEC) infects the higher part of the colon and causes hemorrhagic colitis, a major cause of death in Third World countries. Currently, there is no effective antibiotic therapy for EHEC infections. Therapy is therefore based on relieving the symptoms of EHEC infection and is based on containing dehydration by reintegrating electrolyte and blood loss by transfusion (Takahashi *et al.*, 2004). Alternative therapeutic strategies include administration of probiotics (viable preparations of nonpathogenic bacteria that have beneficial effects on the host health) and plant extracts (Lee and Puong, 1999; Voravuthikunchai *et al.*, 2004). Many laboratories have focused their research on the analysis of modifications in host cell physiology (e.g. production of the cytokine IL-8, actin remodeling, etc.) induced by proteins that are translocated in the host cell by the type III secretion system of EHEC. In order to identify additional components of host response to infection, a genome-wide functional genomics approach is an appropriate strategy. Among the various options available (expression profiling, proteomics and metabolomics) expression profiling using microarray technology is the most practical option (Dopazo *et al.*, 2001).

Microarray technology is an extremely powerful tool to simultaneously investigate the expression pattern of thousands of different genes in a single experiment. Several groups have applied this approach to the analysis of the response of human cells to infection by bacterial pathogens. Examples include the analysis of human cell response to infection with *Helicobacter pylori* (Cox *et al.*, 2001), *Listeria monocytogenes* (Baldwin *et al.*, 2003), *Staphylococcus aureus* (Boldrick *et al.*, 2002), *Salmonella* (Detweiler *et al.*, 2001), and *Yersinia enterocolitica* (Sauvonnet *et al.*, 2002). These studies have confirmed the general role of the NF-κB pathway in the immune innate response to bacterial infection and have provided a hypothesis on a new mechanism of response to infection. Microarrays have also been used for the investigation of the bacterial transcriptional pattern during interaction with host cells. For example, *Salmonella enterica* (Eriksson *et al.*, 2003) and *Mycobacterium tuberculosis* (Talaat *et al.*, 2004) gene expression during infection have been well characterized using this technique.

Very little is known about gene regulation in pathogenic *E. coli* and in host cells during infection. EHEC cells preferentially infect the epithelium of the large intestine (colon). For this reason the majority of groups have used CaCo2 cells (a colon carcinoma cell line that is able to differentiate in polarized epithelia) as an *in vitro* model of infection. In this case study we describe the transcriptional response of CaCo2 cells during infection. Host response has been monitored using a custom-built gene microarray representing the main cellular functions (Stekel *et al.*, 2005).

6.3 Design of the experiment

6.3.1 Case study 1: The response of *E. coli* cells to adaptation to body temperature

This case study describes the transcriptional response of *E. coli* K-12 cells during adaptation from 10°C to 37°C. In simple terms this process mimics the adaptation of *E. coli* during transition from ambient environmental temperatures to the animal/human host environment.

In order to initially characterize the system we measured the effect of the temperature shift on the growth rate. An *E. coli* culture in LB broth grown for several days at 10°C with aeration up to mid-exponential phase (OD_{600} = 0.4) was transferred to 37°C and the optical density was monitored over time. Results are plotted in *Figure 6.1* (open symbols) together with the growth curve of a culture at 37°C (closed symbols). The inset of the figure is a close-up of a similar experiment with more frequent OD measurement time points. It is clear from these data that the temperature shifted culture enters a lag phase of about 10 min (see inset) and, once the lag is over, grows more slowly than the nonshifted culture for the remainder of the exponential phase. These initial results suggest that the adaptation to 'optimum' temperature could be as stressful on bacterial metabolism as the inverse transition. In order to address this hypothesis we designed the following microarray experiment. The experiment aims to compare the transcriptional profile of cells as they adapt from 10°C to 37°C with the transcriptional profile of a control culture that has always been growing at 37°C. The experiment starts at time 0 where both control and experimental cultures are at the same optical density. The changes in the transcriptional profile of these two experimental samples were monitored using single-channel glass microarrays representative of the entire genome of the K-12 strain at 12 time points (0, 2, 4, 8, 16, 24, 32, 40, 56, 64, 72, and 80 min). RNA was extracted and labeled (by direct labeling) using the procedures described in Protocol 2 (p. 231). Three experimental replicates were performed and labeled cDNA was hybridized on three different slides for each time point to take into account the experimental and technical variability.

6.3.2 Case study 2: The response of human intestinal cells to *E. coli* infection

The second case study describes the transcriptional response of host cells to infection by the enterohemorrhagic *E. coli* (EHEC) strain O157. The aim of this study was to identify genes differentially expressed between infected

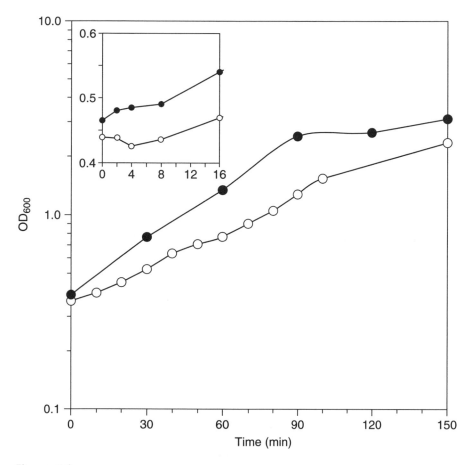

Figure 6.1

Growth curve of control and shift-up *E. coli* MG1655 cells. The *y* axis represents the optical density (OD) of the *E. coli* cells whereas the *x* axis represents the time. Closed circles represent the growth of the control strain at 37°C whereas the open circles represent the growth of strain after temperature shift from 10 to 37°C. The inset reports an identical experiment in which culture was sampled at closely spaced early time points.

and uninfected cells and identify the components of the response that are associated with factors secreted by bacterial cells. *Figure 6.2* outlines the schema of the experiments. Bacterial cells were grown overnight in LB at 37°C. Cells were then diluted 1:100 in DMEM mammalian cell culture media and incubated for 3 h. This incubation is necessary for the bacteria to express the virulence factors necessary for adherence to human host cells. Bacterial cells were then transferred to a monolayer of CaCo2 cells either directly (infection) or on the top of a trans-well semipermeable filter separating the bacteria from the host cells. The filter allows the diffusion of diffusible factors but does not allow direct contact between the pathogen and the host cells. A monolayer of CaCo2 cells (growth in DMEM) represents the uninfected control cells. During the course of the infection (0, 2,

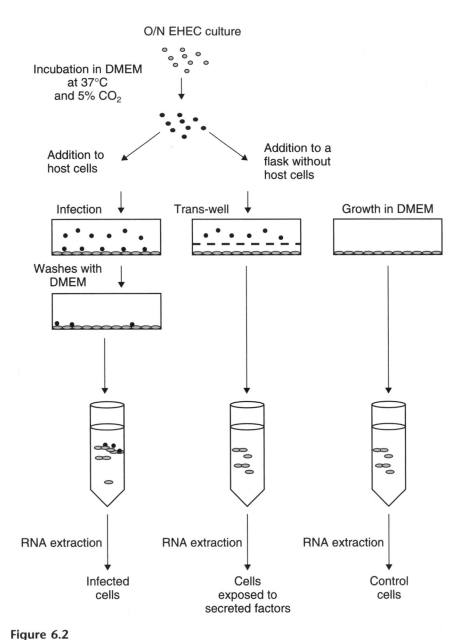

Figure 6.2

Experimental design of the host cell infection experiment (Case study 2).

4, and 6 h) the human cells were recovered and RNA extracted for micro-array analysis. The experiment was performed with two experimental replicates and three technical replicates (a total of six measurements for every gene). The expression of each gene at 2, 4, and 6 h was expressed as a ratio with respect to the reference profile of CaCo2 cells before infection (two-channel microarray experiments).

6.4 A description of data analysis procedures

6.4.1 Data normalization

Microarray data are a numerical integration of an image that is obtained via scanning the slide with a laser scanner (this process is described with some detail in Chapter 1). All other experimental measures of gene expression profiling are subject to systematic errors that need to be corrected before any analysis is performed. In addition to the errors that can be introduced by the molecular biology procedures; microarray technology may also have additional sources of error that are introduced by the robotic equipment for the manufacturing of the arrays and the scanning procedures. The process of removing these errors is called 'normalization'. The process differs greatly between single-channel and double-channel arrays.

One-channel array normalization

The most common procedure used to normalize single-channel array data is called 'quantile normalization' (Bolstad *et al.*, 2003). It is commonly used with Affymetrix Gene Chips (see Chapter 1 for an overview of this technology) but can be applied to any single-channel array design. This procedure assumes that the signal distribution associated with the different arrays is identical in the absence of any experimental error. The procedure averages the ($\log_2$) values by percentiles or quantiles (for example, from averaging all values from the beginning to the next 1%, then from the 1% to 2%, and so on). Data from Case study 1 has been normalized using this procedure.

Two-channel array normalization

In the two-channel array design the experimental sample and the reference sample (for example infected and control cells) are labeled with two different dyes (usually Cy3 and Cy5) and hybridized on the same slide. With this design the control and experimental samples can be directly compared in a single hybridization thereby reducing the measurement error. However, a two-channel array design also introduces sources of variability that can be corrected with appropriate normalization procedures. These are the overall difference in signal between the experimental and reference sample and the differential signal that is associated with Cy3 and Cy5, which can arise in direct labeling due to differential incorporation efficiencies of the dyes. The normalization procedure used with two-channel microarrays is based on the assumption that the majority of the genes are not differentially expressed or that the number of genes up- and downregulated are similar. The procedure, called Lowess normalization (a good introduction to this methodology is available in Stekel, 2003), is based on an M versus A plot where $M = \log_2(R/G)$ is drawn in the vertical axis, $A = \log_2 (R \times G)/2$ is drawn in the horizontal axis, and R and G are the intensities of the red and green channels, respectively. The normalization is performed by fitting a smooth line through the M values along the A values. Then, the new normalized ratios M are computed by subtracting the smoothed line to the original ratio M. When there is spatial variability in the slide, a *print-tip* or *spatial* normalization is performed doing the same procedure as above but performed in each print-tip block independently. Data from Case study 2 has been normalized using this procedure.

6.4.2 Identifying differentially expressed genes

The t-test and the false discovery rate

One of the major goals in performing microarray experiments is to know which genes are differentially expressed between two conditions. In this context, the most common procedure used to test the statistical significance of the gene expression data is to perform a t-test for each individual gene. The objective of the t-test is to assign the probability that the observed difference in two sets of measurements are given by random chance. The lower this p value is the greater the confidence that the difference is indeed genuine. The t-test works in two distinct steps. The first step computes a number that represents the extent of separation between the two sets of measurements. This number is called the t-statistic and can be computed using the following formula:

$$t = (\bar{x}_1 - \bar{x}_2)/\sqrt{s_1^2/n_1 + s_2^2/n_2}$$

where $\bar{x}_1$ and $\bar{x}_2$ are averages, s_1^2 and s_2^2 are variances, and n_1 and n_2 are the number of replicates for the two conditions. In order to compute the p value the computed t-statistic is compared with the distribution of values of t-statistics that is expected by random chance.

Ultimately, the p value will depend on the value of the t-statistics (as the larger the absolute value of the t-statistic the lower the p value) and from parameters that modify the shape of the null distribution (such as the number of experimental replicates).

The t-test is really designed for comparing two sets of measurements. A microarray experiment would involve performing this operation up to tens of thousands of times, however. In this scenario many genes will appear differentially regulated at a given p value threshold by random chance. For example, if we use completely random values, the t-test will consider 5% of genes as significant (for $\alpha = 0.05$). On a dataset with 10 000 genes this could result in 500 genes appearing to be differentially regulated. For this reason, a correction in the p values is commonly performed in large-scale data analysis in an attempt to control for false positives. One of the most widely used and most powerful correction methods is the false discovery rate (FDR) proposed by Benjamini and Hochberg (1995). In this method the correction is performed by multiplying each p value by the total number of tests and dividing by the number of (the remaining) larger p values. The result is defined as a q value and is interpreted as the expected number of false positives. Typical q values reported in the literature range from 0 to 0.1 (10%). The number of replicates needed for identifying differentially expressed genes depends on the experimental and technical variability. As a general suggestion a microarray experiment should be performed with at least three replicates. However, in order to compute the statistical power of a t-test (defined as the probability of rejecting a null hypothesis when it is false) for a given number of replicates one should perform a proper power analysis (Stekel, 2003).

A multivariate empirical Bayes' methodology to identify differentially expressed genes from replicated time course data

Tai and Speed (2005) have described a multivariate empirical Bayes' methodology which provides a solution for the analysis of differential expression in replicated microarray time course data. The method is currently implemented in the statistical programming language R and includes functions for identifying genes of interest from longitudinal (time series) replicated microarray time course experiments with one or more biological conditions. This methodology has the advantage over the use of the traditional F statistic in that it incorporates replicate variances, the correlations among gene expression time point samples, and moderation, borrowing information across genes to reduce the numbers of false positives and false negatives. An alternative method for the identification of differential time course expression profiles has also recently been developed by Conesa *et al.* (2006). Chapter 8 describes this alternative approach.

6.4.3 Data exploration techniques

Hierarchical clustering

Clustering is a procedure that is designed to group similar patterns (e.g. gene expression profiles). The core idea is that patterns within a cluster are more similar to members of the same cluster than to members of different clusters. Several methods to quantify the degree of similarity (or dissimilarity) between different expression patterns have been proposed (for an overview of these measures see Stekel, 2003). Once that the distance (or similarity) of each pair of patterns has been computed (distance matrix), the next task is the generation of clusters. What follows is an introduction for *agglomeration* methods that generate clusters hierarchically (clusters are generated adding-up more members to clusters generated in an earlier steps). In general, the first step in hierarchical clustering is to cluster those two patterns (e.g. gene expression profiles) whose distance is (unambiguously) shorter. The following steps agglomerate new patterns to clusters defined in the previous step using an updated distance matrix computed from the generated clusters and remaining patterns. This process continues until all patterns have been clustered.

Principal component analysis

Principal component analysis (PCA) is a mathematical procedure that is designed to represent a very complex (multidimensional) dataset in two or three dimensions whilst still maintaining most of the information. In statistics, PCA is mainly used as an exploratory tool to detect outliers, to verify clusters or to reveal trends. In more technical terms, PCA transforms a set of uncorrelated components that are linear combinations of the original variables. The coefficients of this linear relation are called *loadings* whose magnitude is related to the degree of contribution of each original variable to the principal components (PCs). The first component by definition explains most of the variability of the original dataset. Typically, only a handful of components explain virtually all variability, thus only a few PCs are considered for further analysis.

6.4.4 Implementation

Normalization, processing, clustering, and differential expression for Case study 2 have been done using the web-based tool GEPAS (Vaquerizas *et al.*, 2005). Quantile normalization, PCA, and time-course analysis have been done in *R* (http://cran.r-project.org). *R* is a powerful statistical programming environment which is very popular in the bioinformatics community, but *R* is designed to be used by an expert analyst. In this and other chapters, we will provide the code needed to perform the analysis in a form that can be easily applied by a computer minded biologist. For more information on *R* the reader may refer to http://cran.r-project.org, http://cran.r-project.org/manuals.html, http://cran.r-project.org/other-docs.html.

6.5 Data analysis tutorials

The procedures described in the previous sections with the exception of PCA have been applied to the case studies as they are implemented in the web-based microarray data analysis toolset GEPAS (Vaquerizas *et al.*, 2005). PCA has been performed using the statistical programming environment *R*. What follows is a series of step-by-step procedures used to generate most of the results summarized in the results section.

6.5.1 Case study 1: The response of *E. coli* cells to adaptation to body temperature

Step 1: Normalizing single-channel microarray data using quantile normalization

The raw data consists of 78 .gpr files (acquired using GenePix® software). These files correspond to 39 experiments for the temperature shift from 10°C to 37°C and include three replicates from 13 time points, and 39 experiments for the growth control at 37°C under the same time points and replicates. To normalize this data we used the quantile normalization included in the *affy* package (http://www.bioconductor.org) under the programming environment *R* (http://cran.r-project.org). This package is designed to process *Affymetrix* chip arrays which are single-channel microarrays. The analysis using this package is therefore equivalent for the single-channel data we used in this case study. Before proceeding with the normalization the data must be prepared to be a suitable input for the *normalize.quantiles* function in *R*. The file must be a text tab-delimited table where rows are the gene identifiers and columns report the expression value for each gene in the experimental samples. Such a table can be constructed in Microsoft Excel by copying the relevant columns from the various GenePix files (usually the column reporting the median value of signal for the pixels in a given spot with local background subtracted is used for this purpose). A copy of the data matrix used for this tutorial can be downloaded from the book's website. The data matrix can be loaded in *R* (in the object *signal*) by typing

```
> signal<-read.delim(table.xt)
```

We can now proceed to the normalization as follows.
Initially the appropriate library must be loaded:

```
> library(affy)
```

and data can be normalized:

```
> signal.qn <- normalize.quantiles(log2(signal))
> dimnames(signal.qn) <- dimnames(signal)
> signal.qn[signal.qn <0] <- 0 # set normalized negative
  values to zero
```

Step 2: Data processing

In order to perform the PCA, we need to obtain a measure for each gene
from the two internal replicates and the three biological replicates. We can
use the gene names and identification fields included in the .gpr files to
recognize gene identities. Then, we can compute an average from all gene
replicates within each array. This procedure can be done as follows.

```
> id <- paste(gNames[-unique(del)],gID[-unique(del)],sep="
  || ")
> sqn.g <- by(1:nrow(signal.qn), factor(id), function(x)
apply(signal.qn[x,,drop=FALSE], 2, mean)) # computes the
mean
> sqn.g <- t(data.frame(lapply(sqn.g,function(x) x))) #
  transform to matrix
```

sqn.g now contains the average expression per gene (rows) in each array
(columns). Now we must average the biological replicates (columns) which
can be recognized by the experiment name (colname). We performed this as
follows.

```
> sqn <- by(1:ncol(sqn.g), factor(substr(colnames(sqn.g), 1,
  8)), function(x)
apply(sqn.g[,x,drop=FALSE], 1, mean)) # compute means
> sqn <- data.matrix(data.frame(lapply(sqn,function(x) x)))
  # join data
> xt <- as.numeric(gsub("m","",substr(colnames(sqn),4,5))) +
ifelse(substr(colnames(sqn),1,2)=="Ct",100,0)
> sqn <- sqn[,order(xt)]
```

Here *sqn* contains a matrix with the average expression for each gene (rows)
for each experimental condition, 13 time points for a time-shift experiment, and 13 for controls.

Step 3: Using principal component analysis to describe the change in the transcriptional state of *E. coli* cells during adaptation to 37°C.

We will now perform PCA using the averaged normalized values. Then, we
will use the first two principal components to describe the relationships

between the molecular state of the cells in the different experimental conditions.

The following code performs the PCA procedure:

```
> pca <- prcomp(t(sqn))
```

Results can be plotted to produce *Figure 6.3*:

```
x> timep <- as.numeric(gsub("m","",substr(colnames(sqn),
4,5)))
> scol <- ifelse(substr(colnames(sqn),1,2)=="Ct","white",
"black")
> symbols(pca$x[,1:2], circles=sqrt((timep+1)/3.14159),
inches=FALSE, bg=scol)
> text(pca$x[,1], 1+pca$x[,2]+sqrt((timep+1)/3.14159),
paste(timep,"min",sep=""), col=1, cex=1.25) # add labels to
sample points
```

The results shown in *Figure 6.3* have been described in the Results section of this chapter (p. 141). However, it is worth noting that the first principal component (the *x* axis of *Figure 6.3*) is representative of the specific dynamics associated with temperature shift.

Step 4: Using cluster analysis to identify waves of gene expression associated with temperature shift

As we have mentioned in the previous step, the early dynamics of the temperature shift is captured by the first principal component. To identify the genes that contribute most to the first principal component, we can use the loadings associated to PC1. We will therefore select genes whose *loading* (coefficients) are higher in absolute value then a defined threshold. *Figure 6.4A* shows a graph summarizing the loading values (*y* axis) for each gene (*x* axis). The two lines, parallel to the *x* axis represent the selection threshold (larger than 0.4 in absolute value) used to select interesting genes. The genes can then be clustered and their expression profiles visualized using a heatmap. The figure, the list of interesting genes and the heatmap representing the clustered data can be generated with the following code.

To plot the loadings:

```
> plot(pca$rotation[,1],type="h",xlab="Genes",ylab="PC1
Loading")
> abline(h=c(-0.04,0.04),col="red")
```

To extract the list of interesting genes:

```
 > load.filter <- 0.04
> pc1 <- sqn[abs(pca$rotation[,1]) >= load.filter,]
```

To generate the heatmap and the clustering:

```
> heatmap(pc1, col= c(gray(0:15/16)), Colv=NA, ColSideColors=
rep(gray(13:1/14),2), cexRow=1.5, cexCol=4, margins=c(20,40),
distfun=function(x) as.dist(1-cor(t(x)))) # plotting genes
```

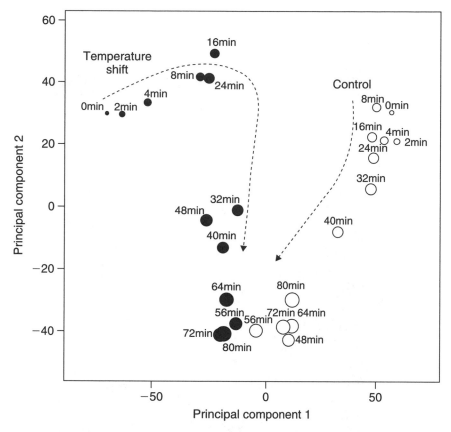

Figure 6.3

Principal component analysis (PCA) analysis of the response of *E. coli* cells to adaptation to 37°C. The plot clearly shows that the adaptation following the shift from 10°C to 37°C follows a trajectory that is parallel to the first component whereas the transcriptional response of cells during growth at 37°C follows a trajectory that is parallel to the second component. The plot also shows that after an initial period of 24 min the cells realign to the reference trajectory.

6.5.2 Case study 2: The response of human intestinal cells to *E. coli* infection

In this case study, the input data contains two replicates for each of the three time points (2 h, 4 h, 6 h) for *Control*, *Infection*, and *Transwell* experimental conditions. The 18 files were obtained using GenePix® software (.gpr extension). To identify the results, the files were named using the experimental condition (ctrl, inf, tw), the time point (2 h, 4 h, 6 h) and the replicate number (r1, r2). The custom-made microarrays were spotted in a grid of 4 × 12 blocks (48 blocks: 4 horizontal by 12 vertical). Each block containing 10 × 8 spots (80 spots: 10 columns by 8 rows). Eight hundred and fifty genes were spotted in triplicate. The reference channel (Cy3 green) was the zero

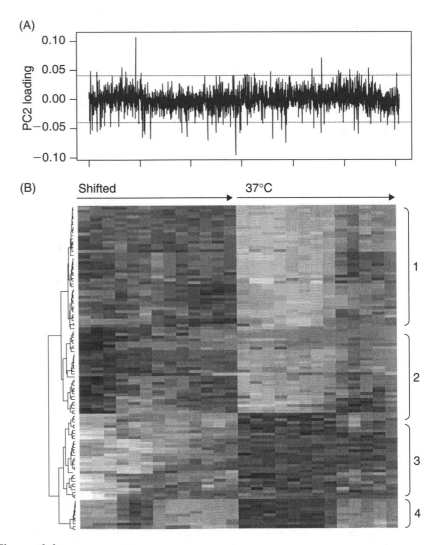

Figure 6.4

Cluster analysis of genes differentially regulated during temperature adaptation. Panel A shows a histogram representing the values of the loadings (*y* axis) on the first PC for all the genes analyzed (gene indexes on the *x* axis). The horizontal lines represent the arbitrary cut off threshold of ±0.04. Panel B shows the results of the cluster analysis. The numbers on the right side of the heatmap defined the four main clusters. The individual genes are listed in Appendix 4.

time-point for each experiment and the red channel (Cy5 red) the experimental sample at 2, 4, and 6 h after infection.

Step 1: Normalizing double-channel microarray data using Lowess normalization

The 18 .gpr files were archived in a unique .zip file using desktop software (equivalent to WinZip®). This format allows multiple file processing in the web-based tool GEPAS (http://dnmad.bioinfo.cipf.es/cgi-bin/dnmad.cgi). For the normalization process, within the web front end we must load the zip file; input the microarray layout, the background correction method, and the normalization approach. In this case, we used **12** rows by **4** columns in the layout–main grid options (48 blocks) and **8** rows by **10** columns (80 spots) for the layout–subgrid options. For the background correction method, we used **background subtraction**. We would like to normalize each block independently for a possible within-slide artefact; therefore we **did not** use **global Lowess**.

The tool produces an extensive set of plots that are very useful for diagnostics purposes. A comprehensive documentation for the interpretation of the graphical output can be found in the tool documentation. Here we focus on the textual output representing the normalized data. The output of the normalization process is the $\log_2$ normalized ratios (M values in GEPAS output as a .txt file) as shown below. Columns are delimited by 'tabs'. First column is gene name (or ID). All lines starting with symbol '#' are commentaries which are not really part of the data. Nevertheless, the line '#name' contains the names of the original .gpr files in columns.

```
#Normalization method: Print-tip loess
#Options for normalization: Use negative flags. Return negative flagged
points
#M values
#names ctrl2h.r1    ctrl2h.r2    ctrl4h.r1    ctrl4h.r2    ctrl
i24.PIG7    -1.04902680277178    0.176908827526412    -0.574741720
m24.BRE    -0.100201752399195    0.969996918053997    -0.094999018
a4.Blank    NA    NA    NA    NA    NA    NA    NA    NA
e4.Blank    NA    NA    NA    NA    NA    NA    NA    NA
```

Step 2: Data processing

Before proceeding to data analysis, we have to consider that some spots were automatically flagged by the microarray scanning software, or manually flagged by user inspection, for being of very low quality or with no associated signal. It is good practise to remove these spots from the data. Furthermore, we may also consider eliminating spots that have too many missing values along the series of samples or genes whose expression profile is not changing across the series of samples.

In order to perform these pre-processing steps, directly from the results page of the normalization in the previous step, we used the option **Send to Preprocess**. Clicking on this option automatically sends the normalized data to the GEPAS Preprocess tool. Because the data has already been transferred we only need to input the parameters for the data processing. We used **1.5** in the option **remove inconsistent replicates** to avoid the use of replicates that were too far from the median. In order to eliminate genes that do not vary in any of the experimental conditions we **filtered flat patterns with at least 1 peak and 0.5 as threshold** (approximately 97% of the genes pass this filter, see histograms in the GEPAS output). We used **0.5** in both, **filter flat patterns by root mean square** and **filter flat patterns by standard deviation** (approximately 90% of the genes pass these filters).

The results, linked in the **Processed data** output, were 737 (86%) genes out of 859 initially included. The output is shown below including the actual filters and options used as commentaries (starting with '#'). Note that unused 'Blank' spots have been removed through commenting those lines.

```
# [Tue Mar 21 11:40:44 2006] Flat Patterns Filtered (Standard deviation
>= 0.25)
# [Tue Mar 21 11:40:44 2006] Flat Patterns Filtered (#Hits 1; Threshold
0.5)
# [Tue Mar 21 11:40:44 2006] Patterns with less than 70% Removed
# [Tue Mar 21 11:40:43 2006] Remove Replicates (Thrs from median 1.5)
# [Tue Mar 21 11:40:43 2006] Check data
#Normalization method: Print-tip loess
#Options for normalization: Use negative flags. Return negative flagged
points a
#M values
#names ctrl2h.r1     ctrl2h.r2     ctrl4h.r1     ctrl4h.r2     ctrl6h.r
i24.PIG7_PIG7 1.049    -0.177 0.575  -0.59   0.96   -1.03  -0.404  0.953
m24.BRE_BRE    0.1     -0.97  0.095  -0.544  0.224   0.145  -0.261  -1.152
#a4.Blank | Blank     NA     NA     NA      NA     NA     NA      NA
#e4.Blank | Blank     NA     NA     NA      NA     NA     NA      NA
```

Step 3: Identifying differentially expressed genes using the web based tool T-Rex.

One of the objectives of this study was to identify genes differentially expressed between human CaCo2 cells that were infected or were uninfected. For this purpose, we can use the web-based tool T-Rex (http://t-rex.bioinfo.cipf.es/cgi-bin/t-rex.cgi) which is part of the GEPAS web suite. To detect differential expression, T-Rex compares 'classes' of experiments in a text file where samples associated to a class are in columns (delimited by tabulator) and genes are in rows. To build a file like this, we have to do some external file editing to generate the class names and the columns for each replicate. An example of the edited file is shown below for control versus infection at 2 h after inoculation.

```
#NAMES   Inf2h.1.1          Ctrl2h.1.1         Inf2h.1.2          Ctrl2h.1.2
#CLASS   Infection2h        Control2h          Infection2h        Control2h
AATK     -0.376  0.941      0.213   0.482      0.317   0.613      0.795   -0.004
ABL1     -0.007  -0.118     -0.641  -0.169     -0.572  -0.380     1.000   0.992
ABS       0.435  -0.751     0.942   -1.081     0.986   -0.694     0.853   0.028
ACTB     -0.660  -0.626     0.835   -1.212     -0.017  -0.459     -0.926  0.504
ACTIVIN A (bA subunit)     -0.255  -0.136     -0.224  -1.103     0.599   -0.837
ACTIVIN B (bB subunit)     -0.317  -0.442     0.709   -0.821     0.628   0.170
ACTIVIN RIA       0.183    -0.507  0.894      -0.431  0.585      0.364   -0.114
ACTIVIN RIB       0.165    -0.743  0.438      -0.755  -0.010     -1.230  0.286
ACTIVIN RIIA      0.708    -0.200  0.737      -0.624  0.819      -0.448  -0.091
ACTIVIN RIIB     -0.260    0.127   0.121      -0.202  0.302      -0.352  -1.311
AD022     0.236  1.917      1.002   0.343      1.725   -0.407     -0.426  0.326
```

In T-Rex all samples labeled with the same class name (Infection2h or Control2h in the text example shown above) are considered as replicates. Finally, we used *t*-test as the differential expression test.

The same procedure was performed for Infection versus Control at 4 and 6 h.

A second objective was to separate the components of the response that were cell–cell contact independent. For this, we performed the same procedure explained for Transwell (noncontacting cells) versus Control at 2, 4, and 6 h. Then, for all six comparisons, the T-Rex output file was saved and analysed. The analysis consisted in counting the number of differential expressed genes at FDR less than or equal to 10% (FDR ≤ 0.10).

Step 4: Cluster analysis of samples

The procedure described in Step 3 has been used to compare the response of infected cells, or the response of cells exposed to soluble factors, with the transcriptional state of control cells at 2 h, 4 h, and 6 h. These comparisons generate six different gene lists. For visualization purposes it is useful to perform cluster analysis on genes that are differentially expressed in at least one of the comparisons. In the previous case study we have shown how to perform cluster analysis in the statistical environment *R*. The web-based toolset GEPAS, however, offers a set of tools which is easier to use when performing cluster analysis. The tool we will be using performs hierarchical cluster analysis and it is called Cluster (http://gepas.bioinfo.cipf.es/cgi-bin/cluster). The input file for the clustering program is a text (tab delimited) file where rows are represented by genes and columns represent different samples. The program T-rex (used in step 3) does not generate the file we need as a result of a single comparison; therefore, in this case, the input file for the clustering program would need to be manually prepared. This can be done using Microsoft Excel or Microsoft Access. Once the file is prepared it can be loaded in the program from the 'data' form in the web front end. The clustering method (average clustering, also called UPGMA) can be selected from the 'Cluster condition' drop-down window and the distance measured can be selected from the 'distance' drop-down window. The graphical output of the program is directly obtained by clicking on the run button.

6.6 Results and discussion

6.6.1 Case study 1: The response of *E. coli* cells to adaptation to body temperature

Exploring the evolution of the transcriptional state of *E. coli* cells during temperature adaptation

Methods that allow the graphical representation of the molecular state of a cell as defined by its full genome transcriptional profile can be extremely useful in understanding the dynamics of a biological process. In this context, PCA has been frequently applied to the analysis of microarray data. In Section 6.4 we have shown that this technique allows the projection of highly dimensional data on a two-dimensional space whilst retaining the majority of the information. We have applied this technique to describe the evolution of the transcriptional state of *E. coli* K-12 cells following temperature shift. For comparison the time course of control cells grown at 37°C in the same growth conditions (comparable cell densities and growth curves) has been included in the analysis. *Figure 6.3* shows a plot representing the first two principal components. The visual inspection of the PCA plot is sufficient to identify that the trajectory describing the evolution of the molecular state of cells readapting to 37°C is very different for the first 24 min to the control cells.

Table 6.1 Cluster 1

Gene	Encoded function
aegA	putative oxidoreductase, Fe S subunit
ansB	periplasmic L-asparaginase II
appB	cytochrome oxidase bd-II, subunit II
appC	cytochrome oxidase bd-II, subunit I
dcuB	anaerobic C4-dicarboxylate transport protein
dmsA	anaerobic dimethyl sulfoxide reductase, subunit A
dmsB	anaerobic dimethyl sulfoxide reductase, subunit B
frdA	fumarate reductase, anaerobic, catalytic and NAD/flavoprotein subunit
frdB	fumarate reductase, anaerobic, Fe-S subunit
fumB	fumarase B (fumarate hydratase class I), anaerobic isozyme
gadA	glutamate decarboxylase A, isozyme
gadB	glutamate decarboxylase isozyme
gldA	glycerol dehydrogenase
hdeA	hypothetical protein
hdeB	hypothetical protein
hyaD	processing of HyaA and HyaB (hydrogenase 2) proteins
hyaF	nickel incorporation into hydrogenase 1
nrfB	formate dependent nitrite reductase
tdcA	transcriptional activator of tdc operon
tdcB	threonine dehydratase, catabolic
tdcC	anaerobically inducible L-threonine/ L-serine permease
tdcD	propionate kinase/acetate kinase II, anaerobic
tdcE	probable formate acetyltransferase 3
tdcG	putative L-serine dehydratase
tdcX	hypothetical protein
yciD	outer membrane protein W; colicin S4 receptor
ydfZ	hypothetical protein
yecH	hypothetical protein
yfcC	putative S transferase
yhbV	hypothetical protein
yhiD	putative Mg^{2+} transport ATPase
yhiU	multidrug resistance lipoprotein
yhiV	multidrug transport protein, RpoS-dependent (RND family)
yhjA	putative cytochrome C peroxidase
yjjI	hypothetical protein
ynfE	putative dimethyl sulfoxide reductase, major subunit
ynfF	putative dimethyl sulfoxide reductase, major subunit
ynfG	putative dimethyl sulfoxide reductase, Fe-S subunit
ynfH	putative dimethyl sulfoxide reductase, anchor subunit
Z1180	hypothetical protein

More precisely, cells growing at 37°C follow a trajectory that is aligned to the second principal component whereas cells adapting to the temperature shift are following a trajectory that is orthogonal to the control cells for the initial 24 min (parallel to the first principal component). After 36 min both control and shifted cells display a very similar molecular profile. Since the trajectory of shifted cells is aligned to PC1 it is possible to use the PC leadings to identify genes that are likely to be important in temperature adaptation (*Figure 6.4A*). Cluster analysis performed on genes selected with this strategy identifies four distinct clusters of genes responding to temperature shift (*Figure 6.4B* and *Tables 6.1, 6.2, 6.3,* and *6.4*). Cluster 1 primarily includes genes that are well expressed in cells exponentially growing at 37°C and decrease their expression level to different extents when approaching the stationary phase. These genes are not expressed in cells exponentially growing at 10°C (with a generation time of about 24 h) and remain essentially nonexpressed upon temperature upshift. However, a subgroup of genes in this cluster appears to transiently increase expression half an hour after the temperature up-shift.

Cluster 2 genes exhibit an expression profile similar to cluster 1 at the initial phase of growth and adaptation (i.e. off at 10°C and shortly after the

Table 6.2 Cluster 2

Gene	Encoded function
adiY	putative ARAC type regulatory protein
appY	DLP12 prophage; transcriptional regulator required for anaerobic and stationary phase induction of genes (AraC/XylS family)
fdhF	formate dehydrogenase H, selenopolypeptide subunit
fdnG	formate dehydrogenase-N, α subunit, nitrate-inducible
fdnH	formate dehydrogenase-N, Fe-S β subunit, nitrate-inducible
fliC	flagellar biosynthesis; flagellin, filament structural protein
garL	α-dehydro-β-deoxy-D-glucarate aldolase–carbohydrate catabolism
hsdR	endonuclease R, host restriction
hybB	putative cytochrome Ni/Fe component of hydrogenase-2
hybC	hydrogenase-2, large subunit
hybD	putative processing element for hydrogenase-2
hybO	hydrogenase-2, small chain
hypA	guanine-nucleotide-binding protein in formate-hydrogenlyase system, functions as nickel donor for HycE of hydrogenlyase 3
hypE	hydrogenase 3 maturation protein
narK	nitrite extrusion protein
nirD	nitrite reductase (NAD(P)H) subunit
nmpC	DLP12 prophage; outer membrane porin, at locus of qsr prophage
srlB	PTS family enzyme IIA, glucitol/sorbitol-specific
tnaA	tryptophanase
tnaL	tryptophanase leader peptide tnaL
treC	trehalose 6-P-hydrolase
ydeN	putative sulfatase
ydhV	hypothetical protein
yfiD	putative formate acetyltransferase
yghJ	predicted inner membrane lipoprotein; transport
yjdE	putative amino acid/amine transport protein, cryptic
yjjW	putative activating enzyme
yliH	putative receptor

Table 6.3 Cluster 3

Gene	Encoded function
adhP	alcohol dehydrogenase
astE	succinylglutamate desuccinylase; amino acid catabolism
cspB	Qin prophage; cold shock protein
cspF	Qin prophage; cold shock protein
cspG	homolog of *Salmonella* cold shock protein
cspI	Qin prophage; cold shock-like protein
cysH	3'-phosphoadenosine 5'-phosphosulfate reductase
deaD	cold-shock DeaD box ATP-dependent RNA helicase
fhuA	outer membrane pore protein, receptor for ferrichrome, colicin M, and phages T1, T5 and phi80
fhuF	ferric hydroxamate transport protein
gmd	GDP-D-mannose dehydratase
mqo	malate:quinone oxidoreductase
nrdF	ribonucleoside-diphosphate reductase 2, β subunit chain
nrdI	hypothetical protein
phoH	PhoB-dependent, ATP-binding *pho* regulon component
proV	ATP-binding component of transport system for glycine, betaine and proline
sodA	superoxide dismutase, Mn
wza	putative polysaccharide export protein
ycdO	hypothetical protein
yceJ	cytochrome b561 homolog 2
ycfJ	hypothetical protein
ydiU	hypothetical protein
yeaR	hypothetical protein
yfdN	CPS-53 prophage; putative transcriptional regulator
yhfC	putative transport protein
yhhQ	hypothetical protein
yncE	putative receptor
yoaG	hypothetical protein
ypeC	hypothetical protein

Table 6.4 Cluster 4

Gene	Encoded function
argT	lysine-, arginine-, ornithine-binding periplasmic protein
astC	acetylornithine δ-aminotransferase; amino acid catabolism
cyoA	cytochrome o ubiquinol oxidase subunit II
cyoB	cytochrome o ubiquinol oxidase subunit I
cyoC	cytochrome o ubiquinol oxidase subunit III
cyoD	cytochrome o ubiquinol oxidase subunit IV
cyoE	protoheme IX farnesyltransferase (heme O biosynthesis)
fadB	4-enzyme protein: 3-hydroxyacyl-CoA dehydrogenase; 3-hydroxybutyryl-CoA epimerase; $\delta(3)$-*cis*-$\delta(2)$-*trans*-enoyl-CoA isomerase; enoyl-CoA hydratase
lldD	L-lactate dehydrogenase
lldR	putative transcriptional repressor for L-lactate utilization (GntR family)

temperature shift-up, but on at 37°C). However, at later times after the shift-up the expression level of these genes tends to increase and to conform to the somewhat decreased level observed at late times in non shifted cultures.

Clusters 3 and 4, by comparison, include genes highly expressed at 10°C and not at 37°C in exponentially growing cells. In cluster 3 the genes remain off or increase their transcript abundance towards the end of the exponential phase in non shifted cultures, whereas in shifted up cells the mRNA abundance tends to decrease. In cluster 4, expression of the genes is activated at the end of the exponential phase whereas the temperature-shifted cells, initially expressing these genes at a high level, quickly conform to the pattern of nonshifted cultures.

Most genes present in these four clusters are directly implicated in bacterial metabolism, from energy production and transport (components of the electron transport chains for both anaerobic and aerobic respiration), to central metabolism (amino acid, nucleotide, and nucleoside catabolism and conversion; sulfur metabolism), membrane receptors, binding, and transport proteins. This can be understood by considering the drastically diverse metabolic conditions of a bacterial cell growing in a rich medium at low temperature (where reaction rates are very slow, generation time very long, and oxygen solubility in the medium is high) compared with 37°C (faster reaction rates, shorter generation times, and lower oxygen solubility). It appears that many of these genes, belonging to cluster 1, do not need to be expressed at 10°C nor to be activated upon temperature shift during the 80 min (approximately three generations) monitored by our experiment, whereas the genes in the other clusters tend to conform to the expression pattern of the control temperature at the late time points.

Only a small group of genes known to encode regulatory functions have been found in clusters 1, 2, and 3 (none in cluster 4). In cluster 1 only *tdcA*, a transcriptional activator for amino acid degradation (and thus implicated in carbon and nitrogen metabolism), is found. Two additional regulators of metabolic operons not expressed at 10°C are found in cluster 2. Interestingly, cluster 3 contains five genes implicated in post-transcriptional regulation of gene expression, namely *cspB*, *cspG*, *cspF*, *cspI*, and *deaD*. Csp proteins belong to a large family of small (approx. 70 amino acids), mostly acidic proteins conserved amongst bacteria and containing the 'cold shock domain', a universally conserved RNA binding fold (Graumann and Marahiel, 1998). Only a subset of the nine homologous Csp proteins present in *E. coli* , namely CspA, CspB, CspG and CspI, are transiently induced upon cold shock. However, the functional role of the different cold-shock proteins seems to overlap since, for example, CspE becomes cold-inducible in a *cspA*, *cspB*, *cspG* triple deletion mutant and these four genes are collectively essential for growth of *E. coli* at low temperatures (Xia *et al.*, 2001). Csp proteins, as well as DeaD, an ATP-dependent RNA helicase, are thought to act as RNA chaperones and are implicated in post-transcriptional processes such as transcription termination, RNA stability and mRNA translation, thus modulating gene expression in diverse environmental conditions.

It would appear from these data that in the 'warm response', as in cold-shock adaptation, changes in gene expression pattern appear to be controlled mostly by post-transcriptional mechanisms, mainly mRNA stability and translatability. However, it should be mentioned that mRNA and cognate protein abundance do not always follow the same pattern and, most importantly, control of transcriptional regulators and their activity might not be primarily at the transcript abundance level and might therefore escape a transcriptomic analysis that selects for genes with the most divergent expression patterns in the two conditions. Careful screening of genes with less drastically diverse expression patterns as well as analysis of temperature up shift of *E. coli* mutants in genes known to be involved in cold adaptation and/or in global regulators of gene expression will be required to identify key regulators of warm adaptation.

Overall, these seminal data suggest that 'warm adaptation' may not be a trivial matter for a bacterial cell as seen by the long time needed by shifted-up cells to recover a growth rate comparable to cells acclimatized to 37°C.

Identifying genes that are differentially expressed in response to temperature shift

The previous analysis provides an example of how relatively simple data exploration techniques, such as PCA, can help identify the general dynamics of the process and select gene subsets that may be representing important components of the overall cell response. Although this approach is very powerful it suffers from some limitations. The main limitation is that it is not very useful unless the transition of interest can be unambiguously associated to one of the principal components. It follows that there is a need for a methodology that is designed to identify genes differentially expressed over time with respect to a control time course. In this chapter we have already described a methodology developed by Tai and Speed (2005) that aims to rank the genes by the degree of differential expression over

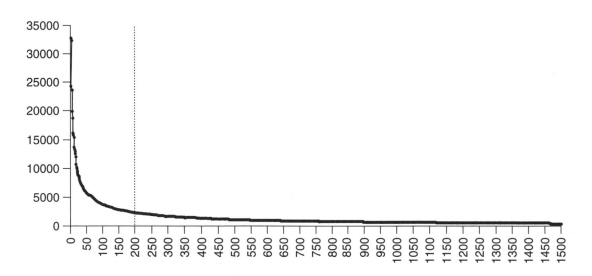

Figure 6.5

Selection of genes differentially expressed according to the Hotelling T^2 statistic. The *x* axis represents the genes ordered in descending order for the value of the Hotelling T^2 statistic whereas the *y* axis represents the value of the statistics. The vertical bar represents the 200 most differentially expressed genes.

time. The method attempts to calculate a p value but ranks the genes on the basis of a multivariate empirical Bayes' log-odds score (the MB statistic) or a Hotelling T^2 statistic. We have applied this methodology to the comparison between the shifted and control time series and selected the 200 most differentially expressed genes (*Figure 6.5*). Interestingly, the large majority of the genes identified with the PCA are also identified as differentially expressed with this methodology. In order to provide an initial biological interpretation of our results we have used the web-based tool GoStat (Beissbarth *et al.*, 2004) to identify biological functions that are significantly associated with differential expression. This web-based tool uses gene ontology to describe the molecular function, biological process and cellular component that are describing a certain gene product. The tool also compares the frequency of appearance of a certain functional term between the group of genes that are differentially expressed and a reference group of nondifferentially expressed genes and associates a p value to this comparison (Chapter 7 provides a detailed overview of the functional analysis of the results of microarray experiments). *Tables 6.5* and *6.6* summarize the results of this analysis. The picture that emerges from this high level functional analysis highlights a general shift in the control of metabolic pathways involved in the regulation of energy metabolism.

Table 6.5 Gene ontology terms (biological processes) significantly enriched in response to temperature shift

Groupcount	Totalcount	Pvalue	GO as name
38	170	2.87E-15	BP (electron transport)
44	231	1.26E-13	BP (generation of precursor metabolites and energy)
68	599	1.24E-06	BP (establishment of localization)
66	576	1.30E-06	BP (transport)
11	23	2.29E-06	BP (cellular respiration)
9	21	8.64E-05	BP (polyol metabolism)
8	17	0.000106	BP (tricarboxylic acid cycle)
8	17	0.000106	BP (coenzyme catabolism)
8	17	0.000106	BP (acetyl-CoA catabolism)
8	17	0.000106	BP (cofactor catabolism)
8	18	0.000173	BP (aerobic respiration)
8	21	0.000651	BP (acetl-CoA metabolism)
14	63	0.000683	BP (energy derivation by oxidation of organic compounds)
11	47	0.00296	BP (main pathways of carbohydrate metabolism)
124	1644	0.0064	BP (cellular physiological process)
4	7	0.00818	BP (galactitol metabolism)
4	7	0.00818	BP (alditol metabolism)
4	7	0.00818	BP (hexitol metabolism)
5	14	0.0192	BP (glycerol metabolism)
124	1678	0.0193	BP (cellular process)
13	87	0.029	BP (alcohol metabolism)
10	58	0.0389	BP (carboxylic acid transport)
97	1243	0.0411	BP (cellular metabolism)
2	2	0.0441	BP (anaerobic electron transport)
2	2	0.0441	BP (disaccharide transport)
8	47	0.0841	BP (carbohydrate transport)

Table 6.6 Gene ontology terms (molecular function) significantly enriched in response to temperature shift

Groupcount	Totalcount	Pvalue	GO as name
45	241	1.28E-13	MF (oxidoreductase activity)
31	162	3.35E-09	MF (iron ion binding)
42	292	6.73E-07	MF (transition metal ion binding)
43	313	1.67E-06	MF (cation binding)
18	90	1.37E-05	MF (iron-sulfur cluster binding)
46	369	1.70E-05	MF (metal ion binding)
5	5	4.31E-05	MF (succinate dehydrogenase activity)
17	75	9.12E-05	MF (4 iron, 4 sulfur cluster binding)
7	18	0.00172	MF (nickel ion binding)
3	3	0.00531	MF (dimethyl sulfoxide reductase activity)
3	3	0.00531	MF (oxidoreductase activity, acting on heme group of donors, oxygen as acceptor)
3	3	0.00531	MF (heme-copper terminal oxidase activity)
3	3	0.00531	MF (cytochrome-c oxidase activity)
3	3	0.00531	MF (oxidoreductase activity, acting on heme group of donors)
6	16	0.00585	MF (oxidoreductase activity, acting on the CH-CH group of donors)
5	11	0.0064	MF (oxidoreductase activity, acting on sulfur group of donors)
4	8	0.0148	MF (3 iron, 4 sulfur cluster binding)
11	65	0.0308	MF (carbohydrate transporter activity)
11	65	0.0308	MF (sugar transporter activity)
11	65	0.0308	MF (sugar porter activity)
5	16	0.033	MF (molybdenum ion binding)
19	153	0.0386	MF (carrier activity)
3	6	0.0447	MF (formate dehydrogenase activity)
3	6	0.0447	MF (oxidoreductase activity, acting on hydrogen as donor, iron-sulfur protein as acceptor)
3	6	0.0447	MF (ferredoxin hydrogenase activity)
10	61	0.0447	MF (carboxylic acid transporter activity)
10	61	0.0447	MF (organic acid transporter activity)
11	78	0.0922	MF (electron transporter activity)
8	48	0.0928	MF (oxidoreductase activity, acting on CH-OH group of donors)

6.6.2 Case study 2: The response of human intestinal cells to *E. coli* infection

The identification of genes differentially expressed in response to the presence of bacteria

The design of this experiment aims to compare gene expression in CaCo2 cells infected by *E. coli* with uninfected control cells at three time points. In Case study 1 we have shown that the analysis of time course data requires the use of particular statistical methods that are designed to identify genes with a different expression profile over time. In this case study, because of the limited number of time points, we have chosen a different strategy to identifying differentially expressed genes. We will compare infected and uninfected samples at a given time point using a *t*-test. We will therefore

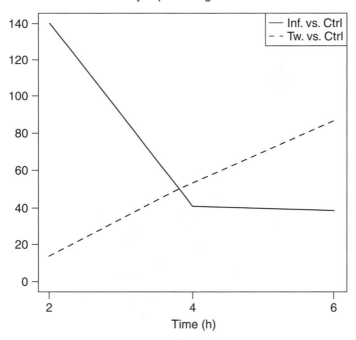

Figure 6.6

Number of differentially expressed genes in response to infection. The plot shows the number of differentially expressed genes (*y* axis) as a function of time. The number of differentially expressed genes is the result of the comparison between infected cells or cells in transwell and control cells.

perform six comparisons (infected host cells versus control cells and host cells exposed to bacteria factors (transwell) versus control cells at 2 h, 4 h, and 6 h). *Figure 6.6* shows the number of genes differentially regulated (at a threshold FDR ≤ 10%) at the different time points. The summary plot shows that infected cells display the largest number of differences at 2 h (140 genes) with a dramatic reduction at 4 and 6 h. Cells exposed to diffusible factors instead show a gradual increase from 2 h (18 genes), 4 h (58 genes) to its maximum at 6 h (80 genes). *Figure 6.7* shows the results of cluster analysis performed using the expression profiles of genes differentially expressed in at least one of the comparisons. A simple visual inspection of the dendrogram classifying the experimental samples shows two very separate clusters. On the left hand side the infected samples cluster together with the transwell samples. In particular it appears that the molecular state of cells at 6 h is most similar to the molecular profile of infected cells at 2 and 4 h suggesting that simple contact of cells with diffusible factors is sufficient to produce a delayed response that is comparable to the acute response to infection. The dendrogram classifying the genes clearly separates genes that are upregulated in response to infection (cluster 1) to downregulated genes (cluster 2). Appendix 4 lists the genes used for the clustering arranged by their position in the dendrogram in *Figure 6.7*.

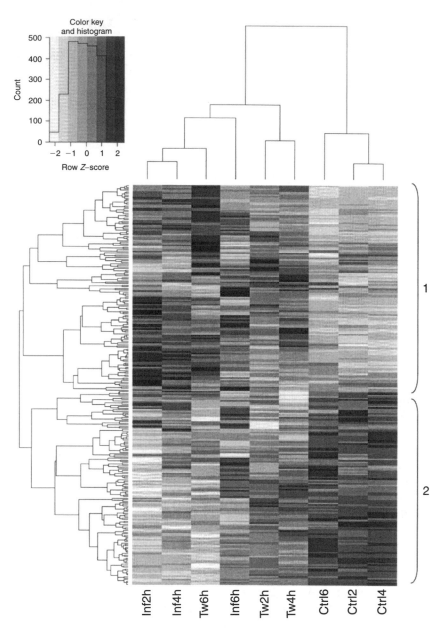

Figure 6.7

Cluster analysis of genes differentially expressed during host response to infection.

Amongst the genes that were highly expressed in infected cells at the 2 h time point were those encoding factors belonging to (or activators of) the NF-κB pathway, such as CXCL2, GRO-β, ICAM1, CCL20, CXCL3, GRO-γ, IRAK2, Myd88, and IL1RI. Myd88 is an adapter protein involved in the Toll-like receptor and IL-1 receptor signaling pathway in the innate immune response. Myd88 acts via IRAK1, IRAK2 and TRAF6, leading to NF-κB activation, cytokine secretion, increase of IL-8 transcription and to the activation of the inflammatory response (Cenni *et al.*, 2003), especially in response to bacterial LPS (Zhang *et al.*, 1999). Myd88 was also upregulated in CaCo2 cells exposed to factors released by EHEC cells in the transwell culture system at 4 and 6 h time points. Other genes related to the NF-κB pathway were also upregulated in the transwell experiment (*CXCL2, TNFAIP2, CXCL3,* GRO-γ, *ICAM1, CXCL1,* Groα, and *IRAK2*) indicating that NF-κB activation is, to some extent, not dependent on the host–pathogen direct physical interaction.

The differentiation of T-cell/NK-cell (mediated by the Notch signaling pathway), the activation of bactericidal processes in macrophages (mediated by Inos) and promotion of leukocyte adhesion and migration (mediated by CX3CL1) are important components of the antibacterial host response. We discovered that factors involved in these important functions are upregulated during infection. Interestingly, Inos and CX3CL1 expression did not change in CaCo2 cells infected by EHEC in the transwell culture system. This may indicate their activation was contact dependent.

In parallel to the activation of the NF-κB pathway and the immune response, EHEC infection resulted in the early upregulation of genes encoding adhesion molecules, such as ICAM1, ALCAM, ITGB6, ITGA2, ICAM-3, ITGAX (2 h time point). At later time points only ICAM1 and ITGA2 expression was maintained at higher levels in CaCo2 cells directly infected by EHEC than in uninfected cells. ICAM proteins are ligands for the leukocyte adhesion LFA-1 protein (integrin alpha-L/beta-2) and for this reason they are involved in leukocyte adhesion and migration processes across endothelia (Amartey *et al.*, 1997). Integrins are important adhesion molecules, and play a pivotal role in the immune response. In particular the complex integrin alpha-X/beta-2 is a receptor for fibrinogen, and mediates cell–cell interaction during inflammatory responses. It is especially important in monocyte adhesion and chemotaxis (Garnotel *et al.*, 2000).

Additionally, many cell adhesion genes were upregulated in the transwell system, such as *ITGA2, ITGAX, ITGA6,* and *ICAM1* (at 4 and 6 h time point), *ITGB2* and *ICAM3* (only at the 4 h), and *ITGB6* (at the 6 h time point). This may indicate that the contribution of integrins to the host inflammatory response was not dependent on direct contact with the bacteria, but it could also be induced by soluble bacterial factors. A number of cell adhesion molecules were downregulated in the infected cells. These were *MCAM* (melanoma cell adhesion molecule) and *ITGA5* (at the 2 h time point in the direct infection condition); ITGA5 (at the 4 h time point in the transwell culture system); *ALCAM* and *Mad-CAM1* (at the 6 h time point in the transwell culture system). MCAM protein is involved in cell adhesion, and in cohesion of the endothelial monolayer at intercellular junctions in vascular tissue (Anfosso *et al.*, 2001), whereas ITGA5 is a receptor for fibronectin and fibrinogen (Ly *et al.*, 2003). *ALCAM* binds to CD6 and may play a role in the binding of T- and B-cells to activated leukocytes (Bowen *et al.*, 1996). These cell adhesion genes have important functions for the host immune response, especially in the recruitment of immune cells to the site of infection. The reduction of their expression may represent a bacterial attempt to counteract the general onset of factors activating or sustaining the immune response, which is triggered in the host cells by interaction with bacteria.

Pro-apoptotic genes were consistently downregulated in response to infection at earlier time points (2 and 4 h) but were upregulated at 6 h. A similar pattern was observed in the transwell culture experiment. Genes involved in cell motility, such as members of the *Wnt* signaling pathway and *AGRIN*, were downregulated consistently in all time points both during infection and in the transwell experiment.

These results suggest that genes involved in many important physiological functions are regulated in the initial phase of infection. These effects on cell physiology seem to be largely independent on cell–cell contact and therefore not under the control of effector proteins linked to the activity of the type III secretion system.

6.7 Conclusions

The two case studies we have described in this chapter show how relatively simple experiments in combination with the appropriate data exploration techniques can reveal a surprisingly large amount of information. Beyond the biological importance of the individual genes, the analysis of *E. coli* cells undergoing temperature up-shift shows how the adaptation process can be described as a rapid transition between two equilibrium states. The methodologies described here are general strategies to apply to the analysis of state transition including eukaryotic cell differentiation. The analysis of host cell response to bacterial infection on the other hand has been a simple example of the comparison of different experimental conditions. The analysis strategy has been based on the identification of genes differentially expressed using a simple *t*-test followed by the visualization of patterns using cluster analysis. This example has shown how interesting conclusions can be drawn by the simple visual inspection of a clustering dendrogram. In particular it has revealed that the response of host cells to bacterial infection is in large part dependent on diffusible factors.

Acknowledgments

The research described in this manuscript is, in part, supported by an IDBA grant from MRC/ EPSRC, by the BBSRC Grant EGA16107, and the Fondazione Cariplo Grant 2005.1076/10.4878. This material is also, in part, based upon work supported by the National Science Foundation under Grant No. 0524331. Any opinions, findings and conclusions or recommendations expressed in this material are those of the authors and do not necessarily reflect the views of the National Science Foundation (NSF). Victor Trevino and Donatella Sarti are recipients of Darwin Trust fellowships.

References

Amartey JK, Parhar RS, al-Sedair ST (1997) The effect of antibodies against cell-surface adhesion molecules (LFA-1 alpha and ICAM-1) on the migration and localization of ^{99m}Tc-labeled leukocytes in acute infection. *Nucl Med Biol* **24**: 603–605.

Anfosso F, Bardin N, Vivier E, Sabatier F, Sampol J, Dignat-George F (2001) Outside-in signaling pathway linked to CD146 engagement in human endothelial cells. *J Biol Chem* **276**: 1564–1569.

Baldwin DN, Vanchinathan V, Brown PO, Theriot JA (2003) A gene-expression program reflecting the innate immune response of cultured intestinal epithelial cells to infection by Listeria monocytogenes. *Genome Biol* **2003:4:R2** Epub 2002.

Barrett T, Suzek TO, Troup DB, Wilhite SE, Ngau WC, Ledoux P, Rudnev D, Lash AE, Fujibuchi W, Edgar R (2005) NCBI GEO: mining millions of expression profiles – database and tools. *Nucleic Acids Res* **33**(Database issue): D562–566.

Beissbarth T, Speed TP (2004) GOstat: find statistically overrepresented gene ontologies within a group of genes. *Bioinformatics* **20**: 1464–1465.

Benjamini Y, Hochberg Y (1995) Controlling the false discovery rate – a practical and powerful approach to multiple testing. *J R Stat Soc B – Methods* **57**: 289–300.

Boldrick JC, Alizadeh AA, Diehn M, Dudoit S, Liu CL, Belcher CE, Botstein D, Staudt LM, Brown PO, Relman DA (2002) Stereotyped and specific gene expression programs in human innate immune responses to bacteria. *Proc Natl Acad Sci USA* **99**: 972–977.

Bolstad BM, Irizarry RA, Astrand M, Speed TP (2003) A comparison of normalization methods for high density oligonucleotide array data based on variance and bias *Bioinformatics* **19**: 185–193.

Bowen MA, Bajorath J, Siadak AW, Modrell B, Malacko AR, Marquardt H, Nadler SG, Aruffo A (1996) The amino-terminal immunoglobulin-like domain of activated leukocyte cell adhesion molecule binds specifically to the membrane-proximal scavenger receptor cysteine-rich domain of CD6 with a 1:1 stoichiometry. *J Biol Chem* **271**: 17390–17396.

Cenni V, Sirri A, De Pol A, Maraldi NM, Marmiroli S (2003) Interleukin-1-receptor-associated kinase 2 (IRAK2)-mediated interleukin-1-dependent nuclear factor κB transactivation in Saos2 cells requires the Akt/protein kinase B kinase. *Biochem J* **376**: 303–311.

Conesa A, Nueda MJ, Ferrer A, Talon M (2006) maSigPro: a method to identify significantly differential expression profiles in time-course microarray experiments. *Bioinformatics* **22**: 1096–1102 (e-pub 15 February 2006).

Cox JM, Clayton CL, Tomita T, Wallace DM, Robinson PA, Crabtree JE (2001) cDNA array analysis of cag pathogenicity island-associated Helicobacter pylori epithelial cell response genes. *Infect Immun* **69**: 6970–6980.

Detweiler CS, Cunanan DB, Falkow S (2001) Host microarray analysis reveals a role for the Salmonella response regulator phoP in human macrophage cell death. *Proc Natl Acad Sci USA* **98**: 5850–5855.

Dopazo J, Zanders E, Dragoni J, Amphlett G, Falciani F (2001) Methods and approaches in the analysis of gene expression data. *J Immunol Methods* **250**: 93–112.

Eriksson S, Lucchini S, Thompson A, Rhen M, Hinton JC (2003) Unravelling the biology of macrophage infection by gene expression profiling of intracellular *Salmonella enterica*. *Mol Microbiol* **47**: 103–118.

Garnotel R, Rittie L, Poitevin S, Monboisse JC, Nguyen P, Potron G, Maquart FX, Randoux A, Gillery P (2000) Human blood monocytes interact with type I collagen through alpha x beta 2 integrin (CD11c-CD18, gp150-95). *J Immunol* **164**: 5928–5934.

Graumann PL, Marahiel MA (1998) A superfamily of proteins that contain the cold-shock domain. *Trends Biochem Sci* **23**: 286–290.

Gualerzi CO, Giuliodori AM, Pon CL (2003) Transcriptional and post-transcriptional control of cold-shock genes. *J Mol Biol* **331**: 527–539.

Lee YK, Puong KY (2002) Competition for adhesion between probiotics and human gastrointestinal pathogens in the presence of carbohydrate. *Br J Nutr* **88**(**suppl 1**): S101–S108.

Ly DP, Zazzali KM *et al.* (2003) De novo expression of the integrin alpha5beta1 regulates alphavbeta3-mediated adhesion and migration on fibrinogen. *J Biol Chem* **278**: 21878–21885.

Narberhaus F (1999) Negative regulation of bacterial heat shock genes. *Mol Microbiol* **31**: 1–8.

Raivio TL, Silhavy TJ (2001) Periplasmic stress and ECF sigma factors. *Annu Rev Microbiol* **55**: 591–624.

Sauvonnet N, Pradet-Balade B *et al.* (2002) Regulation of mRNA expression in macrophages after *Yersinia enterocolitica* infection. Role of different Yop effectors. *J Biol Chem* **277**: 25133–25142.

Stekel DJ (2003) *Microarray Bioinformatics*. Cambridge University Press, Cambridge.

Stekel DJ, Sarti D, Trevino V, Zhang L, Salmon M, Buckley CD, Stevens M, Pallen MJ, Penn C, Falciani F (2005) Analysis of host response to bacterial infection using error model based gene expression microarray experiments. *Nucleic Acids Res* **33**: e53.

Tai YC, Speed TP (2005) Statistical analysis of microarray time course data. In: *DNA Microarrays* (ed U. Nuber). BIOS Scientific Publishers, Oxford, chapter 20.

Takahashi M, Taguchi H, Yamaguchi H, Osaki T, Komatsu A, Kamiya S (2004) The effect of probiotic treatment with *Clostridium butyricum* on enterohemorrhagic *E. coli* O157:H7 infection in mice. *FEMS Immunol Med Microbiol* **41**: 219–226.

Talaat AM, Lyons R, Howard ST, Johnston SA (2004) The temporal expression profile of *Mycobacterium tuberculosis* infection in mice. *Proc Natl Acad Sci USA* **101**: 4602–4607.

Vaquerizas JM, Conde L, Yankilevich P, Cabezon A, Minguez P, Diaz-Uriarte R, Al-Shahrour F, Herrero J, Dopazo J (2005) GEPAS, an experiment-oriented pipeline for the analysis of microarray gene expression data. *Nucleic Acids Res* **33**: W616–W620. [Web issue.]

Voravuthikunchai S, Lortheeranuwat A, Jeeju W, Sririrak T, Phongpaichit S, Supawita T (2004) Effective medicinal plants against enterohemorrhagic *E. coli* O157:H7. *J Ethnopharmacol* **94**: 49–54.

Weber MHW, Marahiel MA (2003) Bacterial cold shock responses. *Sci Prog* **86**: 9–75.

Xia B, Ke H, Inouye M (2001) Acquirement of cold sensitivity by quadruple deletion of the cspA family and its suppression by PNPase S1 domain in *Escherichia coli*. *Mol Microbiol* **40**: 179–188.

Zhang FX, Kirschning CJ, Mancinelli R, Xu XP, Jin Y, Faure E, Mantovani A, Rothe M, Muzio M, Arditi M (1999) Bacterial lipopolysaccharide activates nuclear factor-kappaB through interleukin-1 signaling mediators in cultured human dermal endothelial cells and mononuclear phagocytes. *J Biol Chem* **274**: 7611–7614.

Functional annotation of microarray experiments

7

Joaquín Dopazo and Fátima Al-Shahrour

7.1 Introduction

For many years, molecular biology has addressed functional questions by studying individual genes, either independently or a few at a time. In spite of the intrinsic reductionism of this approach, an important part of our knowledge on functional properties and biological roles of genes and gene products was obtained in this way. However, in only a few years, the advent of high-throughput technologies has drastically changed the scenario. DNA microarrays constitute probably the paradigm among these technologies (Rhodes and Chinnaiyan, 2005) which are characterized for producing large amounts of data, whose analysis and interpretation is far from being trivial. Thus, the possibility of obtaining experimental measurements of thousands of genes in a fast and relatively cheap way has opened up new possibilities in querying living systems at the genome level that are beyond the old paradigm 'one-gene–one-postdoc'. Relevant biological questions regarding genes, gene product interactions or biological processes played by networks of components, etc., can now for the first time be realistically addressed. Nevertheless, genomic technologies are at the same time generating new challenges for data analysis and are demanding a drastic change in the habits of data management and in the biological interpretation of the results. Dealing with this overabundance of data must be approached cautiously because of the high occurrence of spurious associations if the proper methodologies are not used (Ge *et al.*, 2003).

In many cases, traditional molecular biology approaches tended to mix up the concepts of data and information. This was partially due to the fact that researchers had a great deal of information previously available about the data units they used (genes, proteins, etc.). Nevertheless, over the last few years the increasing availability of high-throughput methodologies has amplified by orders of magnitude the potential of data production. One direct consequence of this revolution in data production has been the end of the fictitious equivalence between data and information. Nowadays, data are produced faster than they can be analyzed by traditional means. It is not uncommon when analyzing microarray data to face paradoxes such as finding relevant behaviors for unannotated genes. Microarray experiments allow an explanation to be found at the molecular level after some macroscopic observation (e.g. differences between cases and controls, before and after some stimulus, differences in survival, etc.) Ending up with a list of differentially expressed genes is only a part of this explanation. A detailed understanding of the roles played by these genes is required for the proper interpretation of the results of a microarray experiment. Thus, functional annotation of the data produced in a microarray experiment has become a necessity (Al-Shahrour, *et al.* 2005b; Khatri and Draghici, 2005). In this context, different procedures for extracting terms characteristic of and specific to groups of genes, beyond the classical approach 'one-gene-at-a-time' have recently been developed (Khatri and Draghici, 2005).

Nevertheless, as previously mentioned, when the proper methodologies are not used spurious associations can be found (Ge *et al.*, 2003) which are often considered as evidence of actual functional links, leading to misinterpretation of results. All these characteristics of genomic data must be taken into account for any procedure aiming to properly identify functional roles in groups of genes. Reducing the number of false positives without a drastic loss of sensitivity in the method is also a challenge in functional annotation procedures. Methods such as the FatiGO (Al-Shahrour *et al.*, 2004) and other similar ones (Khatri and Draghici, 2005) provide efficient functional annotation at the desired ratios of false positives.

Moreover, the use of biological information in combination with values of gene expression provides a more sensitive way of detecting moderate (but coordinated) activations or repressions of blocks of genes functionally related which cannot be detected by approaches based merely on the study of the values of gene expression (Mootha *et al.*, 2003; Al-Shahrour *et al.*, 2005a; Subramanian *et al.*, 2005). The use of biological information in innovative algorithms used for detecting differential expression of genes constitutes a new and active field.

7.2 Scientific background

7.2.1 Functional annotation

The problem we are addressing in this chapter is the functional annotation of a microarray experiment. As mentioned above microarray experiments are designed to find the explanation, at the molecular level, of any macroscopic observation. There are different types of biologically or clinically relevant questions that can be posed within this framework. Typical questions are finding the set of genes with different expression levels when comparing cases and controls, or detecting genes with correlated expression values across a number of experimental conditions, etc. Different mathematical and statistical procedures are used to find these genes (Quackenbush, 2001) but this is only the first step in the process of interpretation of the results. Once the list of relevant genes is available then the aim is to understand what particular biological roles they are playing, which account for the macroscopic observation that originated the experiment. To achieve this, many types of available functional information can be used. Perhaps one of the most popular sources of functional information nowadays is gene ontology (GO) (Ashburner *et al.*, 2000). Also, pathways (Kanehisa *et al.*, 2004) or Interpro functional motifs (Mulder *et al.*, 2005) are considered important sources of biologically relevant information. Other approaches involve the use of information automatically extracted from the scientific literature by means of text-mining methodologies (Blaschke *et al.*, 2002).

The inspection of the functional information available for the genes selected is a necessary but not sufficient step for the proper functional interpretation of a microarray experiment. A typical conceptual error is in looking for the functional annotations of the relevant selected genes and then assigning importance to the most abundant functions. For example, let us imagine that 80 genes were found to be differentially expressed between healthy controls and patients with colon cancer, and 24 of them belonged to the GO class *cellular macromolecule metabolism*. Given that 24 out of 80 is a 30%, this seems to point clearly to *cellular macromolecule metabolism* as the molecular function responsible for the macroscopic observation, which, in this case, is the occurrence of colon cancer. This is a typical error because 28.30% of the genes in the human genome have *cellular macromolecule metabolism* as GO annotation. Thus, after observing the background distribution of GO terms we can conclude that *cellular macromolecule metabolism* is completely irrelevant for understanding the molecular basis of the colon cancer (in this fictitious example): genes with this particular GO term were present both among the differentially expressed and among the equally expressed.

A simple exercise of abstraction shows that the most common problems in functional annotation of DNA microarray data can be reduced to the comparison of two groups of genes. The most common uses of DNA microarrays are derived from experimental designs that involve supervised questions (differentially expressed genes when comparing two or more experimental conditions, genes related to survival, etc.) or unsupervised questions (finding clusters of co-expressing genes or subtypes of experimental conditions). Finding genes that account for diseases or traits (Tracey *et al.*, 2002) or the generation of prognostic predictors (van't Veer *et al.*, 2002) are examples or supervised studies. In the case of unsupervised studies, a clear example is the study of gene co-expression, which tends to be evidence of common function (Lee *et al.*, 2004)

In any case we will be interested in finding what makes our set of 'interesting' genes (differentially expressed, co-expressing, etc.) different from the rest of genes regarding its functionality. In any case we must compare the available functional annotations for both sets of genes. There are different ways of performing this type of comparison (see Khatri and Draghici, 2005 for a review) but usually both lists of genes are converted to the corresponding lists of functional terms (GO, or any other) and then a 2×2 contingency table is constructed for each term and any test, commonly a Fisher's exact test for 2×2 contingency tables, is used.

7.2.2 What can be considered a significant functional difference? Statistical approaches and the multiple testing problem

When we test for differences in the distribution of functional terms between two groups of genes and we do not have any a priori hypothesis what we do is to inspect all the available functional terms. This implies, for example, testing over 17 000 terms if we are using GO as source of functional information. Performing such a large number of tests increases the occurrence of apparent asymmetrical distributions of functional terms, that are purely due to chance. *Table 7.1* shows an interesting simulated

Table 7.1 GO terms found to be differentially distributed when comparing ten independent random partitions of 50 gene samples from the complete genome of *E. coli*. See text for explanation

GO term		Percent in random set	Percent in genome	p value	Adjusted p value
Number	Function				
GO:0042592	Homeostasis	4.65	0.40	0.0157944	0.300094
GO:0050874	Organismal physiological process	6.06	2.06	0.151987	1
GO:0008152	Metabolism	59.52	70.32	0.129865	1
GO:0019058	Viral infectious cycle	4.55	0.40	0.016503	0.181353
GO:0016265	Death	4.76	1.36	0.116317	1
GO:0018942	Organometal metabolism	2.33	0	0.0136	0.7406
GO:0050794	Regulation of cellular processes	2.70	0.15	0.0658124	1
GO:0006725	Aromatic compound metabolism	11.90	3.34	0.0138061	0.374662
GO:0046483	Heterocycle metabolism	13.51	2.76	0.00376158	0.259549
GO:0006766	Vitamin metabolism	10.26	2.02	0.00875837	0.604328
GO:0019059	Initiation of viral infection	4.76	0.35	0.0123062	0.459417
GO:0008360	Regulation of cell shape	2.78	0.03	0.0227086	1
GO:0009056	Catabolism	14.63	5.62	0.0276032	1
GO:0007155	Cell adhesion	6.25	1.86	0.122953	1

experiment: we sampled randomly 50 genes from the genome of *E. coli* and compared the occurrence of functional terms in these genes to the rest of the genome. In many cases we can observe GO terms that are apparently (see uncorrected *p* value in the fourth column of *Table 7.1*) significantly over-represented in the random sets of 50 genes (which obviously are not functionally related).

Addressing multiple testing properly is not a trivial problem. Many of the conventional correction methods (e.g. Bonferroni or Sidak) are based on the consideration that a *p* value should be adjusted by multiplying a reasonable significant threshold (e.g. $p < 0.05$) for the number of tests performed to obtain a new threshold. Whenever many thousands of tests are performed the original assumption risks being too conservative.

A better strategy to estimate *p* values is provided by another family of methods that allow less conservative adjustments. Within this context it is more appropriate to control the proportion of errors among the identified functional terms whose differences among groups of genes cannot be attributed to chance instead. The expectation of this proportion is the false discovery rate (FDR). Different procedures offer control of the FDR under independence and some specific types of positive dependence of the tests statistics (Benjamini and Yekutieli, 2001). The fifth column in *Table 7.1* shows the FDR-adjusted *p* values. None of them is significant (as can be expected from random samples).

7.3 Design of the experiment

For the sake of simplicity we will discuss only two case studies of functional annotation of supervised and unsupervised experimental designs: one of them involving an unsupervised analysis and the other one a supervised analysis.

One of the most popular hypotheses in microarray data analysis is that co-expression of genes across a given series of experiments is most probably explained through some common functional role (Eisen *et al.*, 1998). Actually, this causal relationship has been used to predict gene function from patterns of co-expression (Mateos *et al.*, 2002; van Noort *et al.*, 2003). We will study the co-expression of genes in a classical experiment (DeRisi *et al.*, 1997) in which gene expression is measured in seven consecutive time intervals. Our interest is in finding clusters of co-expressing genes and understanding the functional basis of this co-expression. Clustering will be performed with the self-organizing tree algorithm (SOTA) method (Herrero *et al.*, 2001) and functional annotation with the FatiGO program (Al-Shahrour *et al.*, 2004, 2005b).

On the other hand, we will study a case of gene selection between healthy cases and controls. Once the genes have been selected we will study what functional roles are played in the cell. Gene selection will be carried out using a *t*-test, with FDR adjustment for multiple testing, as implemented in the GEPAS package (Herrero *et al.*, 2003, 2004; Vaquerizas *et al.*, 2005) and the functional annotation of the genes selected as differentially expressed will be performed with the FatiGO program (Al-Shahrour *et al.*, 2004, 2005b).

7.4 Theory of data analysis

7.4.1 Testing unequal distribution of terms between two groups of genes

FatiGO (Al-Shahrour *et al.*, 2004) takes two lists of genes (ideally a group of interest and the rest of the genome, although any two groups, formed in any way, can be tested against each other) and converts them into two lists of GO terms using the corresponding gene-GO association table. A Fisher's exact test for 2 x 2 contingency tables is used. For each GO term the data are represented as a 2 x 2 contingency table with rows being presence/absence of the GO term, and each column representing each of the two clusters (so that the numbers in each cell would be the number of genes of the first cluster where the GO term is present, the number of genes in the first cluster where the GO term is absent, and so on). In addition to GO, any other functional label can be used. For example, FatiWise implements InterPro motifs (Mulder *et al.*, 2005), KEGG pathways (Kanehisa *et al.*, 2004) and SwissProt keywords.

The structure of the functional labels has an important impact in the strategy for performing the test. For example, InterPro motifs have a 'flat', unstructured organization with a correspondence of one or more motifs per protein. Terms in GO, on the other hand, have a hierarchical structure called DAG (for directed acyclic graph, where each term can have one or more child terms as well as one or more parent terms). Terms at higher levels, close to the root, of the hierarchy describe more general functions or processes while terms at lower levels are more specific. The level at which a gene is annotated in the GO hierarchy depends on the detail of knowledge about its biological behavior the curator has. Testing terms organized in such a way poses an additional difficulty because in some cases they are not exclusive but only constitute descriptions of the same behavior at different levels of detail (e.g. where is the point in testing apoptosis versus regulation of apoptosis?). To deal with this, FatiGO implements an inclusive analysis (Al-Shahrour and Dopazo, 2005), in which a level in the DAG hierarchy is chosen for the analysis. Genes annotated with terms that are descendants of the term corresponding to the level chosen therefore take the annotation from the parent. If the level corresponding to, for example, apoptosis was selected, any gene annotated as either apoptosis or any child term was considered in the same category (apoptosis) for the test. This increases the power of the test. There are fewer terms, each with more genes, to be tested (Al-Shahrour and Dopazo, 2005; Al-Shahrour *et al.*, 2005b).

7.4.2 Unsupervised approach

Unsupervised clustering comprises a number of techniques that produce arrangements of the data based on a distance function. These methods do not use any external information for constructing groups of similar profiles of conditions or genes. Despite the arsenal of methods used, the optimal way of classifying gene expression data is still open to debate. It is beyond the scope of this chapter to discuss the virtues and pitfalls of different clustering methods, and review can be found elsewhere (Sheng *et al.*, 2005).

Depending on the way in which the data are clustered we can distinguish between hierarchical and nonhierarchical clustering. Hierarchical clustering allows detection of higher order relationships between clusters of profiles whereas the majority of nonhierarchical classification techniques work by allocating expression profiles to a predefined number of clusters, without any assumption on the inter-cluster relationships. Aggregative hierarchical clustering in its different variants (average-linkage, single-linkage, complete-linkage, etc.) (Sneath and Sokal, 1973) is still one of the preferred choices for the analysis of patterns of gene expression. As an alternative to hierarchical clustering, there are nonhierarchical methods, like quality cluster (Heyer *et al.*, 1999) or *k*-means (Hartigan, 1975). These algorithms start with a pre-defined number of clusters and, by iterative reallocation of cluster members, minimize the overall within-cluster dispersion. Common criticisms of these types of algorithms focus on the fact that the number of clusters has to be fixed from the beginning of the procedure.

Unsupervised neural networks, such as self-organizing maps (SOMs) (Kohonen, 1997) or the SOTA (Herrero *et al.*, 2001), provide a more robust framework, appropriate for clustering large amounts of noisy data. Because of their properties, neural networks are suitable for the analysis of gene expression patterns. They can deal with real-world data sets containing noisy, ill-defined items with irrelevant variables and outliers, and whose statistical distributions do not need to be parametric.

Nevertheless, the SOM has some inherent problems. Firstly, the number of clusters is arbitrarily fixed from the beginning, as in *k*-means. In addition, the training of the network (and, consequently, the definition of clusters) depends on the number of items in each cluster. Thus the clustering obtained is not proportional. If irrelevant data (e.g. invariant, 'flat' profiles) or some particular type of profile is over-represented, an SOM will produce an output in which this type of data will populate the vast majority of clusters. Contrary to this, clustering obtained with SOTA is proportional to the heterogeneity of the data, instead of to the number of items in each cluster. Thus, regardless of whether a given type of profile is abundant, all the similar items will remain grouped together in a single cluster and they will not directly affect to the rest of the clustering. This is because SOTA is distribution-preserving while SOM is topology-preserving (Dopazo and Carazo, 1997).

7.4.3 Supervised approach

Contrary to the case of unsupervised analysis, in the supervised approach we use the information available on classes of experiments (e.g. cases versus controls, before and after the administration of a drug, etc.). Typical problems here are class comparison (which often implies finding differentially expressed genes), or class prediction or prognostic prediction.

In class comparison we ask if different classes of experiments differ in their gene expression. Usually, the result is a list of genes ranked by their degree of differential expression between the classes. Conversely, in class prediction or prognostic prediction we aim to predict the class membership of a set of experiments given their gene expression data. Again, the description of the different mathematical procedures used for these purposed is beyond the scope of the chapter and can be found in different reviews (Draghici, 2002; Speed, 2003).

Commonly used methods, involve the use of standard statistical tests (e.g. a t-test for two-class comparisons, ANOVA for multiple class comparisons, Cox models for survival data, etc.), where analyses are carried out gene-by-gene (Cui and Churchill, 2003; Reiner *et al.*, 2003), with appropriate correction for multiple testing, since we can be conducting thousands of hypotheses, via control of the family-wise error rate or the FDR (Reiner *et al.*, 2003).

7.5 Data analysis

Most of the material in this section can be also found on-line in the GEPAS tutorials at http://gepas.bioinfo.cipf.es/cgi-bin/documentation.

7.5.1 Functional annotation of a cluster of co-expressing genes

As previously mentioned, our aim is to find clusters of co-expressing genes and understand the functional basis of this co-expression. Clustering will be performed with the SOTA method (Herrero *et al.*, 2001), which is part of the GEPAS (Herrero *et al.*, 2003, 2004; Vaquerizas *et al.*, 2005) package, and the functional annotation will be carried out using the FatiGO program (Al-Shahrour *et al.*, 2004, 2005b).

Step 1: The data

The **GEPAS** program can be found at http://www.gepas.org. The file can be accessed through the corresponding link within the area of public datasets. The first one is called 'Diaxic shift'. For the sake of simplicity we will skip the steps of normalization and preprocessing and directly use the file already processed through this link: http://gepas.bioinfo.cipf.es/data/diauxic/data.txt. Save the file in your computer.

Step 2: Clustering the data

Go to GEPAS (http://www.gepas.org) open the **Tools** section, open the **Clustering** section and choose **SOTA**. *Figure 7.1* shows the program

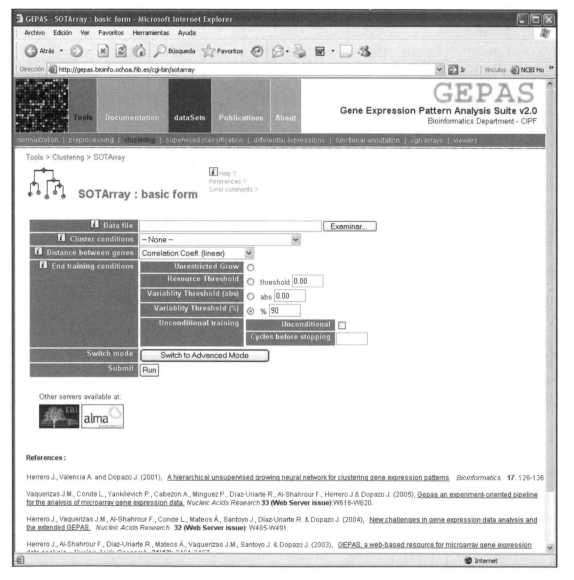

Figure 7.1

Interface to the SOTArray program. (A color version of this figure is available at the book's web site, www.garlandscience.com/9780415378536)

interface. Then upload the file in the **Data file** box and click run with the default options. Once you get the results send them to the tree viewer (click on **Send to Sota Tree**). You will get the interface of the tree viewer. Again click on **run** with the default options. You will get then the resulting tree (see *Figure 7.2*). The circles at the end of the branches represent clusters of genes with similar expression patterns (represented contiguously, on the right side of the cluster).

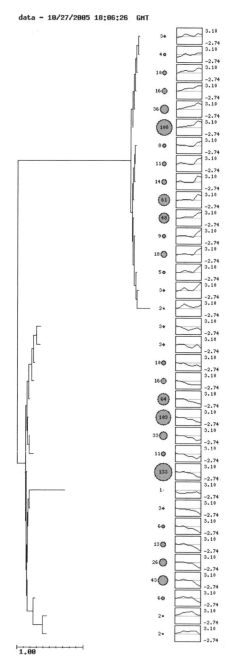

Figure 7.2

Clustering of genes obtained by the SOTA methods.

Step 3: Functional annotation of clusters of co-expressing genes

Click on any cluster and the system will prompt two windows. One of them has the average expression pattern as well as any individual expression pattern and the other one contains the list of genes in the cluster. Using this second window you can send the genes to the annotation tools.

If you click on **Extract this cluster** you will be sent to the cluster extraction window. There you can choose among different annotation tools.

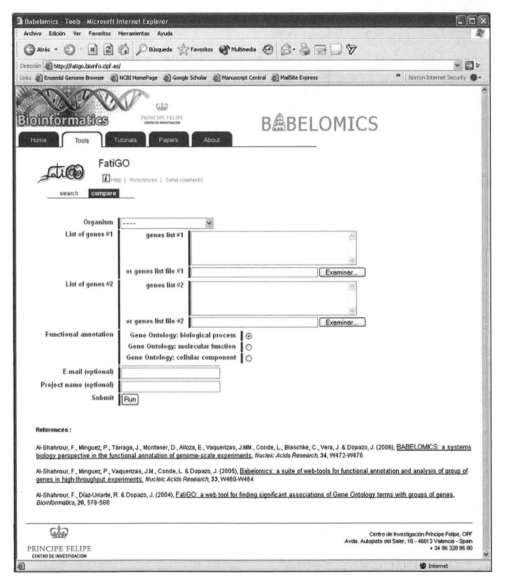

Figure 7.3

Interface to the FatiGo program (http://fatigo.bioinfo.cipf.es). (A color version of this figure is available at the book's web site, www.garlandscience.com/9780415378536)

Click on **Send to FatiGO**. You will get the interface for the FatiGO annotation tool (*Figure 7.3*).

To perform the annotation you have to select the organism, the ontology and the level and then run the program.

The example shown in *Figure 7.4* represents the annotation of the fifth cluster from the top of the tree, annotated in biological process at level 5. **Carbohydrate metabolism** is clearly over-represented in this cluster

Figure 7.4

Results of the FatiGO program. (A color version of this figure is available at the book's web site, www.garlandscience.com/9780415378536)

(41.38%) compared with the background (8.62%) distribution (see second column of adjusted p values). The rest of terms were not significant.

7.5.2 Functional annotation of differentially expressed genes

In the same way you can annotate a supervised experiment. Given a set of genes differentially expressed between two conditions you may be interested in finding what, in functional terms, makes them different from the rest of genes which do not display a different behavior. The **T-Rex** tool from the **GEPAS** (Herrero *et al.*, 2003, 2004; Vaquerizas *et al.*, 2005) package can be used for obtaining the genes differentially expressed.

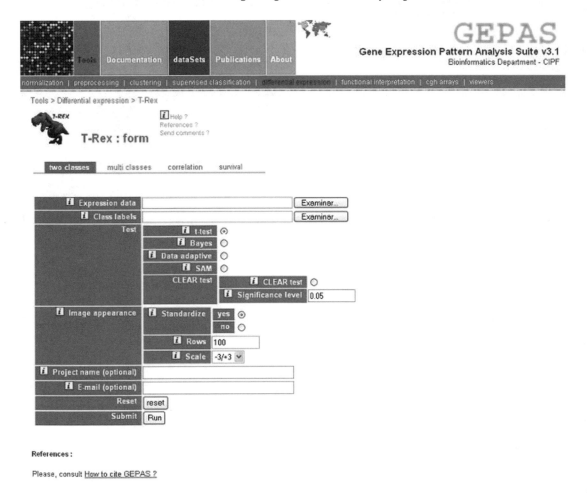

Figure 7.5

Interface to the T-Rex program for finding differentially expressed genes. (A color version of this figure is available at the book's web site, www.garlandscience.com/9780415378536)

Step 1: Finding genes differentially expressed

The **GEPAS** program can be found at http://www.gepas.org. Go to **Tools**, then to **differential expression** and then open **T-Rex**. You will get the interface (see *Figure 7.5*).

Load your data and your labels for classes. See the help pages of the program for additional information (http://t-rex.bioinfo.cipf.es/cgi-bin/docs/t-rex-help). Then run the program. You will get a result similar to what you can see in *Figure 7.6*. Genes differentially expressed are represented color-coded.

Step 2: Functional annotation of the genes

In **T-Rex** you can select a *p* value and send the results to **FatiGO**. The total list of genes will be divided into two groups according this threshold. Similarly to the case previously described you will be sent to the FatiGO interface where you can run the analysis and see if there were any functional terms specific to the differentially expressed genes.

In the tutorials section of **GEPAS** (http://gepas.bioinfo.cipf.es/cgi-bin/tuto), in the **T-Rex** tutorial, you can find several examples in which enrichment in GO terms is found in genes differentially expressed.

7.6 Summary of the results, conclusions, and suggestions for the general implementation of the case study

The importance of using biological information as an instrument to understand the biological roles played by genes targeted in functional genomics experiments has been highlighted in this chapter. Typical functional genomics methodologies permit the arrangement of genes by experimental criteria (co-expression across experiments, differential expression among classes of experimental conditions, etc.). Once genes of potential interest have been detected, the next goal is to determine the significant biological roles that these genes are carrying out, which can explain at the molecular level the macroscopic observation of the experiment. Significance, in the context of genome-scale studies, must be considered with caution because of the high probability of finding spurious associations just by chance (Ge *et al.*, 2003). In addition to FatiGO, which has been described in more detail, there are a number of tools available that allow functional terms (GO and other) to be obtained which significantly characterize a group of genes with common experimental behavior. Actually, the Babelomics (Al-Shahrour *et al.*, 2005b) suite for functional annotation of functional genomic experiments is among the most complete ones available.

Nevertheless, information can be used in other ways than just for obtaining functional labels for groups of genes. There are situations in which the existence of noise and/or the weakness of the signal hamper the detection of real inductions or repressions of genes. Improvements in methodologies of data analysis, dealing exclusively with expression values can to some extent help. Recently, the idea of using biological knowledge as part of the analysis process is gaining in support and popularity. The rationale is similar to the justification of using biological information to understand the biological

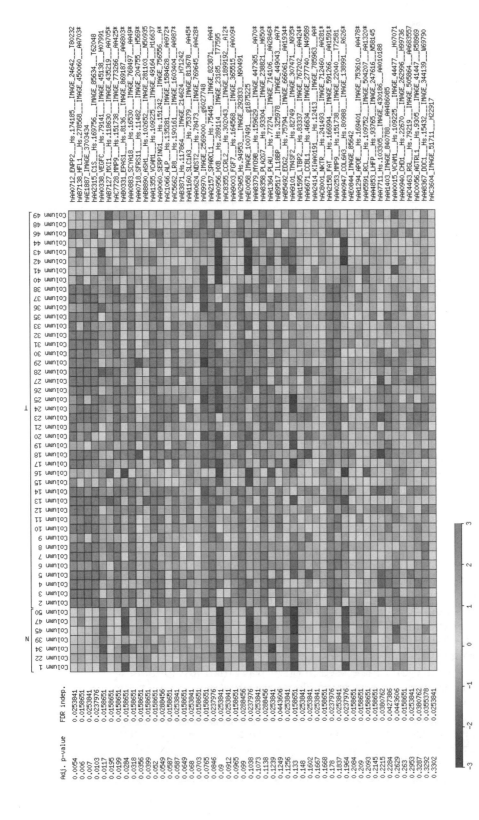

Figure 7.6

Genes differentially expressed as found by the T-Rex program. (A color version of this figure is available at the book's web site, www.garlandscience.com/9780415378536)

roles of differentially expressed genes. What differs here is that genes are no longer the units of interest, but groups of genes with a common function. Let us consider a list of genes arranged according their degree of differential expression between two conditions (e.g. patients versus controls). If a given biological process accounts for the observed phenotypic differences we should then expect to find most genes involved in this process overexpressed in one of the conditions against the other. In contrast, if the process has nothing to do with the phenotypes, the genes will be randomly distributed amongst both classes (for example, if genes account for physiological functions unrelated to the disease studied, they will be active or inactive both in patients and controls). We have recently proposed the use of a sliding window across the list of genes for comparing the distribution of GO terms corresponding to genes within the window against genes outside the window (Al-Shahrour et al., 2005a). Such a procedure has recently been implemented in Babelomics (Al-Shahrour et al., 2005b) as the FatiScan application. If terms (but not necessarily individual genes) were found differentially represented in the extremes of the list, one can conclude that these biological processes are significantly related to the phenotypes. Recently other similar approaches have been proposed (Subramanian et al., 2005).

The application of rigorous definitions of gene function (such as GO) is also likely to play a role in more complex analysis where specific hypotheses on the interaction between different pathways can be tested in the context of statistical models linking molecular signatures to cell physiology. Different creative uses of information in the gene selection process as well as the availability of more detailed annotations will enhance our capability of translating experimental results into biological knowledge.

Acknowledgments

This work is partly supported by grants from NRC-SEPOCT, Fundació La Caixa and Fundacion BBVA and RTICCC from the FIS. Fundación Genoma España supports the Functional Genomics node (INB).

References

Al-Shahrour F, Dopazo J (2005) Ontologies and functional genomics. In: *Data Analysis and Visualization in Genomics and Proteomics* (eds F. Azuaje and J. Dopazo). Wiley, Chichester, pp. 99–112.

Al-Shahrour F, Diaz-Uriarte R, Dopazo J (2004) FatiGO: a web tool for finding significant associations of gene ontology terms with groups of genes. *Bioinformatics* 20: 578–580.

Al-Shahrour F, Diaz-Uriarte R, Dopazo J (2005a) Discovering molecular functions significantly related to phenotypes by combining gene expression data and biological information. *Bioinformatics* 21: 2988–2993.

Al-Shahrour F, Minguez P, Vaquerizas JM, Conde L, Dopazo J (2005b) BABELOMICS: a suite of web tools for functional annotation and analysis of groups of genes in high-throughput experiments. *Nucleic Acids Res* 33: W460–W464.

Ashburner M, Ball CA, Blake JA *et al.* (2000) Gene ontology: tool for the unification of biology. The Gene Ontology Consortium. *Nat Genet* 25: 25–29.

Benjamini, Y, Yekutieli D (2001) The control of false discovery rate in multiple testing under dependency. *Ann Stats* 29: 1165–1188.

Blaschke C, Hirschman L, Valencia A (2002) Information extraction in molecular biology. *Brief Bioinform* **3**: 154–165.

Cui X, Churchill GA (2003) Statistical tests for differential expression in cDNA microarray experiments. *Genome Biol* **4**: 210.

DeRisi JL, Iyer VR, Brown PO (1997) Exploring the metabolic and genetic control of gene expression on a genomic scale. *Science* **278**: 680–686.

Dopazo J, Carazo JM (1997) Phylogenetic reconstruction using an unsupervised growing neural network that adopts the topology of a phylogenetic tree. *J Mol Evol* **44**: 226–233.

Draghici S (2002) *Data Analysis for DNA Microarrays*. Chapman and Hall, London.

Eisen MB, Spellman PT, Brown PO, Botstein D (1998) Cluster analysis and display of genome-wide expression patterns. *Proc Natl Acad Sci USA* **95**: 14863–14868.

Ge H, Walhout AJ, Vidal M (2003) Integrating 'omic' information: a bridge between genomics and systems biology. *Trends Genet* **19**: 551–560.

Hartigan J (1975) *Clustering Algorithms*. Wiley, Chichester.

Herrero J, Valencia A, Dopazo J (2001) A hierarchical unsupervised growing neural network for clustering gene expression patterns. *Bioinformatics* **17**: 126–136.

Herrero J, Al-Shahrour F, Diaz-Uriarte R, Mateos A, Vaquerizas JM, Santoyo J, Dopazo J (2003) GEPAS: A web-based resource for microarray gene expression data analysis. *Nucleic Acids Res* **31**: 3461–3467.

Herrero J, Vaquerizas JM, Al-Shahrour F, Conde L, Mateos A, Diaz-Uriarte JS, Dopazo J (2004) New challenges in gene expression data analysis and the extended GEPAS. *Nucleic Acids Res* **32**: W485–W491.

Heyer LJ, Kruglyak S, Yooseph S (1999) Exploring expression data: identification and analysis of coexpressed genes. *Genome Res* **9**: 1106–1115.

Kanehisa M, Goto S, Kawashima S, Okuno Y, Hattori M (2004) The KEGG resource for deciphering the genome. *Nucleic Acids Res* **32**: D277–D280.

Khatri P, Draghici S (2005) Ontological analysis of gene expression data: current tools, limitations, and open problems. *Bioinformatics* **21**: 3587–3595.

Kohonen T (1997) *Self-organizing Maps*. Springer-Verlag, Heidelberg.

Lee HK, Hsu AK, Sajdak J, Qin J, Pavlidis P (2004) Coexpression analysis of human genes across many microarray data sets. *Genome Res* **14**: 1085–1094.

Mateos A, Dopazo J, Jansen R, Tu Y, Gerstein M, Stolovitzky G (2002) Systematic learning of gene functional classes from DNA array expression data by using multilayer perceptrons. *Genome Res* **12**: 1703–1715.

Mootha VK, Lindgren CM, Eriksson KF *et al.* (2003) PGC-1alpha-responsive genes involved in oxidative phosphorylation are coordinately downregulated in human diabetes. *Nat Genet* **34**: 267–273.

Mulder NJ, Apweiler R, Attwood TK *et al.* (2005) InterPro, progress and status in 2005. *Nucleic Acids Res* **33**: D201–D205.

Quackenbush J (2001) Computational analysis of microarray data. *Nat Rev Genet* **2**: 418–427.

Reiner A, Yekutieli D, Benjamini Y (2003) Identifying differentially expressed genes using false discovery rate controlling procedures. *Bioinformatics* **19**: 368–375.

Rhodes DR, Chinnaiyan AM (2005) Integrative analysis of the cancer transcriptome. *Nat Genet* **37(suppl)**: S31–S37.

Sheng Q, Moreau Y, De Smet F, Marchal K, De Moor B (2005) Advances in cluster analysis of microarray data. In: *Data Analysis and Visualization in Genomics and Proteomics* (eds F. Azuaje and J. Dopazo). Wiley, Chichester, pp. 99–112.

Sneath P, Sokal R (1973) *Numerical Taxonomy*. Freeman, San Francisco.

Speed T (2003) *Statistical Analysis of Gene Expression Microarray Data*. Chapman and Hall, London.

Subramanian A, Tamayo P, Mootha VK *et al.* (2005) Gene set enrichment analysis: A knowledge-based approach for interpreting genome-wide expression profiles. *Proc Natl Acad Sci USA* **102**: 15545–15550.

Tracey L, Villuendas R, Ortiz P *et al.* (2002) Identification of genes involved in resistance to interferon-alpha in cutaneous T-cell lymphoma. *Am J Pathol* **161**: 1825–1837.

van Noort V, Snel B, Huynen MA (2003) Predicting gene function by conserved co-expression. *Trends Genet* **19**: 238–242.

van't Veer LJ, Dai H, van de Vijver MJ *et al.* (2002) Gene expression profiling predicts clinical outcome of breast cancer. *Nature* **415**: 530–536.

Vaquerizas JM, Conde L, Yankilevich P, Cabezon A, Minguez P, Diaz-Uriarte R, Al-Shahrour F, Herrero J, Dopazo J (2005) GEPAS, an experiment-oriented pipeline for the analysis of microarray gene expression data. *Nucleic Acids Res* **33**: W616–W620.

Microarray technology in agricultural research

8

Ana Conesa, Javier Forment, José Gadea, and
Jeroen van Dijk

8.1 Introduction

Similarly to other disciplines in life sciences, the agro-food branch is increasingly applying microarray technology to answer fundamental and applied research questions. The aim of this chapter is to offer a view of the scope of microarray technology in agricultural research through one of the major issues of modern agriculture biotechnology: the development of genetically modified crops. In the following sections we will describe which types of platform are available for functional genomics studies in agriculture and how researchers are developing new tools for specific needs. We will introduce the problems posed by the introduction of transgenic material to food and feed and extensively discuss the role of microarray technology in addressing these problems. Finally, we will illustrate this role through a data analysis example of the differential response of a transgenic plants to environmental stresses.

8.2 Microarray resources in agricultural research

Although *Arabidopsis thaliana* is still widely used as model system to address fundamental biological questions of agricultural research, the growing availability of crop-specific expression analysis platforms is resulting in a progressive shift to the direct use of the target species to address their particular research objectives. These expression platforms have been developed by commercial and public initiatives, frequently in collaboration, and range from cDNA low coverage arrays to genome wide high density oligonucleotide devices.

Table 8.1 gives an overview of available specific gene expression platforms in crops. This list, although not exhaustive, illustrates the diversity of plant systems reached by microarray technology. A more complete impression of the extent to which micorarrays are currently being used in agricultural research can be obtained by browsing the proceedings of the plant and animal genome conferences of the last years (http://www.intl-pag.org/).

8.3 Home-made plant microarrays

Home-made cDNA microarrays are a good alternative for species where commercial microarrays are not yet available, or where flexibility in the

Table 8.1 Plant microarray platforms

Name	Species	Type	Coverage	Reference
GeneChip Arabidopsis	Arabidopsis	Oligonucleotide	24K	Affymetrix (http://www.affymetrix.com)
Arabidopsis 2/3 Oligo Microarray Kit	Arabidopsis	Oligonucleotide	21K–28K	Agilent (http://www.home.agilent.com)
Rice1265 Array	Rice	cDNA	1.2K	Rice Expression Database (http://red.dna.affrc.go.jp/RED/index.html)
GeneChip Rice Genome Array	Rice	Oligonucleotide	51K	Affymetrix (http://www.affymetrix.com)
Rice Oligo Microarray Kit	Rice	Oligonucleotide	22K	Agilent (http://www.home.agilent.com)
NSF Rice Oligonucleotide Array	Rice	Oligonucleotide	20K	TIGR/Iowa state university http://www.ricearray.org/index.shtml
Rice 60K Microarray	Rice	Oligonucleotide	60K	GreenGene BioTech Inc (http://www.ggbio.com/chip2/)
GeneChip Barley Genome Array	Barley	Oligonucleotide	22K	Affymetrix (http://www.affymetrix.com)
Maize GeneChip	Maize	Oligonucleotide	14.8K	Affymetrix (http://www.affymetrix.com)
NSF Maize Oligonucleotide Array	Maize	Oligonucleotide	30K	http://www.maizearray.org/index.shtml
GeneChip Soybean Genome Array	Soybean	Oligonucleotide	37.5K	Affymetrix (http://www.affymetrix.com)
GeneChip Vitis vinifera Genome Array	Grape plant	Oligonucleotide	15.7K	Affymetrix (http://www.affymetrix.com)
GeneChip Wheat Genome Array	Wheat	Oligonucleotide	55K	Affymetrix (http://www.affymetrix.com)
Tom1	Tomato	cDNA	12K	Center for Gene Expression Profiling (http://bti.cornell.edu/CGEP/CGEP.html)
NSF Potato Microarray	Potato	cDNA	10K	http://www.tigr.org/tdb/potato/microarray_comp.shtml
Sunflower Gene Chip	Sunflower	cDNA	3.8K	http://cgb.indiana.edu/genomics/projects/10
Mt16kOLI1	Soybean	Oligonucleotide	16K	http://www.genetik.uni-bielefeld.de/Genetik/legum/arrays.engl.html
Cotton Gene Chip	Cotton	cDNA	10K	http://www.agric.nsw.gov.au/reader/genomics-projects/gene-chips.htm
COM	Cotton	Oligonucleotide	12K	http://cottongenomecenter.ucdavis.edu/microarrays.asp
POP2	Populus	cDNA	25K	PopulusDB (http://www.populus.db.umu.se/project.html)
Sugar Cane GeneChip Genome Array	Sugar Cane	Oligonucleotide	8K	Affymetrix (http://www.affymetrix.com)
µPEACH 1.0	Peach	Oligonucleotide	5K	www.itb.cnr.it/ESTree

design is necessary. In these microarrays, gene-specific probes can be obtained by PCR amplification of cDNA clones and spotted on glass slides. Basically, the steps to produce home-made cDNA microarrays include construction of cDNA libraries, isolation and partial sequencing of cDNA clones, processing and analysis of the sequences, amplification by PCR of a nonredundant set of cDNA clones, and spotting of PCR products on glass slides (*Figure 8.1*). Alternatively, oligonucleotides can be designed for the sequences of interest, purchased and spotted.

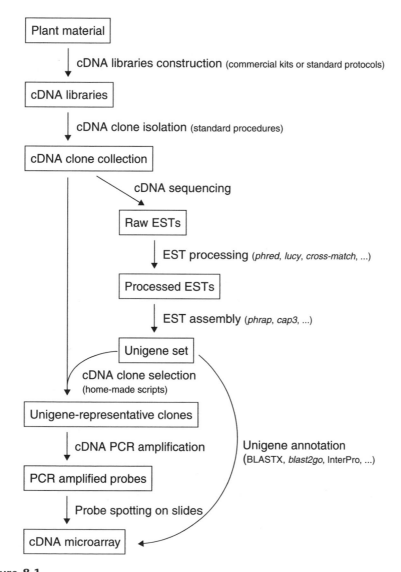

Figure 8.1

Schematic representation of the construction of a home-made cDNA microarray.

8.3.1 Construction of cDNA libraries

The first step in cDNA library construction is the choice of the plant material to be used as RNA source. (URLs for reagents and bioinformatics tools are provided as supplementing material at the chapter site.) cDNA libraries contain a representation of the genes expressed in the particular tissue they are derived from. Therefore, when the aim is an unbiased genome-wide microarray, a good approach is to use as many different sources of plant material as possible (Yamamoto and Sasaki, 1997; Forment *et al.*, 2005; da Silva *et al.*, 2005). These sources of plant material should include different tissues, taken at different developmental stages, and subjected to different biotic and abiotic stress conditions. A single cDNA library can be constructed using a mix of mRNA samples coming from these tissues, developmental stages, and environmental conditions. Alternatively, different single libraries can be generated from each of the different mRNA samples. Although more expensive and laborious, this latter approach offers the possibility of tracking the origin of single cDNA clones, which allows *in silico* gene expression studies (Audic and Claverie, 1997; Greller and Tobin, 1999; Stekel *et al.*, 2000; Qiu *et al.*, 2003; Fei *et al.*, 2004).

Many procedures are available for cDNA library construction, either from commercial kits (e.g. Lambda ZAP®; CloneMiner™, SMART™) or standard protocols (Ying, 2003). Specific manipulations can be introduced to enrich libraries for certain types of molecules in order to increase the diversity and quality of the cDNA collection. For example, special procedures can be used to increase the proportion of full-length clones in the cDNA libraries (Ying, 2003; Ogihara *et al.*, 2004). This type of clone has been claimed to be more convenient than partial clones for microarray construction purposes (Seki *et al.*, 2001). Subtracted libraries, that allow enrichment of mRNAs present in a tissue of interest relative to a control state (Bonaldo *et al.*, 1996; Ying, 2003) would specifically select for potentially interesting expressed genes. Finally, libraries can be normalized in order to increase the proportion of low expressed genes (Bonaldo *et al.*, 1996).

8.3.2 Isolation and partial sequencing of cDNA clones

Once cDNA libraries have been constructed, a number of clones are randomly isolated and partially sequenced to render expressed sequence tags (ESTs) (Adams *et al.*, 1993; Rudd, 2003). Plate formats with 96 or 384 wells are used in subsequent steps to allow high throughput. ESTs are then processed and analyzed to extract their informative content (Dong *et al.*, 2005). Due to the highly data intensive nature of these projects, an efficient laboratory and clone management along the successive rounds of bacterial cultures, plasmid isolation, cDNA sequencing, and PCR amplification is essential in order to avoid clone tracking errors. Both commercial products (e.g. AlmaZen™, CloneTracker™) and open software (e.g. B.A.S.E., Saal *et al.*, 2002) are available for support in this process.

8.3.3 Processing and analysis of EST sequences

The large size of many EST data sets makes manual processing of sequence information difficult or impossible, and computer-based bioinformatics methods are indispensable to address this task. Although some of the available tools can be used from their web servers, locally installed programs are more convenient because: (1) they offer more possibilities of configuration, and can easily be customized for local needs; (2) they can be run from local pipelines that automate the work of repeated processing and analysis; and (3) they offer the possibility of analyzing a higher number of sequences in batch. Many of these computational solutions rely on the open-source Linux operating system, which can be installed on most modern PC processors, and on which a wealth of software has been developed for both bioinformatics and standard computing needs. Although it is also possible to run many of these programs on nonLinux platforms, Linux has become the environment of choice for many bioinformatics researchers and is therefore well supported and rapidly growing in diversity and stability.

ESTs are single-pass DNA sequence reads derived from cDNA clones, and usually contain vector sequences and sequencing errors. cDNA library construction can also inadvertently result in cloning of contaminant DNA, particularly from *Escherichia coli* and its bacteriophages. A first EST processing step is necessary to remove these regions in order to obtain clean, high-quality sequences. Software for base calling from chromatograms (*phred*, Ewing and Green, 1998; Ewing *et al.*, 1998), and for trimming low-quality, vector and contaminant sequences (*lucy*, Chou and Holmes, 2001; *cross-match*) is frequently used.

Due to the random nature of clone isolation, and because of the diversity of mRNA levels for the different genes, more than one cDNA clone per gene is habitually present in the EST collection. The EST data set will therefore have some level of redundancy that can be used to derive better consensus sequences, although these additional clone sequences need to be removed to obtain a set of unique sequences or unigenes. This clustering and assembly step can be done by a number of programs, like *phrap*, CAP3 (Huang and Madan, 1999), CLU (Ptitsyn and Hide, 2004), d2-cluster (Burke *et al.*, 1999), and TGICL (Pertea *et al.*, 2003). These programs use different algorithms to group sequences on the basis of sequence similarity into clusters and to derive assembled consensus sequences from them. The accuracy of EST clustering is affected by various error sources, such as sequencing errors and insufficient overlapping of sequences (Wang *et al.*, 2004). Because of these errors, multiple unigenes can represent the same gene, and a single unigene can involve nonsibling ESTs; therefore, further inspection of the assembly is often necessary.

For identification of the genes represented by the unigene sequences, similarity searches against public or private databases are used. If the organism of interest has already been sequenced, searches against the corresponding genome database will exactly identify genes represented in the collection. For other organisms, cross-species sequence comparisons are used for functional annotation of the unigenes. Conservation of sequences from different species is used to transfer functional annotation from a gene of which the role is known in one species to an otherwise uncharacterized gene in the other. For EST analysis, the most used sequence comparison tool is BLASTX, one of the BLAST family of programs (Altschul *et al.*, 1990), which searches sequence databases (e.g. *nr* nonredundant database at NCBI) to find known proteins with similarity to the potential translation of the

ESTs. Functional annotation using the Gene Ontology controlled vocabulary (The Gene Ontology Consortium, 2004) is becoming a standard. The free software Blast2GO is a user-friendly tool to annotate uncharacterized sequences with Gene Ontology terms (Conesa *et al.*, 2005). Secondary protein databases on protein families, domains and functional sites can also be screened using the integrated database InterPro (Apweiler *et al.*, 2001; Mulder *et al.*, 2005), so that features found in known proteins can be applied to ESTs.

All the information gathered at the EST processing and analysis steps must be stored in relational databases (e.g. MySQL PostgreSQL, Oracle®) for efficient retrieval and updates. Some software packages for EST processing and analysis, and for database creation and management, are publicly available. These include ESTIMA (Kumar *et al.*, 2004), PHOREST (Ahren *et al.*, 2004), ESTAP (Mao *et al.*, 2003), ESTWeb (Paquola *et al.*, 2003), ESTAnnotator (Hotz-Wagenblatt *et al.*, 2003), PipeOnline (Ayoubi *et al.*, 2002), and RED (Everitt *et al.*, 2002). Although they can be useful in some of the steps, they usually do not cover the full process, and are not well suited to the specific needs of every project. Accordingly, locally created EST analysis pipelines are normally created by combining and/or extending some of the available tools.

8.3.4 Generation of the probes by PCR amplification

Probes to spot onto the microarray are generated by PCR amplification of a selected set of cDNA clones representing the genes whose expression is to be monitored. In case a gene is represented by more than one clone in the collection, only one should be used as template for probe amplification. A comprehensive review of microarray probe selection can be found in Tomiuk and Hofmann (2001).

However, clone selection alone does not solve some inherent problems of cDNA arrays such as alternative splicing and cross-hybridization which might mask changes in transcript levels. Each spotted cDNA may still detect a group of closely related sequences rather than a single transcript. The percentage of cross-hybridization depends on the length and the similarity of the related sequences (Evertsz *et al.*, 2001). On the other hand, cross-hybridization has the potential benefit of allowing the use of a single chip design for probing different strains/varieties of the same organism, or even for related species, simply by adjusting hybridization conditions.

To minimize or control these effects, some considerations can be taken into account. First, the relative position of the EST in the cluster assembly is important. For the 3' position most clones have a high probability of including 3'-UTR regions, which are more variable and can be used to discriminate among members of a gene family. On the other hand, when cDNA libraries have been constructed using oligo-dT as a primer, then for the 5' position most clones are probably longer and potentially more convenient (Seki *et al.*, 2001), but they increase the probability of cross-hybridization. Furthermore, the extent of similarity of the EST to the consensus sequence of the cluster should be considered. ESTs with sequence regions of low similarity with the consensus sequence of the unigene are not a good representative of the cluster. Finally, to achieve homogeneous hybridization efficiency for all the probes, clones with special hybridization conditions should be avoided. Accordingly, clones containing high GC/AT content stretches or shorter than a given threshold should not be selected as probes. Since no publicly available programs exist for this task, script programming is of great value to parse clustering and assembly results in order to perform a convenient selection of cluster representatives based on the criteria adopted by each project.

PCR amplification of the cDNA in the representative clones is done using primers flanking the inserts. A single band must be obtained for each clone to ensure the specificity of every probe in the microarray. PCR products are therefore separated by agarose gel electrophoresis to check that single bands representing gene-specific probes are obtained. Reactions producing multiple bands must be discarded, and another representative for that unigene must be selected and amplified until a single band is obtained.

8.3.5 Probe spotting on the glass slides

The spotting step of the gene-specific probes is usually done with commercial microarray printing devices (e.g. Genomic Solutions). Up to 60 000 spots can be printed onto a conventional sized slide using these devices. Different types of glass slides are available from several vendors (e.g. Corning®, ArrayIt®). Although duplicate spots are not a substitute for replicate arrays, they can reveal spatial hybridization defects and are recommended if possible. Positive and negative controls, as well as spike mixes to help in normalization, should also be spotted interspersed with the probes. Random arrangement of duplicate spots, positive and negative controls and spike mixes along the glass slide surface is customary to minimize spatial printing effects.

8.4 Microarrays and genetically modified organisms

8.4.1 Genetically modified crops and their implication for food safety

Genetic modification of plants has proved a powerful tool for the development of desirable traits in plants, which would otherwise have taken long breeding strategies. Furthermore, traits from unrelated plants or bacteria can now also be inserted into the plant genome, resulting already in numerous GM crops, mostly resistant to insects, pests or herbicides. Modifications may also improve industrial processing or the nutritional value of crops. Along with the development of the technique, discussion on potential food safety implications also grew, around the world, but especially in Europe. The discussions between environmentalist groups, consumers, and industry, with the scientific community as yet unable to settle the argument, have caused uneasiness with European consumers as far as GM foods are concerned. Recent outbreaks of diseases in cattle, though not related to the GM issue, have not helped to improve consumers' trust in food safety in general. This has increased the alertness of governments to draw up appropriate legislation concerning food related issues, including the use of GM foods. Apart from national governments, international bodies such as the

Table 8.2 Food safety related internet sites*

Key word	Description/URL
EFSA	European Food Safety Authority
	http://www.efsa.eu.int/
ENTRANSFOOD	European network safety assessment of genetically modified food crops, EU-funded
	http://www.entransfood.com
EU	Food and feed safety internet page of the European Union
	http://europa.eu.int/comm/food/index_en.htm
EU	EU regulation on genetically modified food and feed
	http://europa.eu.int/eur-lex/pri/en/oj/dat/2003/l_268/
	l_26820031018en00010023.pdf
FAO	Food and agriculture organization of the united nations, homepage
	http://www.fao.org/
FSA	Food standards agency of the UK, homepage
	http://www.food.gov.uk/
FSA, GO2	Research project: Safety assessment of genetically modified foods
	http://www.food.gov.uk/science/research/researchinfo/foodcomponentsresearch/
	novelfoodsresearch/g02programme/g02projectlist/
GMOCARE	EU funded -omics project, see workpackage 2 of ENTRANSFOOD
	http://www.entransfood.com
ILSI	International life sciences institute
	http://www.ilsi.org/
OECD	Organization for economic co-operation and development, homepage
	http://www.oecd.org/home/
OECD	Consensus documents for the work on the safety of novel foods and feeds
	http://www.oecd.org/department/0,2688,en_2649_34391_1_1_1_1_1,00.html
TIGR	The institute for genomic research, homepage
	http://www.tigr.org/
WHO	World Health Organization, homepage
	http://www.who.int/en/

*Only nonprofit, academic or governmental sites are listed, in alphabetical order according to key words.

World Health Organization (WHO), Food and Agriculture Organization (FAO), Organization for Economic Co-operation and Development (OECD) and the International Life Sciences Institute (ILSI) have formulated guidelines and regulations for the management of GM crops (*Table 8.2*). In relation to food safety, one of the greatest concerns is the occurrence of unintended effects. As the transgene is being randomly inserted in the host plant genome, the place of insertion may cause erroneously up- or downregulation of endogenous plant genes. This has already spawned investigation into unintended or unpredicted effects in GM crops. While most of such studies are included in dossier information and therefore not publicly available, some examples of unintended effects have been reported in scientific literature (*Table 8.3*) (Kuiper *et al.*, 2001). For these observed effects the question remains whether they pose a risk to human health upon consumption of the foods derived from these GM crops. Few examples of animal studies are available in which the effect of GM food on health has been investigated. Three different studies of GM rice and potato showed no adverse effects in rats (Hashimoto *et al.*, 1999a, 1999b; Momma *et al.*, 1999). One controversial study on GM potato did show indications of

Table 8.3 Unintended effects in genetic engineering breeding* (Kuiper *et al.*, 2001)

Host plant	Trait	Unintended effect	Reference
Canola	overexpression of phytone-synthase	multiple metabolic changes (tocopherol, chlorophyll, fatty acids, phytoene)	(Shewmaker *et al.*, 1999)
Potato	expression of yeast invertase	reduced glycoalkaloid content (−37–48%)	(Engel *et al.*, 1998)
Potato	expression of soybean glycinin	increased glycoalkaloid content (+16–88%)	(Hashimoto *et al.*, 1999a, 1999b)
Potato	expression of bacterial levansucrase	adverse tuber tissue perturbations; impaired carbohydrate transport in the phloem	(Turk and Smeekens, 1999; Dueck *et al.*, 1998)
Rice	expression of soybean glycinin	increased vitamin B6 content (+50%)	(Momma *et al.*, 1999)
Rice	expression pf provitamin A biosynthetic pathway	formation of unexpected carotenoid derivatives (beta-carotene, lutein, zeaxanthin)	(Ye *et al.*, 2000)
Soybean	expression of glyphosphate (EPSPS) resistance	higher lignin content (20%) at normal soil temperature (20°C); splitting stems and yield reduction (up to 40%) at high soil temperatures (45°C)	(Gertz *et al.*, 1999)
Wheat	expression of glucose oxidase	phytotoxicity	(Murray *et al.*, 1999)
Wheat	expression of phosphatidyl synthase	necrotic lesions	(Delhaize *et al.*, 1999)

*Data from publicly available reports.

toxicity in rats (Ewen and Pusztai, 1999), but lack of proper controls and the protein deficiency in the diets caused criticism on this study (Kuiper *et al.*, 1999; Royal Society, 1999). Malatesta *et al.* (2002) showed some ultrastructural changes in the nuclei of hepatic cells of mice fed with a diet supplemented with 14% GM soybeans, compared with mice fed with a diet supplemented with wild-type soybeans, indicating a possible difference in the metabolic activity of the hepatocytes. However, no overall changes were observed in the livers of the mice in the two groups, nor any overall toxicity.

The controversy on the subject has caused hesitance in the European Union (EU) for allowing GM crops to be grown or used for food and feed. EU consumers' demand for a choice in buying GM or nonGM food has resulted in a labeling requirement for all foodstuffs, if an ingredient contains more than 0.9% of a GM crop that is allowed. A recent change in regulation has grouped the evaluation of new GMOs and GMO-derived products for food and feed purposes apart from that of other novel foods. According to the new regulation this evaluation will be performed by the European Food Safety Authority on the basis of the EU guidance document for the risk assessment of GM derived food and feed (*Table 8.2*). Part of this evaluation will be the analysis of a number of compounds, as compiled in the OECD consensus documents on key substances for a number of economically important food crops (*Table 8.2*). A starting point for risk assessment is based on comparative safety assessment, where key components are evaluated for substantial equivalence between the novel food and its nearest comparator, being the nontransformed parent line in the case of GM crops. Any changes found should subsequently lead to targeted toxicological and nutritional testing. Although more information on key compounds for different crops is being gathered, this targeted approach has of course the limitation that not all possible unintended effects may be uncovered. A number of recent reviews elaborate on regulation of GM foods as well as safety assessment and are recommended for further reading (Kuiper *et al.*, 2001, 2003; Kok and Kuiper, 2003; Nordic Council, 2003; Liu-Stratton *et al.*, 2004).

8.4.2 Microarrays as profiling tools for screening of GM crops

Investigating all components of a GM plant and comparing them to their nonGM counterparts would certainly reveal any effects, both intended and unintended. However, such an approach is not possible at present, and will probably never be. Transcriptomics presents a very powerful tool for investigation of unintended effects, despite the fact that changes in the expression level of a certain gene may be less evidently linked to a food safety issue than a change in levels of a certain metabolite (see Section 8.4.4). In fact, it is the only technique so far capable of delivering a full-scale analysis of the GM versus the parental crop, at least at the level of gene expression, given the rapid increase in the number of completely sequenced plant genomes. However, scientific journals do not abound with microarray analyses of GM crops. This is mostly due to the emphasis by the large genome projects, and within their wake the microarray industry, on the human, mouse, and rat genomes. This has forced scientists in food research to develop their own arrays, many of them based on cDNA clones (Section 8.3). Two different projects have coupled development of such arrays with the detection of unintended effects in GM plants, (GMOCARE, FSA GO2, *Table 8.2*). In the GMOCARE study, part of ENTRANSFOOD, three different plant arrays were developed. An *Arabidopsis thaliana* array was spotted containing cDNA libraries enriched for leaf, flower, root, and stressed seedlings. With these, analyses were performed on Arabidopsis plants that had genetically altered flavanoid biosynthesis. Also a tomato array was developed, derived from cDNA libraries enriched for red- or green-specific ripening stages of tomato fruit, and used to evaluate GM tomatoes with elevated lycopene levels. Furthermore, a potato array was constructed making use of a stress-pathway-enriched cDNA library. Several transgenic lines were analyzed with this array. Within a project funded by the Food Standards Agency, UK (*Table 8.2*), this potato array was expanded with cDNA clones specific for plant, tuber, and developing tuber genes. Also, the number of GM lines and the number of wild type potato lines was expanded in this study. Some expression data from these studies are already available (see *Useful Links: Microarray Data Repository* pp. xix and 285), and peer-reviewed results are expected to be published in the near future.

Another reason for the lack of publications may be the size of the experiment needed for the investigation of unintended effects. Unintended modified traits may or may not be apparent in different soil and climatic conditions. Therefore, data from multiple locations and seasons would be more informative than experiments from a single place and season. Furthermore, a thorough investigation of natural variation in traditional crops of a particular species would be needed to place any unintended effects in their proper context and to gain more insight into the potential safety impact of such an effect.

Recently, commercial platforms have started to offer standardized microarrays for major food crops (*Table 8.1*), resulting already in more studies on the transcriptomes of plants, both GM and nonGM. For instance, the availability of full-genome *Arabidopsis* microarrays has already resulted in about a dozen papers in the year since it came on the market. Among them were studies concerning nonGM plants, on auxin responses, and nitrogen and carbon signaling (Armstrong *et al.*, 2004; Palenchar *et al.*, 2004), but also papers on GM plants, affected in certain metabolic routes (Lloyd and Zakhleniuk, 2004; Kristensen *et al.*, 2005; Tohge *et al.*, 2005). These papers focused on pathway analysis rather than food safety issues, particularly the paper by Tohge *et al.*, which elegantly combines transcriptomics with metabolomics to pinpoint the action of genes – already known to be involved in flavanoid biosynthesis – to a specific action in that pathway. Such basic information is much needed for future understanding of unexpected or unintended changes in GM crops. These authors mainly observed changes in predicted pathways, though they did not look specifically for minor, unintended effects. This is contrary to another *Arabidopsis* mutant, investigated by Lloyd and Zakhleniuk. This line was found to harbor a wide array of changes in gene expression, some of which were unexpected. Kristensen *et al.* showed that different mutations involving the same metabolic route can result in dramatic and unexpected metabolic and gene expression changes. In fact, these authors state that targeted introduction of traits by genetic engineering based on a solid knowledge of plant metabolism offers the opportunity to generate cultivars that adhere more strictly to the principle of substantial equivalence than cultivars generated by classical breeding procedures. In any case, food safety studies will benefit from a more basic knowledge of metabolic routes in plants, also at the level of gene expression. The combination of techniques as described above with mutant plants that are affected in a specific metabolic pathway will be of great help in achieving that knowledge.

8.4.3 Technical considerations for plant microarrays

Basically, the techniques for microarray experiments do not differ between different types of samples. In this respect protocols for microarray analysis of GM crops will be the same as for other applications. However, different protocols are available for sample labeling and microarray hybridization. Some of the leading institutes for microarray research already differ in their protocols (*Table 8.4*). Although international efforts have been made to standardize procedures in microarray research, this has not resulted in a rigid consensus protocol, but rather in consensus about the information

Table 8.4 Different microarray hybridization conditions

Source[a]	SSC/SDS[b]	Other components	Blocker
microarray.org	3×SSC, 0.2% SDS	0.02 M HEPES	poly dA, COT-1 DNA
Stanford University	3.3× SSC, 0.3% SDS		poly dA
TIGR	5× SSC, 0.1% SDS	25% formamide	poly dA
National Human Genome Research Institute	3× SSC, 0.15% SDS	1.25, or 2.5× Dernhardt's blocking agent	poly dA, yeast tRNA, COT-1 DNA
Kidney Research Laboratory (Australia)	2× SSC, 0.25% SDS	25% formamide	poly dA, COT-1 DNA

[a] See Useful links, pp. xix–xxi or Appendix 3, pp. 285–286 for relevant internet sites.
[b] SSC: saline sodium citrate; SDS: sodium dodecyl sulfate.

that should be included when sharing information on microarray experiments, such as MIAME (minimum information about a microarray experiment). In general, the first step in a microarray experiment (after RNA isolation) is the choice of whether or not to amplify the RNA. As the amount of RNA is usually ample for crop material, this step is best omitted, to include as few sources of variability as possible. However, with plant material, and especially material rich in sugar chains such as potato tubers or maize kernels, RNA isolation may be harder using phenol-based or column-based isolation methods. Specialized protocols using hexadecyltrimethylammonium bromide (CTAB) in the extraction buffer generally yield less contaminated RNA in those cases (Chang *et al.*, 1993). A plant RNA isolation protocol is available as supplementing material at the chapter site. Apart from numerous in-house protocols, several industrial ready-to-use kits are available for labeling of RNA, and subsequent hybridization to the microarray. On the one hand, these kits will always include some X-factors, or unknown ingredients, something that is generally not well liked by scientists. On the other hand, industrial kits will probably provide if not more rigorous, at least better documented quality controls for their products. This last factor may be of particular importance to GM crop investigation as the results will have to stand up against not only peer review, as is the case for general academic studies, but also to the scrutiny of food companies, environmentalist groups, lawyers, and governments, if they are ever to be used as the basis for regulation and legislation. A further complication is the sheer amount of data that is generated by microarray experiments. This has spawned a complete new science of bioinformatics, dealing with a huge number of variables and a limited number of samples, as opposed to conventional statistics in which the number of samples (far) outnumbers the number of variables. In the next section we will provide an example of data processing in a GMO-related study. However, microarray data analysis methodologies are still an area much in development, and although some general rules are emerging, there is no general analysis strategy and methods are still prone to changes. Therefore, microarray results on unintended effects will still have to be confirmed with other methods. On the transcriptome level this can be achieved with quantitative reverse transcriptase polymerase chain reaction (Q-RT-PCR). However, for food safety issues, confirmation on protein and especially metabolite level seems more desirable, so that transcriptomics may serve as a pointer for more detailed safety assessment. Alternatively, transcriptome analysis could directly lead to targeted toxicological and nutritional testing, when increasing knowledge of links between the different -omics technologies make this possible.

8.4.4 Transcriptomics in relation to other '-omics' techniques

The composition of plants is of such a complexity, ranging from simple ions to macromolecules such as proteins, fibers, and nucleic acids, that no single technique can cover all of them. The compounds most relevant to food safety are macro- and micronutrients, antinutrients, plant toxins, and allergens. Apart from potential allergenic novel proteins, most of these compounds fall in the category of metabolites, and the group as a whole is frequently referred to as the metabolome. The metabolome includes various forms of molecules such as fatty acids, amino acids, polyaromatics, and plant steroids. These compounds vary greatly in size, charge, hydrophobicity etc., so more than one technique is needed for their analysis. Methods such as nuclear magnetic resonance (NMR) and gas or liquid chromatography coupled to mass spectrometry (GC- and LC-MS) have been used for metabolite studies in plants. The combination of these techniques can cover several thousand different metabolites with the current state of techniques, covering only part of the entire metabolome. In this way, metabolite profiling of a GM food can never guarantee that a GM plant is free of unintended effects. However, metabolomics was successful in uncovering unintended effects in GM crops as early as 1998 (*Table 8.3*) (Kuiper *et al.*, 2001; Fernie *et al.*, 2004).

One level up from metabolites are the enzymes that generate them, represented in the proteome. A difference in abundance of a particular protein may indicate a difference in amounts of one or more metabolites, leading to a specific food safety or nutritional issue. The most widely used method for proteome analysis is two-dimensional electrophoresis, a labor-intensive method that requires highly skilled personnel, covering only several hundreds of proteins. Using this technique, no changes in the major proteins were observed when a virus-resistant GM tomato was compared with its wild-type counterpart (Corpillo *et al.*, 2004).

The transcriptome of plants, comprising all transcribed and processed messenger RNA (mRNA) molecules is one step further away still from the (anti-) nutrients of plants, and any changes in the composition of the transcriptome itself are not considered to pose a threat for food safety. Also, effects on the transcriptome and the metabolome or the proteome may not be different only in magnitude, but also distributed in time. Differences in the mRNA level, driving changes in the metabolome may occur during the developmental stage of the crop and may not still be present at the time of harvest, and likewise for the proteome. However, as the technique relies on base-pairing interactions that are common to all nucleic acids, transcriptomics is more likely to deliver uniformly interpretable results. One step higher in the information structure of living cells is DNA, and analysis of the place of insertion of the transgene into the host genome is now routinely asked for as part of the safety assessment procedure according to

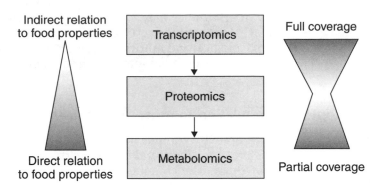

Figure 8.2

Schematic overview of the relation between the different '-omics' technologies, food properties, and coverage of analyses.

EU legislation. However, this is not a profiling technique, though potentially very informative, and will not be discussed in detail here. Recent publications combining metabolomics and transcriptomics have led the way to discover both links and gaps between the two techniques (Kristensen *et al.*, 2005; Tohge *et al.*, 2005; Urbanczyk-Wochniak *et al.*, 2005). In relation to food safety, the future will have to show whether one profiling technique is superior to others, whether they are interchangeable, or whether they should be used in a complementary fashion. Also, future and ongoing studies will have to show the necessity of profiling methods above standard morphological and targeted analysis for food safety. One possible outcome may be the recommendation for a semi-targeted profiling of specific pathways relevant to food safety. *Figure 8.2* shows a scheme of the relation between the different -omics technologies.

8.5 A data analysis example: Time course of gene expression response to stress in transgenic plants

The considerations discussed in the previous sections point to gene expression profiling as a powerful approach to investigate the genome wide effects of genetic manipulation. Microarrays allow, on the one hand, the characterization of the gene expression changes associated with the introduced gene modification and the study of the molecular basis of the target biological process. On the other hand, they have the potential of unraveling possible side effects or unwanted changes introduced by the transformation event in the molecular status of the transgenic plant. As already mentioned, differential gene expression between transgenic and wild-type plants can also be conditional upon external factors.

In this section, we provide an example of data analysis for a gene expression study involving transgenic plants and their interaction with the environment. Unfortunately, publicly available data sets of such types of study are scarce. We have taken for our example a gene expression study of the temporal response to stress in genetically modified tobacco plants.

Resistance to biotic and abiotic stresses is probably one of the most important goals in breeding programs aiming at crop adaptation to diversity of climatic, soil and ecological environments. Stress is a key determinant of plant production and management, and the understanding of the mechanisms that control the stress response and resistance in plants seems crucial for directed crop improvement. Many functional genomics studies have monitored the gene expression response to high salinity, drought, cold and/or heat (abiotic stress), as well as to attack by pests (biotic stress) in plants. Examples of such studies can be found in rice (Rabbani *et al.*, 2003), *Sorghum* (Buchanan *et al.*, 2005), cassava (Lopez *et al.*, 2005), potato (Rensink *et al.*, 2005), tomato (Conh and Martin, 2005), tobacco (Senthil *et al.* 2005), soybean (Moy *et al.*, 2004), barley (Oztur *et al.*, 2001), loblolly pine (Watkinson *et al.*, 2003), cotton (Dowd *et al.*, 2004) and, of course, *Arabidopsis thaliana* (Seki *et al.*, 2002; Narusaka *et al.*, 2004). These approaches have revealed the complexity and abundance of molecular signals involved in stress response and defense mechanisms. Cross-talk in the signaling cascades triggered by the various stress factors but also specificity in the fine-tuning of particular responses comes up as the general conclusion of these studies. Thus, downregulation of basic cell maintenance functions such as transcription or translation processes, metabolism, and photosynthesis seems to be common to different stress responses (Oztur *et al.*, 2001; Okinaka *et al.*, 2002; Lopez *et al.*, 2005). Aquaporins, typically drought-responsive proteins, have also been found to be downregulated upon *Fusarium* infection in cotton (Dowd *et al.*, 2004). Upregulation of plant hormone-responsive, metallotheonine-like, late-embryogenesis-associated (LEA) and Ca-dependent signaling genes is found in both biotic and abiotic stress studies. Other pathways, such as those involving the activation of pathogen-related proteins are more specific for compatible pest–plant interactions that result in a defense response. Chen *et al.* (2002) showed the major role of the plant hormones salicylic acid, jasmonic acid and ethylene in regulating the expression of transcription factors controlling stress response genes. This and other studies demonstrated both the synergic and antagonistic interactions of these hormones when modulating the stress response as well as a clear tissue and stress factor specificity of stress-related transcription factors and responsive genes.

In our data analysis example, the temporal gene expression response to biotic and abiotic stresses was studied in tobacco which had been genetically modified to suppress the function of the salicylic acid induced protein kinase SIPK (Hall, 2005). SIPK is a stress-activated MAP kinase, homologous to AtMPK6 of *Arabidopsis thaliana*, involved in the phosphorylation of the stress-related transcription factor WRKY1 (Menke *et al.*, 2005). Salicylic acid is the key component in the systemic acquired resistance (SAR) in plants, a stress response to compatible plant–pathogen interactions that provides some degree of resistance against unrelated stresses. Alteration of SIPK steady-state mRNA levels in tobacco and *Arabidopsis* has shown to alter the plant sensitivity to oxidative stress (Miles *et al.*, 2005; Samuel *et al.*, 2005). This case study has some of the common features of stress response studies in plants: gene expression is monitored during a more or less long period of time for the biological condition(s) tested and expression differences along the time component are sought. Additionally, in this particular case, the differential behavior of transgenic and wild-type plants is evaluated.

Therefore, in our analysis of the data we will include approaches to both explore the differences in the molecular state between modified and wild-type genotypes, and to find genes with temporal expression changes.

8.5.1 Statistical analysis

Wild-type and SIPK-silenced (RNAi) tobacco suspension cultures were cultivated in the presence of abiotic (xanthine oxidase) or biotic (bacterial harpin or fungal megaspermin) stressors. A 30 min pre-incubation with xanthine was applied to the abiotic stress treatment and its control. For both biotic and abiotic treatments and in both genotypes, three independent cell culture samples were taken after 0, 4, and 6 h of incubation with the stress agent. For each time point, RNA samples from transgenic and wild-type genotypes were randomly paired for hybridization on an 11K cDNA potato microarray using two-color technology. The complete experimental set up is given in *Table 8.5*. Data was Lowess-normalized and replicate spots were averaged. For an introduction to the principles of data normalization and for a step-by-step tutorial see Chapter 6. The original data from this study can be downloaded from http://www.tigr.org/tdb/potato/microarray_comp.shtml.

For the statistical analysis of this data set we will use functions and packages available in the statistical language *R* from the Bioconductor repository (www.bioconductor.org). In this section we will depict the main steps for performing a comprehensive analysis of our case study. Different

Table 8.5 Experiment set up of SIPK-mediated stress response study

Array	Time (h)	Treatment
B_cont_1	0	no_treatment
B_cont_2	0	no_treatment
B_cont_3	0	no_treatment
4h_harp_1	4	harpin
4h_harp_2	4	harpin
4h_harp_3	4	harpin
8h_harp_1	8	harpin
8h_harp_2	8	harpin
8h_harp_3	8	harpin
4h_mega_1	4	megaspermin
4h_mega_2	4	megaspermin
4h_mega_3	4	megaspermin
8h_mega_1	8	megaspermin
8h_mega_2	8	megaspermin
8h_mega_3	8	megaspermin
ab_cont_1	0	xanthine
ab_cont_2	0	xanthine
ab_cont_3	0	xanthine
4h_xox_1	4	xanthine+xanthine oxidase
4h_xox_2	4	xanthine+xanthine oxidase
4h_xox_3	4	xanthine+xanthine oxidase
8h_xox_1	8	xanthine+xanthine oxidase
8h_xox_2	8	xanthine+xanthine oxidase
8h_xox_3	8	xanthine+xanthine oxidase

.txt files are used throughout this example which can be found at the book's website, www.garlandscience.com/9780415378536. An additional 'code.txt' document is also available that contains the *R* code shown here, as well as additional code for the data handling, processing, and plotting used to generate the text files and figures of this example.

The data input for the initial part of the analysis is a text file containing the log-transformed expression ratios where missing values have been replaced with ratios inferred using the KNN approach (Troyanskaya *et al.*, 2001) as rows and where columns represent the different samples. An easy-to-use web-based tool for data processing is contained in GEPAS (Herrero *et al.*, 2003) which is described in Chapter 6. Find the 'data.M.txt' file at the book website and load it into an R session by:

```
> data.M <- read.table("data.M.txt", header = T, row.names
= 1) #load data
```

It is also convenient to create an experimental design table as given in *Table 8.5*. This can be easily done with any spreadsheet and loaded into *R* by:

```
> targets <- read.table("Table 5.txt",header = T) #load
design
```

A first approximation to data analysis is to explore the overall differences that gene expression profiling shows for the various treatments and the two genotypes. Principal component analysis (PCA) can be used for this purpose (see Box 8.1 for statistical background).

```
> PCA <- princomp(data.M, cor = T) # perform principal
component analysis
> cumsum(PCA$sdev^2/sum(PCA$sdev^2)) # show the accumulated
explained variance
```

The first two components explain 70% of the variance and can be used to visualize the variability among samples. The following code will show graphically the result of this PCA analysis:

```
> plot(PCA$loadings[,1], PCA$loadings[,2], col = "grey",
main = "PCA SIPK-silenced stress response data", xlim =
c(-0.24,-0.13))
> text (PCA$loadings[,1], PCA$loadings[,2],
colnames(data.M), col =rep(c(1:8),each = 3))
> legend( -0.13, 0.05, legend =
unique(substring(colnames(dataM), 1,
+ nchar(colnames(data.M))-2), col = c(1:8), xjust = 1,
yjust = 0,lty = 1)
```

The PCA plot (*Figure 8.3*) clearly shows the clustering of samples corresponding to each of the conditions of abiotic stress treatments: this indicates that there are clear gene expression features associated with the differential response of SIPK-silenced and wild-type plants to these treatments which are not present in the 0 h controls. We also observe that the samples corresponding to the xanthine oxidase treatment do not separate from their 0 h controls, and also not from the 0 h biotic control samples. This points to a much milder differential response to this treatment

---

Box 8.1 Statistical background

Principal component analysis (PCA) is a data dimensionality reduction technique. The result of PCA is basically that we obtain a new set of variables that summarizes the information present in the data, representing the main directions of variability. The new variables or principal components are a linear combination of the original variables and are independent of each other. Each principal component is characterized by the amount of data variability that it is able to explain. Frequently, the first two or three PCs explain most of the variance, although higher PCs can also reveal important aspects of the analyzed data. Additionally, the original data can be expressed in the new variable space. The representation of both arrays and genes on the main PCs allows visualizing differences and similarities in the expression profile of the different samples and, eventually, the identification of the major genes associated with each condition. PCA can be done with the functions *princomp* and *prcomp* of the R package *stats*.

Limma R package for differential expression analysis of data arising from microarray experiments. The central idea is to fit a linear model to the expression data for each gene. The method computes a moderated *t*-statistic where posterior residual standard deviations are used in place of ordinary standard deviations. An empirical Bayes' approach is used to borrow information across genes and shrink the estimated sample variance towards a pooled estimate, which results in a more stable inference when the amount of arrays is small.

maSigPro Statistical methodology for the analysis of single and multi series time course microarray data. The method is a two-step regression approach where experimental groups (tissues, treatments, etc.) are defined by binary variables. In the first step, a global regression model is fit for each gene and the multiple test corrected *p* value associated to the *F* statistic is used to select significant time.

---

between transgenic and wild-type plants (this would account for less than 30% of the variation present in the data set).

The same technique can be used to explore more general, stress response independent, differences in the gene expression state of wild-type and transgenic plants. For this analysis we would need the independent signal intensities to both genotypes along the whole experiment. Normalized single-channel signal intensities can be generated from two-color microarray data by normalizing the *A*-values to have the same distribution across the arrays. The R package Limma provides functions to perform such analysis. Limma is a comprehensive package for micorarray data analysis based on linear methods and empirical Bayes (Smyth, 2005; Box 8.1). It also contains a wide range of functions for data handling, visualization and normalization. Single channel normalization of two-color microarray data is explained in detail in Chapter 9 of the Limma tutorial. Find the single-channel normalized data file 'data.sc.txt' at the book's website. Load this file into your R session and perform a PCA analysis as in the previous example:

```
> data.SC <- read.table("data.SC.txt", header = T,
row.names = 1) # load single channel data
> PCA.sc <- princomp(data.SC, cor = T) # perform
principal component analysis
```

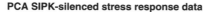

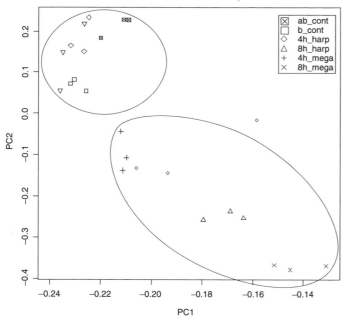

Figure 8.3

PCA plot of SIPK-silenced stress experiments. Conditions (slides) are shown for their values at the first and second principal components. (A color version of this figure is available on the book's website, www.garlandscience.com/9780415378536)

In this case, we observe that the first PC alone explains 80% of the variability which is associated with the individual arrays. The third PC provides a good separation between wild-type and SIPK-silenced signals (*Figure 8.4*). Once more we observe little difference between xanthine oxidase samples and 0 time controls.

So far, this analysis has given a general view of how gene expression features are particular to each condition of the study and can spot relationships between samples. We could be also very interested in identifying the particular genes which are more associated with the observed differences. This can be done by analyzing the scores of the PCA analysis. These give the value that each initial variable (genes) takes in the new dimensional space. Then, by selecting genes with extreme values in the PCs that discriminate samples, we will obtain those genes contributing more to these differences. In this case, genes taking high values on the third PC will be genes associated with the wild-type whereas genes taking low values in this component will be related to the SIPK-silenced signal. This variable selection is usually done by defining an interval on the standard deviation of the discriminant PC (dPC), normally $\pm 2.5 \times SD_{dPC}$, and taking genes with scores outside this interval. This can also be visualized graphically on the PCA biplot (*Figure 8.5*) where both samples and genes are displayed in the PC space.

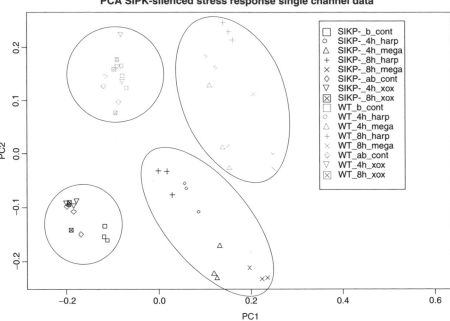

PCA SIPK-silenced stress response single channel data

Figure 8.4

PCA plot of single channel data in SIPK-silenced stress experiments. Values on the second and third principal components are shown. (A color version of this figure is available on the book's website, www.garlandscience.com/9780415378536)

```
> biplot(PCA.sc, choices=c(1,3)) # shows PCA plot
> wt.genes <- rownames(data.SC)[PCA.sc$scores[,3] >
2.5*PCA.sc$sdev[3]] # get WT genes
> sipk.genes <- rownames(data.SC)[PCA.sc$scores[,3] <
-2.5*PCA.sc$sdev[3]] # get SIPK genes
```

This approach produces 190 genes associated with the differences between wild-type and transgenic plants. The method, however, does not provide a statistical assessment for the differential expression between both genotypes. For obtaining a list of significant differentially expressed genes, a means comparison approach is more appropriate. There are several *R* packages to perform such analyses. We illustrate here how to use Limma functions.

```
> fit <-lmFit(data.M[,1:3]) # linear fit on biotic control
data
> fit <- eBayes(fit) # empirical Bayes adjustment of
significance
> result <- topTable(fit, adjust.method="fdr",number=nrow
(data.M)) # statistical results
```

The object 'result' is a table containing the values of different statistics for all genes present in the data set. We will define as significant genes those having an adjusted *p* value lower than 0.05, a positive log-odds for

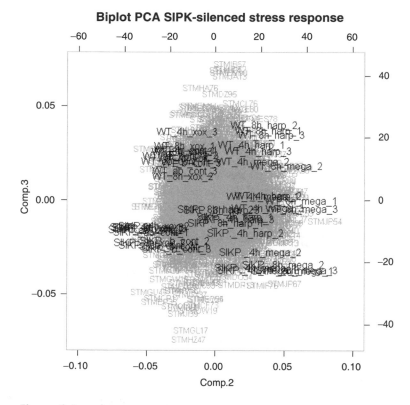

Figure 8.5

PCA biplot of single channel data in SIPK-silenced stress experiments. The values of both variables (genes) and conditions (RNA samples) in the second and third principal components are shown. (A color version of this figure is available on the book's website, www.garlandscience.com/9780415378536)

differential expression (*B*-statistics) and at least two-fold differential expression between wild-type and SIPK-silenced strains (*M*-value >1):

```
> result <- result[(result["P.Value"] <0.05) & (result["B"]
> 0) & (abs(result["M"]) > 1),]
```

This gives a list of 288 significant genes. *Figure 8.6* shows a scatter plot of the mean single channel $\log_2$ intensities of the three biotic control hybridizations. Note that both induction and repression in gene expression can be found in the SIPK-silenced plants with respect to the wild-type, and that although there is a substantial overlap in the genes detected by both analytical methods, the two approaches did not produce exactly the same result. The biological interpretation of these results will be discussed in the next section.

Until now, we have only considered overall gene expression differences between transgenic and wild-type plants and among experimental groups. One might want to go a step further and ask how gene expression evolved differently between wild-type and transgenic plants in response to stress stimuli. This type of question can be answered by studying the temporal

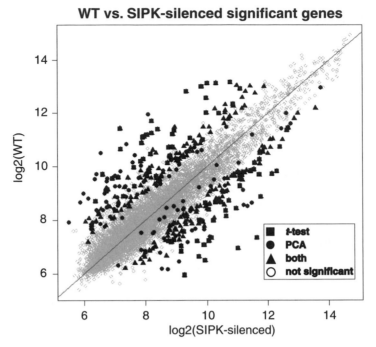

Figure 8.6

Scatter plot of average single-channel intensities of unstressed WT and SIPK-silenced plants. Significant features obtained with different statistical approaches are highlighted. (A color version of this figure is available on the book's website, www.garlandscience.com/9780415378536)

expression profiles for the different treatments and the differences between them. The *R* package maSigPro can be used for this type of analysis. maSigPro is a two-step regression approach to find genes with statistically significant time dependent expression changes or with significant profile differences between experimental groups (Conesa *et al.*, 2006) (Box 8.1). We need to create an experimental design object (edesign) where array assignment to *time*, *replicates*, and *experimental groups* are specified. You can define this object directly in *R* or create a .txt document with the information and import it into your working session.

```
> library(maSigPro) # load maSigPro package
> edesign.h <- read.table("edesign.h.txt",header = T) #load
edesign
> design.h <-make.design.matrix(edesign.h, degree = 2) #
create quadratic regression model
> fit.h <- p.vector(data.M, design.h, Q = 0.01,
nvar.correction = T) # first regression step
> tstep.h <-T.fit(fit.h, alfa = fit.h$alfa) # variable
selection regression step
> sigs.h <- get.siggenes(tstep.h, vars = "groups", rsq =
0.7) # select significant profiles
> sigs.h$summary # show lists of significant genes
```

In this example, the first experimental group (harpin) is taken as reference. This implies that time-dependent gene expression changes are calculated for this group and differences between experimental groups take harpin as the comparison group. To evaluate expression changes for the megaspermin and xanthine oxidase treatments, we simply need to set the first experimental group position of the design object to each of these conditions,

```
> edesign.m <- edesign.h[,c(1,2,4,3,5)] # edesign matrix
for megaspermin as reference
> edesign.x <- edesign.h[,c(1,2,5,3,4)] # design matrix
for xanthine oxidase as reference
```

and perform the rest of the analysis as described above.

Using this approach we obtained 616, 797, and 168 significant genes for the harpin, megaspermin, and xanthine oxidase treatments, respectively. This result is in agreement with our observations in the PCA analysis, where most gene expression differences could be associated to the biotic stress treatments. *Figure 8.7* shows heatmap representations of these data. This statistical approach provides, additionally, a gene-wise analysis on significant profile differences between experimental groups. We will discuss this point in the last part of this section.

The object generated by the get.siggenes function contains different sources of information on the selected genes. One of them is the significant coefficients of the regression models. In this case, a polynomial of degree two was chosen to construct models, and therefore intercept, linear, and quadratic coefficients are present. We can use these to typify responses. For example, genes showing a continuous induction pattern along time will have a positive significant coefficient at the linear term and a nonsignificant coefficient in the quadratic term; genes showing a transitory induction response will be significant and positive on the linear term and significant and negative in the quadratic term. Here it is shown how to select genes on the different possible types of kinetics:

```
> h.coeffs <- sigs.h$sig.genes$harp$coefficients # find
significant coefficients
> harp.up <-rownames(h.coeffs)[h.coeffs[,"betaTime"] > 0 ]
# induced genes at t = 4h
> harp.up.up <-rownames(h.coeffs)[h.coeffs[,"betaTime"] > 0
& h.coeffs[,"betaTime2"] == 0]
# continuous induction along the time component
> harp.up.down <- rownames(h.coeffs)[h.coeffs[,"betaTime"]
> 0 & h.coeffs[,"betaTime2"] <0]
# transitory induction
> harp.down <- rownames(h.coeffs)[h.coeffs[,"betaTime"] < 0
] # repression at t = 4h
> harp.down.down <- rownames(h.coeffs)[h.coeffs[,"betaTime"]
< 0 & h.coeffs[,"betaTime2"] == 0] # continuous repression
along the time component
> harp.down.up <- rownames(h.coeffs)[h.coeffs[,"betaTime"]
< 0 & h.coeffs[,"betaTime2"] > 0]
# transitory repression
```

(A)

(B)

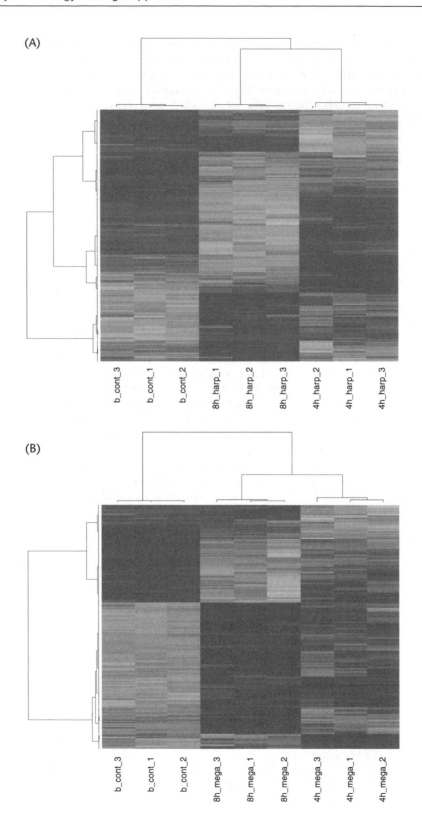

(C)

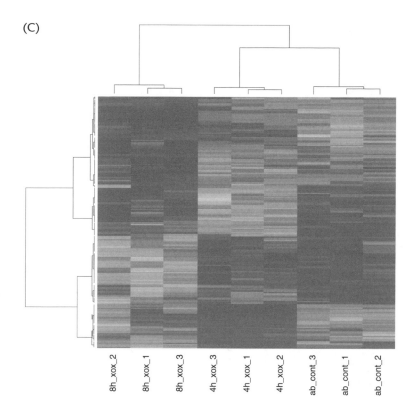

Figure 8.7

Heatmap representation of differentially expressed genes. (A) Harpin treatment, (B) megaspermin treatment, (C) xanthine oxidase treatment. (A color version of this figure is available on the book's website, www.garlandscience.com/9780415378536)

Having found the genes that have a significant gene expression response to the different treatments, we might like to know the similarities and differences between them. For example, genes showing a shared response to both biotic stress treatments will be those common to both treatments and display a similar expression pattern. We can find these genes by imputing the intersection between both gene lists and selecting genes with positively correlated expression profiles (*Figure 8.8*). This analysis revealed a total of 274 genes showing a similar gene expression response to both harpin and megaspermin stressors. Applying a similar approach, we find out that only 15 genes share a common profile for all three stress conditions.

Finally, significant differences in gene expression trends related to the treatments considered in this experiment can be studied by analyzing the comparison slots of the get.siggnes object. The slot megavsharp of sigs.h gives the genes which have significant different expression patterns between the megaspermin and harpin groups, in this case 637, whereas xoxvsharp displays 724 genes as significant for the xanthine oxidase versus harpin comparison. Biologically relevant differences can best be obtained

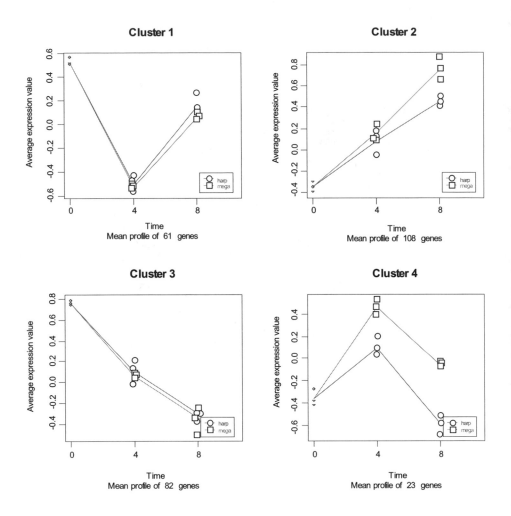

Figure 8.8

Temporal expression profiles of common significant genes of the harpin and megaspermin treatments. Genes are divided into four distinct clusters and the average cluster profile is shown.

by adding a low correlation filter on these statistical differences, resulting in 406 and 673 genes, respectively. Also here, analysis of the significant regression coefficients for each gene will provide the means of identifying distinctive pattern differences. The package function see.genes can be used to easily visualize the profile differences between experimental groups and identify groups of genes with distinctive expression patterns (*Figure 8.9*); for example, cluster 3 is characterized by containing genes upregulated exclusively by the harpin stressor.

```
> megavsharp <- as.character(sigs.h$summary[c(1:637),2])
# genes with significant different profiles for
megaspermin vs harpin comparison
```

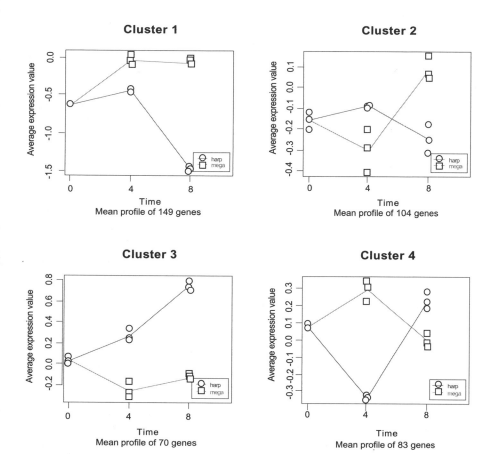

Figure 8.9

Temporal expression profiles of megaspermin vs. harpin differential responsive genes. Genes are divided into four distinct clusters and the average cluster profile is shown.

```
> corr.d <- diag(cor(t(data.M[megavsharp,c(1:9)]),
t(data.M[megavsharp,c(1:3,10:15)])))
> diffs.h.m <- megavsharp[corr.d <0.7] # selects low
correlated genes
> see.dis <- see.genes(as.matrix(data.M[diffs.h.m,
c(1:15)]), edesign = edesign.h.m, groups.vector =
make.design.matrix(edesign.h.m)$groups.vector, k = 6)) #
graphical display
> only.harp.up <- diffs.h.m [see.diffs$cut==3] # selects
genes in Cluster 3
```

8.5.2 Biological interpretation of gene expression results

The methodological approaches described in the previous section provided an evaluation of the gene expression alterations associated to the different factors present in the study, and also generated specific lists of significant genes for each of the questions posed to the statistical analysis (transgenic versus wild-type, kinetics of response to stress, gene expression features for the different stressors, etc.). In this section we will address the problem of giving a biological meaning to these results. In the first part, we will show how explorative statistical tools can offer global views of the biological message behind a microarray result. In the second part, we will illustrate how gene-function-based analysis can be addressed in our specific data example. We do not aim here to provide a comprehensive biological interpretation of our study but to give some examples of available tools and approaches to that purpose.

PCA is a useful technique to spot the molecular state of the samples and reveal relationships between treatments, genotypes and genes. The spatial distribution of control and stress samples displayed in the PCA plot (*Figure 8.3*) shows us how gene expression reflects the action of the different treatments. Megaspermin and harpin, clearly separated from control samples, are most effective in provoking a differential response between SIPK-silenced and wild-type plants, whereas the xanthine oxidase treatment does not have this strong effect. This is later corroborated by the number of significant genes obtained by each treatment and is very much in agreement with the proposed biotic stress response specificity of the jasmonic acid signaling pathway, which is altered in the SIPK-silenced plants. Also PCA can reveal the gene expression information associated with the genotypes used in this study and shows differentiation of the two strains irrespective of stress conditions (*Figure 8.4*).

A widely used approach for the evaluation of gene functions in a microarray result is to analyze the abundance of specific of Gene Ontology (GO) terms within the list of significant genes. Chapter 7 is a thorough review of this issue. For most agricultural and nonmodel species, traditional GO tools are not suitable since these rely on public databases and standard gene identifiers. However, other more dedicated tools, have become recently available. Here we will show how to use the software Blast2GO (www.blast2go.de) for exploration of functional significance in our analysis example. Blast2GO (B2G) is a GO annotation and analysis tool (Conesa *et al.*, 2005). The application uses similarity searches to find putative GO assignment to unannotated sequences and selects the most reliable terms based on some user-adjustable criteria. Additionally, the software includes a number of statistical and visualization tools for GO term based analysis of gene lists. In our case, GO term associations for the potato chip are already available at the NSF functional genomics potato site, so we can directly feed B2G with these data (some simple file formatting would be required for this, details can be found in the B2G tutorial).

We can start our functional analysis by studying the functional categories altered in the SIPK-silenced versus wild-type comparison. For this, we first need to select this gene subset by uploading a .txt file containing the gene list (menu *Extras -> select subset*). Different graphical methods are available in B2G to visualize the joined GO annotation of a group of genes. The function *Make Combined Graph* at the *Extras* Menu displays the annotation of the selected subset in the context of the GO graph highlighting the most relevant nodes. This representation is useful for spotting significant information, exploring relationships between terms and viewing GO categories within their biological context. The application provides different ways to adjust the size and information content of this graph. For this example, we will display Biological Process using default parameters except for *Seq Filter* and *Node Score Filter* that will be set to 3. We observe highlighting of categories such as protein biosynthesis, sucrose biosynthesis, mevalonate-independent isopenthenyl diphosphate biosynthesis, regulation of transcription, ribosome biosynthesis, and response to stress (*Figure 8.10*). Additionally, molecular function categories such as oxidoreductase activity, transferase activity, and hydrolase activity also appear abundant within the differentially expressed genes. These results indicate alteration of both general and specific processes in the SIPK-silenced plants.

Visualization of GO term distributions provides a means of exploring the functional information contained in your result and, when the number of significant genes is small, can be the only effective way to use this information resource. However, a proper evaluation of the importance of a given functional category in a set of differentially expressed genes would need to rely on the statistical assessment of the corresponding GO term abundance. Tests on frequency differences are appropriate for this purpose (see Chapter 7 for more information on this topic). We can use the *Enrichment Analysis* function in Blast2GO for this kind of analysis. For example, to study the GO categories statistically more abundant in the set of upregulated genes of the harpin treatment, we would upload this gene list in *Extras -> Enrichment Analysis -> Make Fisher's Exact Test -> Select Test-Set* and run this function (not selecting a Reference set will use the entire loaded annotation file, i.e. the potato chip annotation, as comparison group in the Fisher Exact's Test). The application will return a list of GO terms with a significance value for category enrichment. Typically, we would choose as significant those terms with a robust false discovery rate (Blüthgen *et al.*, 2004) value below 0.05. Results can be visualized both as bar charts (*Figure 8.11*) or a highlighted GO graph. Again, here the protein biosynthesis machinery clearly emerges through significantly enriched terms such as *ribosome biogenesis, protein biosynthesis, cytosolic small ribosome subunit*, and *RNA methylation*. In this last category, the *fribillari* gene has recently been shown to be involved in stress response in tomato (Fung *et al.*, 2005). *Cysteine biosythesis*, reported as induced in stress resistant transgenic tobacco plants and the related *sulfite reductase activity* appear also as significant, as does *caffeoyl-CoA 3-O-methyltransferase activity*, a function of the lignin biosynthesis, which is a known response to parasite attack. Finally, other significant terms are *geranyltransferase activity, dimethylallyltranstransferase activity*, and *farnesyl diphosphate activity*, involved in the biosynthesis of isopreonoids, which are major structural, environment responsive plant components. *Figure 8.12* offers an overview of the enriched functional categories for different gene lists derived in the previous section.

8.6 Concluding remarks

The arrival of biotechnology and high-throughput approaches in agricultural sciences has brought new and exciting perspectives for studying the molecular basis of plant production and for opening new possibilities to

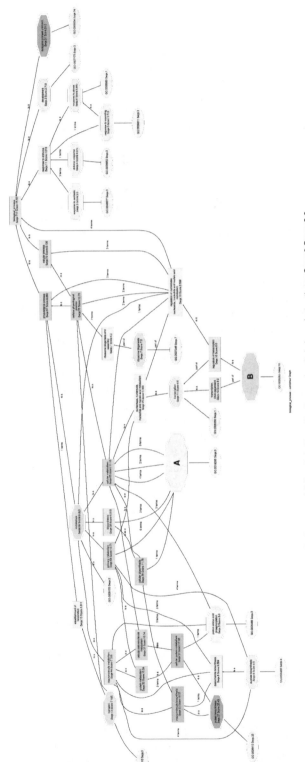

Figure 8.10

Directed acyclic graph representation of relevant Gene Ontology categories present within genes showing a significant differential expression between wild-type and SIPK-silenced plants. (A color version of this figure is available on the book's website, www.garlandscience.com/9780415378536)

A = isopentenyl diphosphate biosynthesis, mevalonate-independent pathway Seqs:3 Score:3.0

B = regulation of transcription, DNA-dependent Seqs:10 Score:10.0

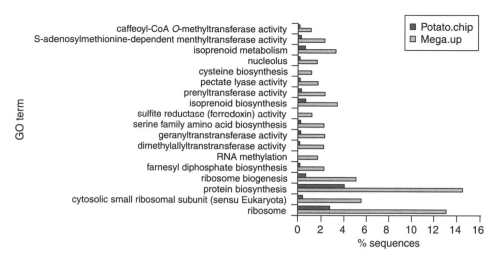

Figure 8.11

Functional category enrichment analysis of harpin response significant genes. Bar length indicates the percentage of genes belonging to a given GO category for both the reference set (potato array) and the test set (group of significant genes).

crop improvement. The agricultural research community has responded to this challenge and is actively developing tools to take full advantage of these research strategies. However, this has also led to concerns with the public about the safety of these genetically modified crops. Analysis of the plant transcriptome through microarray technology, can not only help in unraveling gene regulation mechanisms behind most important crop physiological processes, but offers also a high coverage screening methodology to evaluate wanted and unwanted effects of genetic modification.

	SIPK vsWT	harp	mega	xox	common stress	common biotic	differ. h.m	harp.up	mega. up	xox.up	SIPK vsWT.up	harp. down	SIPKvs WT.down
regulation of transcription. DNA-dependent											S		
regulation of transcription							S				S		
translation elongation factor activity							S						
FK506 binding							S						
translation factor activity			S										
AMP binding	S	S											
RNA methylation	S	S				S							S
ATP binding			S						S				
cytosolic large ribosomal subunit (sensu Eukaryota)			S						S				
protein biosynthesis	S	S				S	S	S	S	S			S
ribosome biogenesis	S	S	S	S		S	S	S	S	S			S
cytosolic small ribosomal subunit (sensu Eukaryota)		S	S			S	S	S	S	S			
structural constituent of ribosome			S	S				S	S	S			
protein folding								S	S				
cysteine biosynthesis						S			S				
regulation of translational initiation									S				
lysyl-tRNA aminocyclation			S						S				
lysine-tRNA ligase activity			S						S				
voltage-gated ion-selective channel activity							S		S				
farnesyl diphosphate biosynthesis		S				S		S	S				
cafeoyl-CoA *O*-methyltransferase activity		S		S	S					S			
lignin biosynthesis			S	S	S	S	S		S	S			
methyltransferase activity			S			S	S		S				
dimethylallyltranstransferase activity		S	S			S		S	S				
geranyltranstransferase activity		S	S			S		S	S				
transferase activity, transferring hexosyl groups	S										S		
proteolysis and peptidolysis												S	
threonyl-tRNA aminocyclation		S				S						S	
threonine-tRNA ligase activity		S		S		S						S	
trehalose biosynthesis			S										S
trehalose phosphate activity			S										S
diphosphomevalonate decarboxylase activity									S				
pectate lyase activity	S					S			S				S
CTP synthase activity	S		S										S
phosphoethanolamine *N*-methyltransferase activity	S												S
sucrose biosynthesis	S										S		
mevalonate-independent isopentenyl diphosphate	S										S		
glutamate-ammonia ligase activity	S										S		
peroxidase activity	S	S	S			S							
mevalonate kinase activity							S		S				
anion transport									S				
protein kinase activity												S	
glutathione disulfide reductase activity												S	
cinnamyl-alcohol dehydrogenase activity				S		S			S				
sulfite reductase activity					S	S							
oxidoreductase activity, acting on the aldehyde or oxo group of donors							S				S	S	
long-chain-3-hydroxyacyl-CoA dehydrogenase activity											S		
formate dehydrogenase activity	S										S		
aconitate hydratase activity	S			S							S	S	
acid phosphatase activity	S										S		
seed germination	S										S		
response to antibiotic	S										S		
response to external stimulus											S	S	
response to pathogen						S							
microtubule motor activity				S									

Figure 8.12

Summary of functional category enrichment analysis for different significant gene lists derived from the statistical analysis. GO categories are given in rows and gene sets are given in columns. The shaded cells indicate significant enrichment of the category for the given gene set.

References

Adams MD, Soares MB, Kerlavage AR, Fields C, Venter JC (1993) Rapid cDNA sequencing (expressed sequence tags) from a directionally cloned human infant brain cDNA library. *Nat Genet* **4**: 373–380.

Ahren D, Troein C, Johansson T, Tunlid A (2004) PHOREST: a web-based tool for comparative analyses of expressed sequence tag data. *Mol Ecol Notes* **4**: 311–314.

Altschul SF, Gish W, Miller W, Myers EW, Lipman DJ (1990) Basic local alignment search tool. *J Mol Biol* **215**: 403–410.

Apweiler R, Attwood TK, Bairoch A *et al.* (2001) The InterPro database, an integrated documentation resource for protein families, domains and functional sites. *Nucleic Acids Res* **29**: 37–40.

Armstrong JI, Yuan S, Dale JM, Tanner VN, Theologis A (2004) Identification of inhibitors of auxin transcriptional activation by means of chemical genetics in Arabidopsis. *Proc Natl Acad Sci USA* **101**: 14978–14983.

Audic S, Claverie JM (1997) The significance of digital gene expression profiles. *Genome Res* **7**: 986–995.

Ayoubi P, Jin X, Leite S *et al.* (2002) PipeOnline 2.0: automated EST processing and functional data sorting. *Nucleic Acids Res* **30**: 4761–4769.

Blüthgen N, Brand K, Cajavec B, Swat M, Herzel H, Beule D (2005) Biological profiling of gene groups utilizing Gene Ontology. Genome inform. ser. workshop. *Genome Inform* **16**: 106–115.

Bonaldo MF, Lennon G, Soares MB (1996) Normalization and subtraction: two approaches to facilitate gene discovery. *Genome Res* **6**: 791–806.

Buchanan CD, Lim S, Salzman RA *et al.* (2005) *Sorghum bicolor's* transcriptome response to dehydration, high salinity and ABA. *Plant Mol Biol* **58**: 699–720.

Burke J, Davison D, Hide W (1999) d2_cluster: a validated method for clustering EST and full-length cDNA sequences. *Genome Res* **9**: 1135–1142.

Chang S, Puryear J, Cairney J (1993) A simple and efficient method for isolating RNA from pine trees. *Plant Mol Biol Reporter* **11**: 113–116.

Chen W, Provart NJ, Glazebrook J *et al.* (2002) Expression profile matrix of Arabidopsis transcription factor genes suggests their putative functions in response to environmental stresses. *Plant Cell* **14**: 559–574.

Chou HH, Holmes MH (2001) DNA sequence quality trimming and vector removal. *Bioinformatics* **17**: 1093–1104.

Cohn JR, Martin GB (2005) *Pseudomonas syringae* pv. tomato type III effectors AvrPto and AvrPtoB promote ethylene-dependent cell death in tomato. *Plant J* **44**: 139–154.

Conesa A, Gotz S, Garcia-Gomez JM, Terol J, Talon M, Robles M (2005) Blast2GO: a universal tool for annotation, visualization and analysis in functional genomics research. *Bioinformatics* **21**: 3674–3676.

Conesa A, Nueda S, Ferrer A, Talón M (2006) maSigPro: a method to identify significant differential expression profiles in time-course microarray experiments. *Bioinformatics* **22**: 1096–1102.

Corpillo D, Gardini G, Vaira AM, Basso M, Aime S, Accotto GP, Fasano M (2004) Proteomics as a tool to improve investigation of substantial equivalence in genetically modified organisms: the case of a virus-resistant tomato. *Proteomics* **4**: 193–200.

da Silva FG, Iandolino A, Al-Kayal F *et al.* (2005) Characterizing the grape transcriptome. Analysis of expressed sequence tags from multiple vitis species and development of a compendium of gene expression during berry development. *Plant Physiol* **139**: 574–597.

Delhaize E, Hebb DM, Richards KD, Lin JM, Ryan PR, Gardner RC (1999) Cloning and expression of a wheat (*Triticum aestivum* L.) phosphatidylserine synthase cDNA overexpression in plants alters the composition of phospholipids. *J Biol Chem* **274**: 7082–7088.

Dong Q, Kroiss L, Oakley FD, Wang BB, Brendel V (2005) Comparative EST analyses in plant systems. *Methods Enzymol* **395**: 400–418.

Dowd C, Wilson IW, McFadden H (2004) Gene expression profile changes in cotton root and hypocotyl tissues in response to infection with *Fusarium oxysporum* f. sp. *vasinfectum*. *Mol Plant Microbe Interact* **17**: 654–667.

Dueck TA, Van Der Werf A, Lotz LAP, Jordi W (1998) *Methodological Approach to a Risk Analysis for Polygene – Genetically Modified Plants (GMPs): A Mechanistic Study*. AB note 59. Research Institute for Agrobiology and Soil Fertility (AB-DLO), Wageningen, the Netherlands.

Engel KH, Gerstner G, Ross A (1998) *Investigation of Glucoalkaloids in Potatoes as Example for the Principle of Substantial Equivalence*. Federal Institute of Consumer Health Protection and Veterinary Medicine, Berlin, pp. 197–209.

Everitt R, Minnema SE, Wride MA, Koster CS, Hance JE, Mansergh FC, Rancourt DE (2002) RED: the analysis, management and dissemination of expressed sequence tags. *Bioinformatics* **18**: 1692–1693.

Evertsz EM, Au-Young J, Ruvolo MV, Lim AC, Reynolds MA (2001) Hybridization cross-reactivity within homologous gene families on glass cDNA microarrays. *Biotechniques* **31**: 1182–1192.

Ewen SW, Pusztai A (1999) Effect of diets containing genetically modified potatoes expressing *Galanthus nivalis* lectin on rat small intestine. *Lancet* **354**: 1353–1354.

Ewing B, Green P (1998) Base-calling of automated sequencer traces using phred. II. Error probabilities. *Genome Res* **8**: 186–194.

Ewing B, Hillier L, Wendl MC, Green P (1998) Base-calling of automated sequencer traces using phred. I Accuracy assessment. *Genome Res* **8**: 175–185.

Fei Z, Tang X, Alba RM, White JA, Ronning CM, Martin GB, Tanksley SD, Giovannoni JJ (2004) Comprehensive ET analysis of tomato and comparative genomics of fruit ripening. *Plant J* **40**: 47–59.

Fernie AR, Trethewey RN, Krotzky AJ, Willmitzer L (2004) Metabolite profiling: from diagnostics to systems biology. *Nat Rev Mol Cell Biol* **5**: 763–769.

Forment J, Gadea J, Huerta L *et al.* (2005) Development of a citrus genome-wide EST collection and cDNA microarray as resources for genomic studies. *Plant Mol Biol* **57**: 375–391.

Fung RW, Wang CY, Smith DL, Gross KC, Tao Y, Tian M (2005) Characterization of alternative oxidase (AOX) gene expression in response to methyl salicylate and methyl jasmonate pre-treatment and low temperature in tomatoes. *J Plant Physiol* **21**: [Epub ahead of print]

Gertz JM, Vencill WK, Hill NS (1999) Tolerance of transgenic soybean (*Glycine max*) to heat stress. In: *Proceedings of the 1999 Brighton Crop Protection Conference*: *Weeds*. British Crop Protection Council, Farnham, **3**: 835–840.

Greller LD, Tobin FL (1999) Detecting selective expression of genes and proteins. *Genome Res* **9**: 282–296

Hall H (2005) Functional analysis of MAPkinase-dependent stress response signalling in tobacco. MSc thesis dissertation. University of British Columbia.

Hashimoto W, Momma K, Katsube T, Ohkawa Y, Ishige T, Kito M, Utsumi S, Murata K (1999a) Safety assessment of genetically engineered potatoes with designed soybean glycinin: compositional analyses of the potato tubers and digestibility of the newly expressed protein in transgenic potatoes. *J Sci Food Agric* **79**: 1607–1612.

Hashimoto W, Momma K, Yoon HJ, Ozawa S, Ohkawa Y, Ishige T, Kito M, Utsumi S, Murata K (1999b) Safety assessment of transgenic potatoes with soybean glycinin by feeding studies in rats. *Biosci Biotechnol Biochem* **63**: 1942–1946.

Herrero J, Al-Shahrour F, Díaz-Uriarte R, Mateos A, Vaquerizas JM, Santoyo J, Dopazo J (2003) GEPAS, a web-based resource for microarray gene expression data analysis. *Nucleic Acids Res* **31**: 3461–3467.

Hotz-Wagenblatt A, Hankeln R, Ernst P, Glatting KH, Schimdt ER, Suhai S (2003) ESTAnnotator: a tool for high throughput EST annotation. *Nucleic Acids Res* **31**: 3716–3719.

Huang X, Madan A (1999) CAP3: a DNA sequence assembly program. *Genome Res* **6**: 829–845.

Kok EJ, Kuiper HA (2003) Comparative safety assessment for biotech crops. *Trends Biotechnol* **21**: 439–444.

Kristensen C, Morant M, Olsen CE, Ekstrom CT, Galbraith DW, Moller BL, Bak S (2005) Metabolic engineering of dhurrin in transgenic Arabidopsis plants with marginal inadvertent effects on the metabolome and transcriptome. *Proc Natl Acad Sci USA* **102**: 1779–1784.

Kuiper HA, Noteborn HP, Peijnenburg AA (1999) Adequacy of methods for testing the safety of genetically modified foods. *Lancet* **354**: 1315–1316.

Kuiper HA, Kleter GA, Noteborn HP, Kok EJ (2001) Assessment of the food safety issues related to genetically modified foods. *Plant J* **27**: 503–528.

Kuiper HA, Kok EJ, Engel KH (2003) Exploitation of molecular profiling techniques for GM food safety assessment. *Curr Opin Biotechnol* **14**: 238–243.

Kumar CG, LeDuc R, Gong G, Roinishivili L, Lewin HA, Liu L (2004) ESTIMA, a tool for EST management in a multi-project environment. *BMC Bioinformatics* **5**: 176–185.

Liu-Stratton Y, Roy S, Sen CK (2004) DNA microarray technology in nutraceutical and food safety. *Toxicol Lett* **150**: 29–42.

Lloyd JC, Zakhleniuk OV (2004) Responses of primary and secondary metabolism to sugar accumulation revealed by microarray expression analysis of the Arabidopsis mutant, pho3. *J Exp Bot* **55**: 1221–1230.

Lopez C, Soto M, Restrepo S, Piegu B, Cooke R, Delseny M, Tohme J, Verdier V (2005) Gene expression profile in response to *Xanthomonas axonopodis* pv. *manihotis* infection in cassava using a cDNA microarray. *Plant Mol Biol* **57**: 393–410.

Malatesta M, Caporaloni C, Gavaudan S, Rocchi MB, Serafini S, Tiberi C, Gazzanelli G (2002) Ultrastructural morphometrical and immunocytochemical analyses of hepatocyte nuclei from mice fed on genetically modified soybean. *Cell Struct Funct* **27**: 173–180.

Mao C, Cushman JC, May GD, Weller JW (2003) ESTAP – An automated system for the analysis of EST data. *Bioinformatics* **19**: 1720–1722.

Menke FL, Kang HG, Chen Z, Park JM, Kumar D, Klessig DF (2005) Tobacco transcription factor WRKY1 is phosphorylated by the MAP kinase SIPK and mediates HR-like cell death in tobacco. *Mol Plant Microbe Interact* **18**: 1027–1034.

Miles GP, Samuel MA, Zhang Y, Ellis BE (2005) RNA interference-based (RNAi) suppression of AtMPK6, an Arabidopsis mitogen-activated protein kinase, results in hypersensitivity to ozone and misregulation of AtMPK3. *Environ Pollut* **138**: 230–237.

Momma K, Hashimoto W, Ozawa S, Kawai S, Katsube T, Takaiwa F, Kito M, Utsumi S, Murata K (1999) Quality and safety evaluation of genetically engineered rice with soybean glycinin: analyses of the grain composition and digestibility of glycinin in transgenic rice. *Biosci Biotechnol Biochem* **63**: 314–318.

Moy P, Qutob D, Chapman BP, Atkinson I, Gijzen M (2004) Patterns of gene expression upon infection of soybean plants by *Phytophthora sojae*. *Mol Plant Microbe Interact* **17**: 1051–1062.

Mulder NJ, Apweiler R, Attwood TK *et al.* (2005) Interpro, progress and status in 2005. *Nucleic Acid Research* **33**: (Database issue): D201–205.

Murray F, Llewellyn D, McFadden H, Last D, Dennis ES, Peacock WJ (1999) Expression of the *Talaromyces flavus* glucose oxidase gene in cotton and tobacco reduces fungal infection, but is also phytotoxic. *Mol Breed* **5**: 219–232.

Narusaka Y, Narusaka M, Seki M, Umezawa T, Ishida J, Nakajima M, Enju A,

Shinozaki K (2004) Crosstalk in the responses to abiotic and biotic stresses in Arabidopsis: analysis of gene expression in cytochrome P450 gene superfamily by cDNA microarray. *Plant Mol Biol* **55**: 327–342.

Nordic Council (2003) *Use of the cDNA Microarray Technology in the Safety Assessment of GM Food Plants*. TemaNord 2003: 558. Nordic Council of Ministers, Copenhagen.

Ogihara Y, Mochida K, Kawaura K *et al.* (2004) Construction of a full-length cDNA library from young spikelets of hexaploid wheat and its characterization by large-scale sequencing of expressed sequence tags. *Genes Genet Syst* **79**: 227–232.

Okinaka Y, Yang CH, Perna NT, Keen NT (2002) Microarray profiling of *Erwinia chrysanthemi* 3937 genes that are regulated during plant infection. *Mol Plant Microbe Interact* **15**: 619–629.

Oztur ZN, Talame V, Deyholos M, Michalowski CB, Galbraith DW, Gozukirmizi N, Tuberosa R, Bohnert HJ (2001) Monitoring the expression pattern of 1300 Arabidopsis genes under drought and cold stresses by using a full-length cDNA microarray. *Plant Cell* **13**:61–72.

Palenchar PM, Kouranov A, Lejay LV, Coruzzi GM (2004) Genome-wide patterns of carbon and nitrogen regulation of gene expression validate the combined carbon and nitrogen (CN)-signaling hypothesis in plants. *Genome Biol* **5**: R91.

Paquola AC, Nishiyama Jr MY, Reis EM, da Silva AM, Verjovski-Almeida S (2003) ESTWeb: bioinformatics services for EST sequencing projects. *Bioinformatics* **19**: 1587–1588.

Pertea G, Huang X, Liang F *et al.* (2003) TIGR Gene Indices clustering tools (TGICL): a software system for fast clustering of large EST datasets. *Bioinformatics* **19**: 651–652.

Ptitsyn A, Hide W (2004) CLU: a new algorithm for EST clustering. *BMC Bioinformatics* **6(suppl 2)**: S3.

Qiu F, Guo L, Wen T-J, Liu F, Ashlock DA, Schnable PS (2003) DNA sequence-based 'bar-codes' for tracking the origins of expressed sequence tags from a maize cDNA library constructed using multiple mRNA sources. *Plant Physiol* **133**: 475–481.

Rabbani MA, Maruyama K, Abe H *et al.* (2003) Monitoring expression profiles of rice genes under cold, drought, and high-salinity stresses and abscisic acid application using cDNA microarray and RNA gel-blot analyses. *Plant Physiol* **133**: 1755–1767.

Rensink WA, Iobst S, Hart A, Stegalkina S, Liu J, Buell CR (2005) Gene expression profiling of potato responses to cold, heat, and salt stress. *Funct Integr Genomics* **5**: 201–207.

Royal Society (1999) Review of data on possible toxicity of GM potatoes http://www.royalsoc.ac.uk/templates/statements/statementDetails.cfm?StatementID=29.

Rudd S (2003) Expressed sequence tags: alternative or complement to whole genome sequences? *Trends Plant Sci* **8**: 321–329.

Saal LH, Troein C, Vallon-Christersson J, Gruvberger S, Borg A, Peterson C (2002) BioArray Software Environment: a platform for comprehensive management and analysis of microarray data. *Genome Biol* **3**: software0003.1–0003.6.

Samuel MA, Walia A, Mansfield SD, Ellis BE (2005) Overexpression of SIPK in tobacco enhances ozone-induced ethylene formation and blocks ozone-induced SA accumulation. *J Exp Bot* **56**: 2195–2201.

Seki M, Narusaka M, Abe H, Kasuga M, Yamaguchi-Shinozaki K, Carninci P, Hayashizaki Y, Shinozaki K (2001) Monitoring the expression pattern of 1300 Arabidopsis genes under drought and cold stresses using a full-length cDNA microarray. *Plant Cell* **13**: 61–72.

Seki M, Narusaka M, Ishida J *et al.* (2002) Monitoring the expression profiles of 7000 Arabidopsis genes under drought, cold and high-salinity stresses using a full-length cDNA microarray. *Plant J* **3**: 279–292.

Senthil G, Liu H, Puram VG, Clark A, Stromberg A, Goodin MM (2005) Specific and common changes in *Nicotiana benthamiana* gene expression in response to infection by enveloped viruses. *J Gen Virol* **86(pt 9)**: 2615–2625.

Shewmaker CK, Sheehy JA, Daley M, Colburn S, Ke DY (1999) Seed-specific overexpression of phytoene synthase: increase in carotenoids and other metabolic effects. *Plant J* **20**: 401–412.

Smyth GK (2005) Limma: linear models for microarray data. In: *Bioinformatics and Computational Biology Solutions using R and Bioconductor*, (eds R. Gentleman, V. Carey, S. Dudoit, R. Irizarry and W. Huber). Springer, New York, pp. 397–420.

Stekel DJ, Git Y, Falciani F (2000) The comparison of gene expression from multiple cDNA libraries. *Genome Res* **10**: 2055–2061.

The Gene Ontology Consortium (2004) The Gene Ontology (GO) database and informatics resource. *Nucleic Acids Res* **32**: D258–D261. [Database issue]

Tohge T, Nishiyama Y, Hirai MY *et al.* (2005) Functional genomics by integrated analysis of metabolome and transcriptome of Arabidopsis plants overexpressing an MYB transcription factor. *Plant J* **42**: 218–235.

Tomiuk S, Hofmann K (2001) Microarray probe selection strategies. *Brief Bioinformatics* **2**: 329–340.

Troyanskaya O, Cantor M, Sherlock G, Brown P, Hastie T, Tibshirani R, Botstein D, Altman RB (2001) Missing value estimation methods for DNA microarrays. *Bioinformatics* **17**: 520–525.

Turk SCHJ, Smeekens SCM (1999) Genetic modification of plant carbohydrate metabolism In: *Applied Plant Biotechnology* (eds V.L. Chopra, V.S. Malik and S.R. Bhat). Science Publishers, Enfield, pp. 71–100.

Urbanczyk-Wochniak E, Baxter C, Kolbe A, Kopka J, Sweetlove LJ, Fernie AR (2005) Profiling of diurnal patterns of metabolite and transcript abundance in potato (*Solanum tuberosum*) leaves. *Planta* **221**: 891–903.

Wang J-PZ, Lindsay BG, Leebens-Mack J, Cui L, Wall K, Miller WC, dePamphilis CW (2004) EST clustering error evaluation and correction. *Bioinformatics* **20**: 2973–2984.

Watkinson JI, Sioson AA, Vasquez-Robinet C *et al.* (2003) Photosynthetic acclimation is reflected in specific patterns of gene expression in drought-stressed loblolly pine. *Plant Physiol* **133**: 1702–1716.

Yamamoto K, Sasaki T (1997) Large-scale EST sequencing in rice. *Plant Mol Biol* **35**: 135–144.

Ye X, Al-Babili S, Kloti A, Zhang J, Lucca P, Beyer P, Potrykus I (2000) Engineering the provitamin A (beta-carotene) biosynthetic pathway into (carotenoid-free) rice endosperm. *Science* **287**: 303–305.

Ying S-Y (ed.) (2003) *Generation of cDNA Libraries*: *Methods and Protocols*. Humana Press Inc., New Jersey.

Protein microarrays

Nigel J. Saunders

9

9.1 Introduction

Microarrays can be used to study any type of interaction in which the arrayed material can be suitably deposited and the binding reagent hybridized and directly or indirectly labeled, or its binding otherwise detected. Microarrays were initially applied to equivalents of Southern and northern blot experiments, but they have considerable potential in the study of other interactions, including between proteins, and proteins and other substrates. Indeed, protein-based microarray applications were first established fairly soon after the DNA-based microarrays were introduced, but their subsequent development has been significantly slower. It was always my intention to develop protein microarrays within the context of our functional genomics research. However, like many we focused initially on nucleic acid-based microarrays, believing that these were more established and easier to perform. Once we finally moved onto proteins, it was immediately apparent that they are certainly no more problematic than nucleic acid-based studies and, indeed in several respects, are actually easier to do. On reflection, there was no need to put this area of work off in the early stages, and those with microarray experience should readily consider working in this area.

Indeed, it is worthy of note that despite a common rhetoric and dominance of those with a different world view (and at least one very large book almost entirely devoted to it), much of the primary work that lay the foundations for microarrays of all types originated in the field of protein microarrays, rather than of PCR products or oligonucleotide microarrays. For those who would appreciate a different view, readers are directed to the articles of Ekins and Chu (1999) and Ekins (1998). For those of us who come to this area of research from DNA microarrays, it is good to realize that we are actually returning to where the whole process in several respects actually began.

In common with all technologies there are practical issues that have to be recognized, some of which will be addressed in this chapter, which consists of an initial introduction to the subject, and then describes some of the specific characteristics of these tools with reference to illustrative examples from one experiment.

9.2 The uses and application of protein microarrays

After the initial hype associated with DNA microarrays the next area to receive significant attention, especially from commercial companies, was protein arrays. This enthusiasm was perhaps initially more muted than following the introduction of DNA arrays and the associated perception was that these had not delivered in the ways that some had promoted that they would. However, the perspectives with regard to protein (and DNA) microarrays are now far more sensible, and this is an area in which great potential exists, and I am confident will ultimately be realized. The important thing to accept is that, like DNA microarrays, researchers need time and experience to determine the strengths and weaknesses of these tools, to identify the technical problems that need to be solved, and to determine the research contexts in which these tools are most appropriate and informative. Protein-based microarrays also have significant additional impediments, compared with those made with readily synthesized probe substrates, related to generating the raw materials needed for their fabrication particularly in high-throughput screening when the array is to be fabricated from complete proteins.

As functional genomics (or systems biology) develops, the whole organism's enmeshed network of internal responses and component interactions is increasingly recognized as central to understanding cellular behavior, and the lack of understanding and tools for studying protein–protein interactions is recognized as a major impediment, and the area is in need of development. Protein microarrays offer one route from which these issues can be addressed. One of the earliest papers in this area, by MacBeath and Schreiber (2000), actually performed a number of tests showing that this approach could be applied to several aspects of protein function, including protein–protein interactions using high affinity interactions such as protein G and immunoglobulin, as well as weaker interactions using NF kappa B and I kappa B, protein–small molecule interactions using the dependence of FKBP immunophillin binding to FRAP upon the presence of rapamycin, and even enzymatic activity testing using three different kinase–substrate pairs. Similarly, Ge (2000) also showed that several types of interactions could be obtained on membrane spotted protein microarrays. Considering that these important papers are now more than 5 years old, it is surprising that progress in this field and using similar tools has not been greater or more widely adopted than it has.

Initially, the range of proteins available is/was frequently limited, and when protein microarrays are used to address focused problems this is not necessarily an impediment. However, the attraction to many, particularly in the fashionable context of high-throughput screening for therapeutic targets and biomarkers, is that it should be possible to assess hundreds, or thousands, of interactions simultaneously to address areas including protein–antibody, protein–protein, receptor–ligand, drug–drug target, enzyme–substrate, and others. Once sufficient numbers of recombinant proteins become available (and in some cases they already are), this approach will progressively expand, and significant progress has been made in this area for some systems.

In many ways the most exciting, but amongst the hardest and most expensive to generate, are microarrays with large catalogs of species specific proteins. These can be used in a variety of ways. For example, a (membrane based) microarray including 1690 *Arabidopsis* proteins has been generated from a cDNA expression library and used to identify mitogen-activated protein kinase in the presence of radioactive ATP. Using this approach, 48 potential substrates for MAPK3, and 39 for MAPKa6 were identified – 26 of them common to both kinases (Feilner, 2005). This follows a similar study in which the production of proteins in an expression system was assessed in a microarray format using immunodetection of a terminal tag in a slide-based format, followed by a kinase activity detection to identify targets for barley casein kinase IIα (Kramer *et al.*, 2004).

The first example of a proteome-scale investigation using protein microarrays, containing about 80% of the yeast proteome, was published by Snyder and his team in 2001 (Zhu *et al.*, 2001). This microarray was successfully used to identify protein ligands for calmodulin, and different phosphoinositide lipids, identifying previously known and novel ligands for these substrates. The range of applications for which microarrays are applicable is often only limited by the imagination of the experimentalist.

A different approach using native and site-directed mutants can be used to perform more specific studies of protein–protein interactions and activation. For example, a study of the influence of specific residues and co-factors on the binding of an *E. coli* phosphodiesterase was studied in this way – using a fluorescently labeled portion of the normally bound ligand for detection (Sasakura, 2005). However, although this paper illustrates some principles of what can be done in a microarray format, the particular study did not have the highly parallel features that really necessitated the use of this approach in this particular experiment. Nevertheless, it does illustrate a good approach, especially when the investigations can be performed with a readily labeled known protein to detect the interaction of interest.

The most obvious next horizon for protein microarrays (at least in the author's opinion) is for coordination to be established between the protein production facilities and programs currently ongoing for structural biology programs, and the fractionation expertise which is rapidly developing in the field of proteomics, with protein microarray-based projects.

9.2.1 Antigen–antibody interactions and immunoassays

The first area in which the protein microarray format was seen to have potential was probably in immunoassays, and performing multiple tests on small experimental or clinical samples. While it is possible to look into the future at the potential of protein-based microarrays in a range of 'systems biology' applications, they are already a well established tool for use in the study of antigen–antibody interactions. Despite this fact, their use in this context has remained relatively restricted, although some centers have advanced expertise in this area. The original work in the area of immunoassays was laid out by Ekins from early in the 1980s. The fact that, when used in certain assay designs, the fractional antibody binding site occupancy is independent of both the antibody concentration and the sample volume laid the groundwork for quantitative microarray-based immunoassays. This is notable particularly because it demonstrates an early case in which the use of two labels is required for proper quantitative assessments to be made. In this case, providing a measure of the arrayed antibody, and the amount of bound target.

One of the most direct applications of microarrays to antibody assessments has been in the area of assessments of antibodies in autoimmunity, with an early key paper in this field being published by Thomas Joos in 2000. Although the presented assay was not refined to the point of a diagnostic tool *per se*, this study demonstrated all of the characteristics necessary for the suitability of this approach for both research and diagnostic applications. Studies in this area to-date have been extended to several diseases of this type (see Robinson *et al.*, 2003), and specific studies have productively been applied to autoantibody profiling studies in diseases including: a range of connective tissue diseases (Robinson *et al.*, 2002), rheumatoid arthritis (Hueber *et al.*, 2005), experimental autoimmune encephalomyelitis (Fontoura *et al.*, 2004), and systemic lupus erythematosus (Graham *et al.*, 2004). A similar approach has been adopted for the identification of non-host antibodies involved in allergic responses (see Harwanegg *et al.*, 2003). A notable feature of these studies, and the power of this application for antibody profiling, is that these studies consistently have a substantial increase in sensitivity over traditional ELISA-based methods, and that the useful detection range is linear and wide.

Another field of disease in which auto-antibodies are important and potentially represent biomarkers for disease diagnosis or monitoring of progression is cancer, in which altered presentation, presentation of sequestered antigens, or altered modification of self-proteins can lead to the generation of antibody responses to tumor-associated antigens. One of the first microarray studies of this type, which was also an important paper because it demonstrated how high-resolution proteomic fractionation can be linked to protein microarray applications, demonstrated that specific protein fractions from prostate cancer materials were immunoreactive with antibodies present in serum obtained from patients with prostate cancer, but not from healthy controls (Yan *et al.*, 2003). Another example of the utility of this approach in the same disease using a similar fractionation-based approach, from a parallel and presumably independent study, was described later in the same year (Bouwman *et al.*, 2003). However, working with fractions adds a whole spectrum of additional problems to those associated with protein microarrays, most notably associated with denaturation and dilution effects.

It is one research goal of protein arrays to extend their use to proteomic profiling. Any assay involving antibodies has inherent issues relating to the fact that different antibodies can have substantially different binding affinities to the antigens to which they are directed. One way around this is to read each data point independently, and to label each antibody used to construct the microarray. However, one of the simplest, and consequently most elegant, solutions is to use a protein microarray in an exactly analogous way to a dual channel transcriptional microarray experiment. In dual-channel experiments a large degree of non-uniformity in the printing, concentrations, and affinity of the probes is overcome through the use of a comparative assay in which each of these parameters affects both channels of the experiment equally (or at least that is how it works in theory). The first paper describing this approach, and one to which anyone interested in this type of assay should still refer, was published by Haab *et al.* in 2001, in which the microarray was used to assess serum proteins. A similar method was used to assess expression of cultured cells in a study of responses to radiation by Sreekumar and colleagues (2001) –

demonstrating that this approach can be applied to focused experimental questions. It may be that protein microarrays themselves hold the key to creating sets of antibodies with appropriately matched, or at least suitably similar binding affinities, since they can be used for the high throughput screening of antibody reagents (e.g. de Wildt *et al.*, 2000; Michaud *et al.*, 2003; Poetz *et al.*, 2005).

In addition to being used for specific experimental assessments, this type of antibody array can also be used for focused proteomic profiling. This was demonstrated in a study, which was in some ways complementary to the cancer-directed antibody detection studies for prostate cancer, in which the sera from patients with prostate cancer were compared with sera from normal controls (Miller *et al.*, 2003). In addition to demonstrating the feasibility of this approach, it also demonstrated the primary limitation that the available antibodies represent, although five 'serological marker' proteins were identified, it is unlikely that any of them will be particularly specific for prostate cancer. A similar study has recently been reported in which a panel of antibodies to 84 serum components that were thought to be potentially useful for the diagnosis of lung cancer were selected and used in an antibody microarray (Gao, 2005). Again, this study was technically successful, in that four markers were identified that might contribute to diagnosis of lung cancer. However, none of the identified markers were new with respect to their known association with the disease being investigated, none of the markers were specific for the different types of lung cancer included in the study, and some of them are raised in several other conditions. What this clearly indicates is that a sufficient bank of specific probe reagents is needed, and that currently the available repertoire of antibodies for most researchers is insufficient for reliable biomarker discovery (but it is always possible to get lucky).

However, it is important to recognize that we are still at the early days of biomarker discovery, and that at present most markers from most studies (other than those looking for specific antibodies) are more readily able to determine whether the patient is unwell and perhaps the general type of illness, rather than what specific disease they might have (partly due to the nature of the currently prevalent experimental designs).

9.2.2 Phage display libraries and protein microarrays

The use of phage display libraries has been one approach to addressing the protein presentation/library issues when specific libraries of whole proteins are unavailable. These naturally have the limitations that all expressed peptides are not necessarily representative of the whole repertoire of naturally occurring proteins, that the peptides will not represent tertiary structures involving remote protein components, and not all peptides will be representative of the naturally occurring reading frames. However, this approach has the considerable advantage that the proteins that are being arrayed, the probes, are produced in a high-throughput fashion that does not incur the high costs of a gene-by-gene, protein-by-protein strategy. There are also additional limitations of scaling, such that libraries rapidly become too large to be effectively arrayed in full diversity, especially with randomly generated libraries. Rather than relying upon random libraries, this approach is also applicable to using libraries prepared from specific-organism cDNA, so that a degree of targeting and the use of larger protein components are possible. Given that antibody interactions are amongst the easiest to assay in a microarray format, that some epitopes are linear and contained within short regions of protein sequence, and antibodies are of interest because they can be used directly in the development of immunoassays, the use of phage display libraries for the identification of 'antibody signatures' is an obvious thing to do, and has been implemented successfully.

An example of this approach is the use of a phage display library used to screen sera from over 100 prostate cancer patients and controls (Wang *et al.*, 2005). In this case the phage display library was generated from cDNA from prostate cancer tissue, and five rounds of biopanning. Using this approach a set of 22 peptides were identified that could be used as a detector that performed better than the currently used test for prostate specific antigen (PSA) – although it should be noted that this current test has a far from ideal sensitivity and specificity, so neither test is optimal for this disease. Also, it is worth noting that only four of these peptides are derived from what are currently considered to be in-frame named coding sequences, so it is open to question whether the recognized peptides represent, or mimic, the actual antigenic targets. However, this study does show the potential and one application for this technique. Strangely, this recent paper did not reference the previous highly relevant protein microarray studies using fractionated proteins of Yan or Bouwman referred to previously in this chapter, which indicated the likely success of a project of this type in this particular disease.

A similar study has been performed for the identification of diagnostic markers for non-small cell lung cancer (Zhong *et al.*, 2005). In this study a notable feature was that the T7 phage display library that was used was enriched with four rounds of biopanning for 4000 initial candidate proteins using the sera from cancer patients to prepare the probe set that was interrogated, from which 212 were then selected for use in the test microarray. Using just 20 patient and control samples with 212 proteins this study was able to identify seven good markers, the combined use of five of which was able to generate an assay with close to 90% diagnostic accuracy.

9.2.3 Use of protein microarrays to assess arrayed samples

In the examples given so far, the model has been that a capture reagent, or something with which the sample would interact is arrayed, and the sample under investigation is then assessed using the microarray. It is also possible to perform protein microarrays 'the other way around'. In this context a microarray format is useful if the materials are relatively precious or only available in small volumes, and if a large number of similar tests are required. In this form, for example, a series of samples are arrayed, and then probed with a common test reagent. This approach is more frequently associated with 'tissue microarrays', but it is not limited to this application. One aspect of the experiment described in the latter section of this chapter addresses one example of this approach – but this is a potentially useful and perhaps underused approach to using protein microarrays.

9.2.4 Focused functional assays on protein microarrays

The main limitation of this approach is that (at least with current models) it is not possible to assess functions that depend upon multi-component complexes. On the other hand, this is an advantage when the experimental goal requires differentiation of direct interactions, from those that are dependent or are bridged by other proteins, which cannot necessarily be distinguished using other methods, such as yeast two-hybrid systems. While it is not inconceivable that proteins could be deliberately prepared and arrayed as complexes in the future, this is not something that has been developed (as far as the author is aware) to date.

While the greatest promise (or at least the greatest 'hype') is for the study of very large and highly diverse protein interactions in 'systems-level' studies, the most immediate return and much of the progress using microarrays has been in relatively focused/targeted investigations. In these studies, particular classes of interaction are studied, and this process is greatly facilitated when the metabolic process can be specifically reported – for example, through the incorporation of a detectable agent, or where the goal is to detect the binding of a directly labeled DNA or protein ligand.

A good example of the use of a protein array format to study interactions in a targeted fashion was one performed to investigate the range of dimerization interactions between members of the human bZIP transcriptional regulators. In these assays 2400 pairwise candidate, pairwise interactions were studied between 49 different members of this protein family, and many novel interactions were identified, which correlated well with validation tests and previous reports (Newman and Keating, 2003).

9.3 The practicalities of protein microarrays

A protein microarray consists of a tool in which the proteins are bound to a solid phase. The surfaces used include glass, membranes, plastics (including microtiter wells), mass spectrometer plates, and also beads and other particles that are used in 'liquid phase' microarrays (e.g. see Waterboer et al., 2005). The assays would normally be multiplexed, or there is usually little advantage to using the microarray format. For studying single pairs of protein interactions of particular interest, surface plasmon resonance and atomic force microscopy would probably be more suitable and provide more detailed information on individual interactions in most cases. The amounts of protein used as the bound substrate, and the amounts of material needed for binding to the microarray are typically very small, which can provide significant cost savings and allow for assessment of multiple substances in a single assay using a small sample. The whole justification for microarrays are ones of materials and scalability – if neither of these is an issue, unless you have a particular need for the characteristics of binding associated with microarrays, then you would probably be better served using different techniques.

9.3.1 Slide substrates

Most protein microarrays, and those that will be addressed later in this chapter, are based upon glass slides – although initially (as with DNA-based microarrays) they were primarily membrane based. There are debates about the best surfaces for use in protein-based microarrays, and there are not always clear answers to which this is. The bottom line is that different applications are more dependent upon having a particular surface than others. The most simple is to use a modified silane-coated slide, with a modification such as amine, aldehyde or epoxy, to which the arrayed material can bind. However, this sort of surface is dry and is not specifically designed to provide an environment favoring naturally occurring three-dimensional protein structures. To address this, some manufacturers produce surfaces onto which the proteins are arrayed in a way in which they are supposed to retain more of their natural tertiary structure. These surfaces include coatings such as polyacrylamide – which may be modified or not, agarose, other polymers, or nitrocellulose. The hydrated/gel surfaces are better suited to noncontact printing than using traditional pin-based contact printing which can damage the surface coating (although this can be worked around with sufficiently large pins), which is currently a less common microarray printing infrastructure. There are also surfaces specifically made for anchoring tagged proteins to the surface, and therefore facilitate molecular orientation of the protein on the slide surface (although even with this approach a significant proportion of proteins are actually likely to be present in other orientations) such as Ni-NTA, streptavidin, and avidin-coated microarray slides. We have used several different slide surfaces, and find each to have its particular strengths and weaknesses. However, one should note that it is not necessary to wait or store microarrays for long periods before using them – although the fact that this is possible with nucleic acid-based microarrays is highly useful and convenient. 'Storability' is not one of the primary reasons to adopt a microarray format, nor a particular strength of these tools when working with proteins. It is therefore not necessary to allow your proteins to dry on the slides, even when using a non-matrix coated slide. In the early study of MacBeath and Schreiber (2000), the proteins were arrayed in 40% glycerol so that the nanodrops did not dry on the (aldehyde modified silane) slide surface. Time was allowed for the proteins to bind to the slide surface (through amine–aldehyde interactions) and then the microarrays were used in assays that did not involve drying the surfaces. To be able to do this, a microarray printing facility that is close to and able to be used in a way that facilitates the experiment is obviously required (although this is not always the case).

One thing that anyone approaching this area of work should be aware of is that the method of binding of the probes to the slide surface can affect more than simply its presentation. The binding process can also have effects on the amount of protein that is available for study, or the consistency of the material bound in different spot features. In a microarray with 6000 expressed yeast proteins it was found that the quantities of the bound proteins varied between 10 and 950 fg (Zhu et al., 2001). One advantage of using antibody capture of common tags, for example binding with anti-His antibodies, is that this not only orientates the protein, and possibly favors native protein conformations, but it also leads to more uniform quantities of protein being bound to each spot location (Sasakura, 2005).

Generally, we prefer to use the simplest method possible for any given assay, and attempt to use methods that require a minimum of microarray postprocessing and handling prior to hybridization, because in general, every additional step in microarray processing is an additional opportunity to add potentially avoidable non-uniform background signal and artefacts to the assay.

9.3.2 Detection/labeling systems

Detection systems have progressively changed, as they have for DNA-based microarrays, from radiolabel-based to the use of fluorescent dyes, which have the advantage of being suited to multiplexing and the inclusion of internal controls. If single color experiments are to be performed, it is generally preferable to use a red label, because the general tendency of unlabeled protein spots is to generate a green signal. This 'green' signal is particularly visible prior to blocking on dry-surfaced microarray slides, and can be useful to assess the quality of array printing – particularly because spot morphology can vary quite dramatically with variations in protein concentrations – which, for example, commonly occurs when using supernatants obtained from various expression systems. For those with resonance light scatter (RLS) scanning facilities available, this detection method may also be very useful. The full optimization of scanning for dual-channel experiments using RLS is unfortunately not facilitated by the most widely available commercial system (or at least not as it was supplied to the author) and until independent scans at optimal settings for each channel are facilitated by the software then this system cannot be considered optimal. Also, the provision of materials and clear protocols necessary for protein labeling using this methodology is not provided by the suppliers (or at least not to us following several attempts). However, when a system is developed (e.g. Haab *et al.*, 2005) in which antibodies pre-labeled with the nanoparticles can be used, this detection method has been proven to be highly useful in protein binding detection.

9.3.3 Data extraction and analysis

As long as compatible labeling systems are used, there are no additional hardware requirements for those groups in which DNA-based microarrays are already established. Either fluorophores or RLS systems are readily applicable to protein-based applications, and the fluorescent dye option is probably the easiest and best selection for most. (The RLS-based systems are very sensitive, and the detection system used in transcription studies is actually based upon labeled antibodies. However, until the companies supplying these systems make the protocols for labeling readily available, and sort out the software such that a dual channel scan can be performed in which both labels can be assessed at their own optimal settings (rather than as a compromise between the two), we avoid using this system except when other methods are not applicable.)

Microarrays are scanned in the usual way, and standard signal quantification tools for expression analysis or genomotyping are applicable. Indeed, the signal-to-noise ratio is typically so high that most algorithms will give excellent results, likewise good spot morphology is relatively easy to obtain, which also assists in data extraction and analysis. It is perhaps surprising that not many specific tools exist for protein microarrays, but the market does not seem to have been large enough to encourage most software providers to include additional tools for protein microarray assessment. One recent exception is the addition of protein tools in the Perkin Elmer QuantArray software package, which includes tools for calibration curves and quantification. These are not particular handicaps, since such analysis can be easily performed in Microsoft Excel, but improvements in this area should be forthcoming as the methodology is more widely used. As ever, the best tool for data extraction from the microarray, and hence its form for subsequent analysis, is determined by the particular nature of the experiment. Sometimes, particularly if using larger pins or arrayed volumes, the larger spot sizes associated with some protein arrays can be problematic for some image extraction systems.

9.3.4 DNA–protein studies

This section is an aside, to perhaps save others possibly unproductive effort as we have had an interest in this area, and have done a number of pilot studies. To cut a long story short, there are significant problems with the binding affinity and specificity of these interactions when the DNA is arrayed – probably because most DNA binding proteins have a low/intermediate affinity for DNA in general, such that it is very difficult to titrate the binding and washing affinities to give clear signals in the absence of large numbers of false positives. However, this approach can work in reverse, such that proteins bound to the microarray can be interrogated with the DNA targets. There are three published studies illustrating this approach to which those interested in this field should refer (Boutell *et al.*, 2004; Hall *et al.*, 2004; Kersten *et al.*, 2004).

9.3.5 An experimental example of the use of protein microarrays

Some of the issues and practical steps necessary to establish a protein microarray system will now be illustrated using a series of simple experiments. The study to be used for illustration in this chapter has two broad components that address common features of protein-based microarrays. One aspect is for the implementation of an antigen–antibody interaction assessment, and the second is for direct analysis of receptor–ligand interactions. Specifically, in these experiments we wished to assess a series of CD200 mutants with respect to the epitopes that had been altered (epitope mapping), and to be able to assess the interactions of CD200 with its receptor. This study and the results of the investigation are published in detail elsewhere (Letarte *et al.*, 2005), and a series of color figures illustrating each of the key steps and types of microarray described here can be found in this article (which is Open Access, so all readers will be able to access/view them). Before moving to the experimental questions we had to assess the methods being used and their applicability to our research problem.

Determining the labeling efficiency

Initially, we wished to determine the labeling efficiency for antibodies in our hands, and the most appropriate binding slide substrate. We therefore labeled the proteins with fluorophore, and printed these labeled proteins onto a variety of different microarrays.

Findings (labeling)
Using the Alexa labeling kit from Molecular Probes we determined that in our hands antibodies were labeled with one to four dyes per antibody.

Findings (slides)
We found little difference in the performance of a wide range of different microarray slide substrates from different suppliers and with different surface modifications, and found Genetix Epoxy-coated microarray slides to be the most cost-effective option giving good consistent protein binding and very low background. The spot morphology was generally reasonable, with the least hydrophobic surfaces generally giving the best looking spots.

Antibody capture and antigen detection

We were concerned that the number of labels per antibody is substantially fewer than those typical of transcription or genomotyping microarrays. We were also concerned that antibodies bound to the slide surface, particularly a dry slide surface, would not retain good antigen binding properties. Some people have emphasized the need for orientation of protein binding to slide surfaces, using specific tags and slide-coating ligands, but this is not a feature of these microarrays, so we did not know the proportion of antibodies or antigens that would be orientated suitably for detection/binding. We therefore set up two assays to test these aspects of the microarrays: to assess the ability of bound antibody to capture antigen, and to determine the direct detection of antigen on the slide. We used two different antibodies to human protein CD4. The specific antibodies used were OX68 (Brown and Barclay, 1994) and W3/25 (White *et al.*, 1978). In the second assay, CD4 was spotted directly onto the microarray, and the monoclonal antibodies used for detection. The spotted proteins were prepared in serial dilutions and printed using 300 µm solid aQu pins on a QArray mini microarray printer (Genetix) at a final concentration of 1× Genetix protein spotting solution. In the first assay, fluorescently labeled CD4 was used for detection, in the second labeled antibody.

Findings: antibody capture
Spotted antibody was readily able to bind detectable amounts of CD4, so a sufficient proportion of the arrayed material on the Epoxy surface was still able to bind antigen.

Findings: antigen detection
The spotted CD4 was readily detected using labeled antibodies.

An additional observation: pin cleaning
The use of phosphate-based printing solutions, at least when working with proteins, leads to reduced quality spot morphology if the pins are not regularly washed. Similarly, when setting up protein printing protocols ethanol washes should only be used after far more rigorous water washes

than are needed for spotting DNA (or omitted completely). This is because the ethanol will precipitate the protein onto the microarray pins.

Antibody binding

We next wished to check that the antibody binding was fully specific and normal with regard to the epitopes being bound. To do this we bound each antibody in serial dilutions to different subgrids, then bound CD4 to the arrayed antibodies, and then used each labeled antibody to detect the bound CD4.

Findings: binding specificity

There was strong binding, in proportion to the serial dilutions of the first arrayed antibody for each antibody, as detected by binding of the labeled second antibody (which address different epitopes), and there was near zero signal in the subgrids in which the same antibodies were being used to both bind and detect the CD4. This indicates that the binding in this assay system was fully directional and the presentation of the protein was entirely determined by the binding by the first arrayed antibody.

Mapping the epitope binding sites

Next we wished to perform the first part of our intended experiment: To map the epitope binding sites of three anti-CD200 monoclonal antibodies. First, we arrayed a series of CD4-CD200 fusion proteins prepared from transient expression supernatants and detected them with fluorescently labeled anti-CD4 OX68 which binds to an epitope of CD4 included in the fusion protein. This revealed that there was a wide dynamic range of protein concentration that were deposited in the microarrayed spot features – probably reflecting primarily the wide range of different concentrations obtained from the protein expression system (because the proteins and solutions were otherwise very similar). We therefore decided on using a dual color methodology for the purposes of the epitope-mapping experiment. So we labeled the respective antibodies under investigation with Cy-5 and the OX68 antibody with Cy-3, and after arraying the fusion proteins co-incubated the microarrays with both antibodies in an equal concentration.

Findings: dual color epitope mapping

We were able to identify the areas of CD200 to which the selected monoclonal antibodies bound, and the dual color labeling strategy enabled us to overcome the significant differences in the amounts of protein that were arrayed.

Comparison of differences in receptor–ligand binding affinity

We wished to determine whether we could compare differences in receptor–ligand binding affinity in a microarray format. Initial experiments failed to detect binding, which we hypothesized was due to the relatively high 'off-rate' associated with monomeric binding, and which we were aware of as a feature of previous experiments on a BiaCore protein interaction analysis system. We therefore sought to decrease the 'off-rate' by multimerizing the ligand in solution, by binding the protein to fluorescent FITC-coated beads. This was successful, and the binding could be related to the relative affinity of the mouse, rat, and human CD200R used in the experiment.

Findings: receptor–ligand interactions
By multimerizing the ligand in the liquid phase, thereby reducing the 'off-rate' it is possible to maintain quantitative binding of sufficient stability to allow for postprocessing/washing and to detect low affinity interactions between receptors and ligands in a microarray format.

9.3.6 Some summary points

(i)   When working with, or interpreting results generated using, protein microarrays one must always ask questions about consistency and internal controls. With respect to quantitative assays in particular, either very high reproducibility in protein deposition (probably beyond what is routinely obtained using contact printers) and/or an additional internal reporter is required for quantitative analysis. Therefore, if possible, it is preferable to work with a dual (or more) labeling strategy in which variables in the amount of substrate arrayed and its integrity can be quantified and/or controlled for.

(ii) When lower affinity interactions are addressed, the binding is not likely to be sufficiently robust to allow for detection and/or sufficiently stringent washing. The key to this is to understand that while the 'on rate' for binding is equal in monomeric and polymeric interactions, the 'off rate' is significantly less in monomeric interactions which are typical of the way in which materials are printed in a microarray. One solution to this is to make the substance binding to the solid phase polymeric. The way in which this is best achieved will depend upon the nature of the assay and interactions being addressed.

(iii) Generally, protein microarrays should be used as soon after manufacture as practically possible. Generally their performance deteriorates substantially, and largely unpredictably, over time when stored.

9.4 Overall concluding remarks

While the example experiment described can be used as a guide, we recommend that a similar (if abbreviated) series of tests are done that are specifically applicable to other experimental questions. We do not use the same printing conditions (or microarray printer) for all protein microarray experiments, and, as with all microarray-based applications, it is important to determine how an assay performs in your own hands. Also, the performance is likely to improve with some practice. Equally, different microarray printers and printing pin types perform differently. Although they were not necessary in the described study, the use of hydrated matrix slides, in which the protein is not dried out and probably retains it secondary and tertiary structure more than on dry slides, may be necessary or preferable for some applications. Sometimes, using arrayed material that does not actually dry onto the slide surface between printing and use may be necessary.

What this chapter has attempted to do is to introduce this area to those who are not familiar with it, and to use a relatively straightforward illustrative example to demonstrate that this technology should not be avoided because of false impressions that it is particularly difficult or problematic, although it is necessary to assess the technical performance of each part of the experimental process. More complex studies are largely simply issues of increased scale, but the basic technology is sound and we would encourage those interested in this area to explore it further. If one reads the literature, there are plenty of descriptions of the difficulties of working with protein microarrays: the lack of orientation on most surfaces, drying, improper folding, cross-reactivity, lack of multi-protein complexes, post-translational modification issues, and so on. These concerns delayed our pursuing this area of work for some time, but it is probably wiser to consider the advantages of whether it does work, and to try it.

References

Boutell JM, *et al*. (2004) Functional protein microarrays for parallel characterisation of p53 mutants. Regulation of gene expression by a metabolic enzyme. *Proteomics* 4: 1950–1958.

Bouwman K, *et al*. (2003) Microarrays of tumor cell derived proteins uncover a distinct pattern of prostate cancer serum immunoreactivity. *Proteomics* 3: 2200–2207.

Brown MH, Barclay AN (1994) Expression of immunoglobulin and scavenger receptor superfamily domains as chimeric proteins with domains 3 and 4 of CD4 for ligand analysis. *Prot Eng* 7: 515–521.

de Wildt RM, *et al*. (2000) Antibody arrays for high-throughput screening of antibody-antigen interactions. *Nat Biotech* 18: 989–994.

Ekins RP (1998) Ligand assays: from electrophoresis to miniaturized micoarrays. *Clin Chem* 44: 2015–2030.

Ekins R, Chu FW (1999) Microarrays: their origins and applications. *Trends Biotechnol* 17: 217–218.

Feilner T (2005) High throughput identification of potential *Arabidopsis* mitogen-activated protein kinases substrates. *Mol Cell Proteomics* 4: 1558–1568.

Fontoura P, *et al*. (2004) Immunity to the extracellular domain of Nogo-A modulates experimental autoimmune encephalomyelitis. *J Immunol* 173: 6981–6992.

Gao WM (2005) Distinctive serum protein profiles involving abundant proteins in lung cancer patients based upon antibody microarray analysis. *BMC Cancer* 5: 110.

Ge H (2000) UPA, a universal protein array system for quantitative detection of protein-protein, protein-DNA, protein-RNA and protein-ligand interactions. *Nucleic Acids Res* 28: e3.

Graham KL (2004) High-throughput methods for measuring autoantibodies in systemic lupus erythematosus and other autoimmune diseases. *Autoimmunity* 4: 269–272.

Haab BB, *et al*. (2001) Protein microarrays for highly parallel detection and quantitation of specific proteins and antibodies in complex solutions. *Genome Biol* 2: 2.

Haab BB, *et al*. (2005) Immunoassay and antibody microarray analysis of the HUPO Plasma Proteome Project reference specimens: systematic variation between sample types and calibration of mass spectrometry data. *Proteomics* 5: 3278–3291.

Hall DA, *et al.* (2004) Regulation of gene expression by a metabolic enzyme. *Science* **306**: 482–484.

Harwanegg C, *et al.* (2003) Microarrayed recombinant allergens for diagnosis of allergy. *Clin Exp Alergy* **33**: 7–13.

Heuber W, *et al.* (2005) Antigen microarray profiling of autoantibodies in rheumatoid arthritis. *Arthritis Rheum* **9**: 2645–2655.

Joos TO (2000) A microarray enzyme-linked immunosorbent assay for autoimmune diagnostics. *Electrophoresis* **21**: 2641–2650.

Kersten B, *et al.* (2004) Protein microarray technology and ultraviolet crosslinking combined with mass spectrometry for the analysis of protein-DNA interactions. *Anal Biochem* **331**: 303–313.

Kramer A, *et al.* (2004) Identification of barley CK2alpha targets by using the protein microarray technology. *Phytochemistry* **65**: 1777–1784.

Letarte MD, *et al.* (2005) Analysis of leukocyte membrane protein interactions using protein microarrays. *BMC Biochem* **6**: 2.

MacBeath G, Schreiber SL (2000) Printing proteins as microarrays for high-throughput function determination. *Science* **289**: 1760–1763.

Michaud GA, *et al.* (2003) Analyzing antibody specificity with whole proteome microarrays. *Nature Biotech* **21**: 1509–1512.

Miller JC, *et al.* (2003) Antibody microarray profiling of human prostate cancer sera: Antibody screening and identification of potential biomarkers. *Proteomics* **3**: 56–63.

Newman JR, Keating AE (2003) Comprehensive identification of human bZIP interactions with coiled-coil arrays. *Science* **300**: 2097–2101.

Poetz O, *et al.* (2005) Protein microarrays for antibody profiling: specificity and affinity determination on a chip. *Proteomics* **5**: 2402–2411.

Robinson WH, *et al.* (2002) Autoantigen microarrays for multiplex characterization of autoantibody responses. *Nat Med* **8**: 295–301.

Robinson WH, *et al.* (2003) Protein arrays for autoantibody profiling and fine-specificity mapping. *Proteomics* **3**: 2077–2084.

Sasakura Y (2005) Investigation of the relationship between protein-protein interaction and catalytic activity of a heme-regulated phosphodiesterase from Escherichia coli (Ec DOS) by protein microarray. *Biochemistry* **44**: 9598–9605.

Sreekumar A, *et al.* (2001) Profiling of cancer cells using protein microarrays: Discovery of novel radiation-regulated proteins. *Cancer Res* **61**: 7585–7593.

Wang X, *et al.* (2005) Autoantibody signatures in prostate cancer. *N Engl J Med* **353**: 1224–1235.

Waterboer T, *et al.* (2005) Multiplex human papillomavirus serology based on in situ-purified glutathione s-transferase fusion proteins. *Clin Chem* **51**: 1845–1853.

White RA, *et al.* (1978) T-lymphocyte heterogeneity in the rat: separation of functional subpopulations using a monoclonal antibody. *J Exp Med* **148**: 664–673.

Yan H, *et al.* (2003) Protein microarrays using liquid phase fractionation of cell lysates. *Proteomics* **3**: 1228–1235.

Zhong L, *et al.* (2005) Using protein microarray as a diagnostic assay for non-small cell lung cancer. *Am J Respir Crit Care Med* [E-pub ahead of print. 18 August 2005]

Zhu H, *et al.* (2001) Global analysis of protein activities using proteome chips. *Science* **293**: 2101–2105.

Protocol 1: Printing oligonucleotide microarrays

STANDARD OPERATING PROCEDURE FOR OPERON OLIGONUCLEOTIDE ARRAYS USING CORNING PRONTO™ SPOTTING BUFFER

A. Oligonucleotide resuspension from new oligo set

1. Oligonucleotides are supplied as 600 pmol. Centrifuge plates at 1000 × g for 5 min in a swing-out rotor centrifuge before carefully removing the adhesive film covering the plates. Discard adhesive film. Resuspend contents of each well in 15 μL of PRONTO SPOTTING BUFFER, for a 40 μM concentration.

2. Cover the plates carefully with sticky plastic film to prevent cross-contamination between wells, and centrifuge at 1000 × g for 5 min. Place on a shaking table moving at 60 rpm overnight at 4°C.

3. When not in use store plates at –20°C.

B. Printing from –20°C stored resuspended oligos

4. Remove the 384-well oligo plates from the freezer, and allow them to equilibrate to room temperature. Centrifuge plates at 1000 × g for 5 min in a swing-out rotor centrifuge before carefully removing the adhesive film covering the plates. Discard adhesive film.

C. Preparation of Biorobotix MicroGrid TAS II array spotter (day 1)

5. Filter-sterilize 10 L of high-purity water (18.2 MΩ) using a 0.22 μM tower filter (Sarstedt or Millipore-compatible with Duran bottles). Dispense water into prewashed bottles.

D. Preparation of Biorobotix MicroGrid TAS II array spotter (day 2)

6. Drain the main wash reservoir of the MICROGRID TAS II array spotter, and introduce 5–6 L of HPSFW (high purity sterile filtered water). Empty out the 2 × 2 L bottles that feed the pre-wash stations; rinse with HPSFW and refill with at least 2 L of HPSFW. Reseal caps carefully, after checking the 'O'-rings are in place, and keep tightly closed.

7. Drain water from the sonication station, clean out with an AZOWIPE and rinse several times with HPSFW. Replace water to the fill mark with fresh HPSFW.

8. Sonicate the 48-pin print head for at least 1 min, moving the pins through the water. Place pin holder in machine.

9. Clean out plate trays with a brush and remove any dust etc. from the inside of the printer using a microfiber electrostatic cloth. Clean trays with AZOWIPES.

10. Load slides. CORNING GAPS II BARCODED or ULTRAGAPS BARCODED. Mark slide boxes with batch number, expiry date, and order slides were placed in the printer.

When handling the slides, always wear powder-free NITRILE gloves and avoid exposure to dust!

11. Remove the 384-well oligo plates from the shaking table. Centrifuge plates at $1000 \times g$ for 5 min in a swing-out rotor centrifuge before carefully removing the adhesive film covering the plates. Discard adhesive film. NB The plates should have removable lids to prevent evaporation of the sample in the wells, whilst printing.

OPTIONAL: IT IS RECOMMENDED THAT A PIN TEST IS CARRIED OUT BEFORE PRINTING THE ARRAY SET

Use a 384-well plate containing PRONTO SPOTTING BUFFER either containing oligo or CyDye™, and program robot to fill at one source visit, followed by 400 spots. Check for missing spots. Clean/sonicate blocked pins.

12. Load 384-well plates into the biobank. Start loading plate 1 at the top left; plate 2, top right; plate 3, second row left; plate 4, second row right etc. MAKE SURE THE LIDS CAN BE FREELY REMOVED, AND PUSH PLATES TO THE BACK OF THE BIOBANK. Load biobank into holder and replace cover.

13. Start program.

Programming

Washes

• WS1	10 s + wiggle	
• WS2	10 s + wiggle	} Repeat three times
• MWS	Wash and dry as per standard	

This program is run 3× (total of 27 washes) prior to printing, and once BEFORE EACH NEW SOURCE VISIT.

Other print parameters

FILL – soft touch   SPOT – soft touch

PROTOCOL FOR CORNING ULTRAGAPS SLIDES PRINTED WITH PRONTO: POST-PRINTING TREATMENT

When handling the slides, ALWAYS wear powder-free gloves and avoid exposure to dust!

Treatment after printing the array

1. Remove slides from printer, and put them back into the slide boxes in the same order they were loaded.

2. Vacuum dessicate the slides for 24–48 h.

3. Crosslink slides as detailed below.

PROTOCOL FOR CORNING GAPS II SLIDES: POST-PRINTING TREATMENT

3. Crosslink slides (Barcode side face up) in a UV crosslinker at 600 mJ.

4. Scan barcodes of slides into an Excel file using a barcode scanner.

5. Store slides in a sealed box containing silica gel desiccant in the dark until required.

Protocol 2: Extraction of *E. coli* RNA

NOTE: READ THE QIAGEN RNeasy MANUAL BEFORE COMMENCING EXPERIMENTS. Always wear gloves and take precautions to minimize RNase contamination of samples.

The maximum RNA binding capacity of RNeasy spin columns is 100 μg (Mini column). Remember that *E. coli* grown in LB yields approximately 120 μg RNA per 7.5×10^8 cells. For MG1655(seq) grown in LB use 4.0 mL of cells OD_{600} 0.6. *E. coli* grown in minimal media yields approximately 40 μg RNA per 7.5×10^8 cells. For MG1655(seq) grown in minimal media use 10.0 mL of cells OD_{600} 0.6. When cells are grown in LB to stationary phase 6–8 mL of cells are required.

For OD_{600} values, calibrate spectrophotometer readings (OD_{600}) to cell numbers derived by plate counts for your own strain.

Equipment and reagents to be supplied by user

- QIAGEN
  Cat. No. 76506 – RNAprotect bacteria reagent, (2 × 100 mL) +
  Cat. No. 74104 - RNeasy Mini kit (50)

  or

  Cat. No. 74542 – RNeasy Protect Bacteria Mini kit (50)

- QIAGEN Cat. No. 79254 – RNase-free DNase set

- SIGMA Cat. No. L-7651 – Lysozyme, from chicken egg white

- SIGMA Cat. No. M-7154 – β-mercaptoethanol 14.3 M, electrophoresis reagent

- Ethanol (96–100%) (for enzymatic lysis protocol)

- TE buffer – 10 mM Tris-HCl 1 mM EDTA pH8.0
  Tris-HCl
  EDTA

General requirements:

- RNase free filter tips - p10, p20, p100, p200 and p1000

- Sterile RNase-free - 2 mL, 1.5 mL, and 0.5 mL microcentrifuge tubes

- Disposable gloves

- Greiner tube (sterile) - 50 mL and 15 mL

A. GROWTH AND HARVESTING OF *E. COLI*

1. Subculture a colony onto a fresh LB agar plate and incubate at 37°C overnight.

2. Inoculate a single colony in 50 mL LB broth into 250 mL flask and incubate at 37°C with shaking.

3. Approximately 2.5 h later check OD_{600}. When OD_{600} reaches 0.6–0.8 or the desired OD is obtained the cells are harvested.

4. Add the required volume of bacterial culture to the Greiner tube containing 2× culture volumes of RNAprotect. Mix immediately by vortexing for 5 s. Incubate for 5 min at room temperature.

5. Centrifuge for 10 min at 5000 × *g* or 3600 rpm. For a greater cell recovery use Eppendorf tubes and spin at 13 000 rpm for 2–3 min.

6. Decant supernatant and remove residual liquid by inversion over a paper towel. Cells can be stored at –70°C for up to 4 weeks.

B. RNEASY MINI PROTOCOL FOR TOTAL RNA ISOLATION FROM *E. COLI*

7. Prepare TE buffer [10 mM Tris-HCl, 1 mM EDTA pH 8.0] containing lysozyme at 1 mg mL^{-1}.

8. Prepare RLT buffer by adding 10 µL of β-ME per 1 mL of RLT. RLT is stable for 1 month after addition of β-ME.

9. Prepare RPE buffer (supplied as a concentrate) by adding four volumes of ethanol (96–100%).

10. Add appropriate volume of TE buffer containing lysozyme to cell pellets. (See table below.)

11. Mix by vortexing for 10 s. Incubate at room temperature for 5 min. Vortex for at least 10 s every 2 min during incubation.

12. Add appropriate volume of RLT to the sample and vortex vigorously. If particulate material is visible it should be pelleted by centrifugation and only the supernatant should be used in subsequent steps.

Number of bacteria	Rneasy spin column	TE + lysozyme	Buffer RLT	Ethanol
5×10^6 to 7.5×10^8	Mini	200 µL	700 µL	500 µL

13. Add the appropriate volume of ethanol (96–100%) to the lysate. Mix by pipetting. **DO NOT CENTRIFUGE**.

14. Apply the lysate containing ethanol, including any precipitate that may have formed, to an RNeasy Mini column placed in a 2 mL collection tube (supplied). The maximum loading volume is 700 µL. Close the tube lid carefully, and centrifuge for 15 s at ≥8000 × *g* (≥10 000 rpm). Discard the flow through.

15. If the volume exceeds 700 μL, load aliquots successively onto the RNeasy column and centrifuge as above.

16. Pipette 350 μL of buffer RW1 into the RNeasy mini column, and centrifuge for 15 s at $\geq 8000 \times g$ ($\geq 10\,000$ rpm) to wash. Discard the flowthrough.

17. Add 10 μL DNase I stock solution to 70 μL of buffer RDD. Mix by gently inverting the tube. **DO NOT VORTEX**. Centrifuge briefly to collect residual liquid from the sides of the tube.

18. Pipette the 80 μL DNase I incubation mix directly onto the RNeasy silica-gel membrane, and place on the bench (20–30°C) for 15 min.

19. Pipette 350 μL buffer RW1 into the RNeasy column. Leave on bench for 5 min and then centrifuge for 15 s at $\geq 8000 \times g$ ($\geq 10\,000$ rpm). Discard flowthrough and collection tube.

20. Transfer the RNeasy column to a **new** 2 mL collection tube (supplied). Pipette 500 μL of buffer RPE onto the RNeasy column. Close the tube carefully, and centrifuge for 15 s at $\geq 8000 \times g$ to wash the column. Discard the flow through. NB RPE is supplied as a concentrate to which four volumes of ethanol (96–100%) should have been added prior to use.

21. Pipette a further 500 μL of buffer RPE onto the RNeasy column. Close the tube carefully, and centrifuge for 2 min at $\geq 8000 \times g$ ($\geq 10\,000$ rpm) to pass through the column, and for the column to dry.

22. Place the RNeasy column in a new RNase-free 2 mL collection tube (not supplied in kit) and discard the old collection tube containing the flowthrough. Centrifuge at full speed for 1 min in a microcentrifuge to remove residual RPE from column. (Carryover of ethanol from the RPE will interfere with subsequent manipulations of the RNA.)

23. To elute RNA transfer the RNeasy column to a new 1.5 mL collection tube (supplied with the kit). Pipette 30 μL of RNase-free water (supplied) DIRECTLY ONTO the RNeasy silica-gel membrane. Close the tube gently, and centrifuge for 1 min at $\geq 8000 \times g$ ($\geq 10\,000$ rpm) to elute. Take a 2 μL aliquot for analysis of RNA samples using the Bioanalyzer or on an RNA gel or spectrophotometer and store at −70°C until required.

Optional: If the expected RNA yield is > 30 μg, repeat the elution step (step 23) as described, with a second volume of RNase-free water. ELUTE INTO THE SAME COLLECTION TUBE. To obtain a higher total RNA concentration, the second elution step may be performed using the first eluate (from step 23). The yield will be 15–30% less than the yield obtained using a second volume of RNase-free water, but the final concentration will be higher.

Protocol 3: Probe labeling

Equipment and reagents to be supplied by user

- Amersham Biosciences Cat. No. RPK3164 – Lucidea Universal ScoreCard Refill
- Amersham Biosciences Cat. No. RPN56660 – CyScribe Post-Labelling Kit
- 2.5 M NaOH

  1 g NaOH in 10 mL water

  Filter sterilize

  Store in dark for up to 1 month

  Note: Metal contamination or prolonged storage in glass vessel exposed to light will result in a solution that would degrade Cy5.
- 2 M HEPES

  4.77 g HEPES-free acid in 10 mL water

  Filter sterilize

  Store at room temperature for up to 3 months
- 0.1 M Na_2CO_3

  0.53g in 50 mL water
- 0.1 M sodium bicarbonate pH 9.0

  4.2 g of $NaHCO_3$ in 500 mL water (pH should be about 8.5)

  Adjust pH to 9.0 by adding 0.1 M Na_2CO_3 (25–30 mL)

  Filter-sterilize the buffer

  Note: Recommend making fresh, although it may be dispensed into aliquots and stored at –20°C for up to 2 months.
- Ethanol

  96–100%

  80%
- 4 M hydroxylamine hydrochloride

  Sigma H2391 27.8 g in 100 mL water
- PCR machine/heating block
- Microcentrifuge that can accommodate GFX columns

General requirements

- RNase free filter tips - p10, p20, p100, p200, and p1000
- Sterile RNase-free - 2 mL, 1.5 mL, and 0.5 mL microcentrifuge tubes
- Disposable gloves

A. PRIMER ANNEALING

1. Add the following components to a 0.5 mL microcentrifuge tube on ice

For prokaryotic cells:

Total RNA (20 µg) – maximum volume 9 µL	X µL
Test/reference RNA from Lucidea kit	2 µL
Random nonamers	1 µL
Water	Y µL
Total volume	12 µL

For eukaryotic cells:

Total RNA (20 µg) – maximum volume 9 µL	X µL
Test/reference RNA from Lucidea kit	2 µL
Oligo dT (0.5 µg µL^{-1})	1 µL
Water	Y µL
Total volume	12 µL

2. Mix gently by pipetting up and down.

3. Incubate reaction at 70°C for 5 min.

4. Cool reaction at room temperature for 10 min to allow the primers and mRNA template to anneal.

5. Spin down reaction for 15 s in a microcentrifuge.

B. Extension reactions

6. Place the cooled annealing reaction on ice and add the following components to it, making sure to add the enzyme last:

5× CyScript buffer	4 µL
0.1 M DTT	2 µL
Nucleotide mix	1 µL
aa-dUTP	1 µL
CyScript reverse transciptase (RT)	1 µL
Total volume	20 µL

7. Mix by gently pipetting or by stirring with a pipette tip and spin for 15 s in a microcentrifuge. Note: Reverse transcriptase is sensitive to denaturation at air–liquid interfaces, therefore avoid foaming.

8. Incubate the reaction at 42°C overnight.

9. Store the amino-allyl-modified cDNA on ice for immediate purification or place at –20°C for storage. Do not store in a frost-free freezer.

C. DEGRADATION OF mRNA

10. Add 2 μL 2.5 M NaOH to each cDNA reaction.

11. Mix reaction by vortexing and spin for 15 s in a microcentrifuge.

12. Incubate reaction at 37°C for 15 min.

13. Add 10 μL 2 M HEPES to each cDNA reaction.

14. Mix reaction by vortexing and spin for 15 s in a microcentrifuge

15. The cDNA reaction is ready for purification or can be stored at –20C.

D. PURIFICATION OF cDNA WITH CYSCRIBE GFX PURIFICATION KIT

16. For every cDNA reaction to be purified, place one CySCribe GFX column into a clean collection tube. Add 500 μL of capture buffer to each CyScribe GFX column.

17. Briefly spin down unpurified cDNA products and transfer them into each CyScribe GFX column and mix the cDNA by gently pipetting up and down five times.

18. Centrifuge each column in a microcentrifuge at 13 000 rpm for 30 s.

19. Remove the CyScribe GFX column and discard liquid at the bottom of each collection tube, return each CyScribe GFX column into the used collection tube.

20. Add 600 μL of 80% ethanol to each column and centrifuge in microcentrifuge at 13 000 rpm for 30 s.

21. Remove the CyScribe GFX column and discard liquid at the bottom of each collection tube, return CyScribe GFX column into the used collection tube.

22. Repeat wash step twice for a total of three washes. After the final wash, return CyScribe GFX column into the used collection tube.

23. Centrifuge each column in microcentrifuge at 13 000 rpm for 10 s to remove all wash buffer in the tip of column. Discard the collection tube.

24. Transfer each CyScribe GFX column to a fresh amber 1.5 mL microcentrifuge tube and add 60 µL 0.1 M sodium bicarbonate pH 9.0 to the top of the glass fiber matrix in each CyScribe GFX column.

25. Incubate the CyScribe GFX column at room temperature for 1–5 min.

26. Centrifuge each column in microcentrifuge at 13 000rpm for 1 min to collect the purified cDNA.

27. Repeat elution steps (23–25) once more.

28. Measure the absorbance of the purified cDNA on a spectrophotometer at 260 nm (use 2 µL on the Nanodrop).

E. LABELING OF AMINO ALLYL-MODIFIED cDNA WITH CYDYE

29. Add amino allyl-modified cDNA (in 0.1 M sodium bicarbonate pH 9.0) directly into an aliquot of CyDye NHS ester and resuspend NHS ester by pipetting several times. Place the mixture of aminoallyl-purified cDNA (in 0.1 M sodium bicarbonate pH 9.0) and CyDye NHS ester in amber 1.5 mL Eppendorf tube.

30. Centrifuge at 13 000 rpm for 1 min to collect the liquid at the bottom of the tube.

31. Mix by stirring and incubate at room temerature in the dark for 2–4 h and mix once half-way during the incubation.

32. Add 15 µL 4 M hydroxylamine to each coupling reaction.

33. Mix by pipetting up and down several times and incubate at room temperature in the dark for 15 min.

34. Purify the CyDye-labeled cDNA immediately as described below (Section F).

F. PURIFICATION OF LABELED cDNA WITH CYSCRIBE GFX PURIFICATION KIT

35. Pre-warming the elution buffer to 65°C may increase the labeled cDNA yield by approx. 5%.

36. For every cDNA reaction to be purified, place one CyScribe GFX column into a clean collection tube. Add 500 µL of capture buffer to each CyScribe GFX column.

37. Briefly spin down unpurified cDNA products, transfer into each CySCribe GFX column and mix the cDNA by gently pipetting up and down five times.

38. Centrifuge each column in microcentrifuge at 13 000 rpm for 30 s.

39. Remove the CyScribe GFX column and discard liquid at the bottom of each collection tube. Return each CyScribe GFX column into the used collection tube.

40. Add 600 µL of wash buffer to each column and centrifuge in microcentrifuge at 13 000 rpm for 30 s.

41. Remove the CyScribe GFX column and discard liquid at the bottom of each collection tube, return each CyScribe GFX column into the used collection tube. Repeat wash step twice for a total of three washes. After the final wash, return each CyScribe GFX column into the used collection tube.

42. Centrifuge each column in microcentrifuge at 13 000 rpm for 10 s to remove all wash buffer in the tip of column. Discard the collection tube.

43. Transfer each CyScribe GFX column to a fresh 1.5 mL microcentrifuge tube and add 60 µL of elution buffer (prewarmed at 65°C) to the top of the glass fiber matrix in each CyScribe GFX column.

44. Incubate the CyScribe GFX column at room temperature for 1–5 min. Centrifuge each column in microcentrifuge at 13 000 rpm for 1 min to collect the purified labeled cDNA.

45. Measure the absorbance of the labeled cDNA on a spectrophotometer (1–2 µL in Nanodrop) at 550, 650, and 260 nm. Store at –20°C until required. Then, the amounts of Cy3, Cy5, and cDNA (in picomoles) are given by:

$$Cy3 = \frac{A_{550}}{150} \times V \times 10^3$$

$$Cy5 = \frac{A_{650}}{250} \times V \times 10^3$$

$$cDNA = \frac{A_{260} \times V \times 37 \times 1000}{324.5}$$

where A_{550}, A_{650}, and A_{260} are the absorbances at 550, 650, and 260 nm and V is the volume (in microliters). The values 37, 1000 and 324.5 refer to ng mL^{-1}, pg ng^{-1}, and pg pmol^{-1}, respectively.

46. Calculate the nucleotides/dye ratio and the required volume for 80 pmol of both Cy5- and Cy3-labeled cDNA:

$$\text{nucleotides/dye ratio} = \frac{\text{pmol cDNA}}{\text{pmol Cydye}}.$$

47. It may be necessary to dry Cy3- and Cy5-labeled cDNA depending on the size of your array and coverslips being used. See Protocol 4 for total volume restriction for probe preparation.

Protocol 4: Hybridization and washing

Equipment and reagents to be supplied by user

- Formamide: AnalaR BDH Cat. No.103266T

- 20× SSC
  175.3 g NaCl and 88.2 g sodium citrate in 800 mL of distilled H_2O
  Adjust the pH to 7.0 with a few drops of 1 M HCl
  Adjust the volume to 1 L with additional distilled H_2O
  Sterilize by autoclaving
  All solutions are to be filter sterilized before use

- BSA, fractionV: ICN Cat. No.160069

- 10% SDS

- Wash buffer I: 2× SSC, 0.1% SDS
  20 mL 20× SSC
  2 mL 10% SDS
  in final volume of 200 mL

- Wash buffer II: 0.1× SSC, 0.1% SDS
  2 mL 20× SSC
  2 mL 10% SDS
  in final volume of 200 mL

- Wash buffer III: 0.1× SSC
  2 mL 20× SSC
  in final volume of 400 mL

- Salmon sperm DNA (5 mg mL^{-1})

- 100× Denhardt's
  1 g Ficoll 400, 1 g polyvinylpyrrolidone, and 1 g BSA in a final volume of 50 mL
  aliquot 500 µL per Eppendorf tube
  store at –20°C
  avoid repeated freeze–thawing

- Pronto! Background Reduction Kit: Corning Cat. No. 40027

- Ethanol
  96–100%

- HPLC-grade water: VWR Cat. No. 23596320

- Cy3- and Cy5-labeled probes

- Hybridization chambers: Corning

- Lifterslips (ERIE Scientific Cat. No. 24 × 60IS-2-4795 or 25 × 60I-M-5439-001-LS)

- Moist box
  Parafilm
  Wet paper towels
  Aluminum foil

- Plastic black troughs (×4): VWR Cat. Nos 406023600 and 406023700

- Glass troughs (×3): VWR Cat. No. 406023010

- Scanner: Axon 4000B

General requirements

- Filter tips: p10, p20, p100, p200, and p1000

- Sterile 2 mL, 1.5 mL, and 0.5 mL microcentrifuge tubes

- Disposable, powder-free gloves

- Microcentrifuge

- Benchtop centrifuge

- 50 mL centrifuge tubes

- Incubator at 42°C

A. HYBRIDIZATION OF SLIDES

1. Pre-warm overnight at 42°C

   Duran bottle: 250 mL

   Measuring cylinders: 100 mL and 250 mL

   Glass trough with cover: 200 mL

   Coplin staining jar with cover: 2 × 50 mL

   Pronto! Universal Pre-Soak Solution: 50 mL per five slides

   Pre-hybridization solution containing 50 mL 20× SSC and 130 mL HPLC-grade water

2. Next day, switch on 42°C shaking water bath and then dissolve in pre-hybridization solution while stirring:

- BSA (fraction V) 0.2 g

  10% SDS   2 mL

- Make up volume to 200 mL with HPLC-grade water

Note: it is important to add the components in the order they are listed to avoid precipitation.

3. Filter sterilize all the pre-hybridization solution with 0.22 μm filter into the pre-warmed glass trough containing a slide holder and place back at 42°C until ready (minimum 1 h; see step 8).

4. In a 100 mL glass cylinder containing a magnetic flea, dilute 500 μL sodium borohydride into 50 mL of Universal Pre-Soak Solution pre-warmed to 42°C. Swirl gently to mix. Immediately proceed to step 5.

5. Immerse arrays in glass slide containers filled with approx 50 mL of solution from step 4 above and incubate in shaking water bath at 42°C and 100 rpm for 20 min.

Do not incubate for longer than 20 min.

6. Transfer the slides to a black plastic trough, with a lid, containing wash buffer II and wash the slides in wash buffer II for 30 s with vigorous shaking.

7. Repeat step 6 above using a second plastic trough, with a lid, containing fresh wash buffer II.

8. Transfer slides to the glass trough containing the pre-warmed pre-hybridization solution from step 3. Seal the glass trough with parafilm and incubate in a shaking water bath at 42°C and 100 rpm for a minimum of 120 min (the longer, the better).

Meanwhile, if the labeled probe is required to be vacuum dried in order to reach the desired volume then it can be done during this time.

9. Transfer the slides to black plastic trough, with a lid, containing wash buffer II and wash the slides in wash buffer II for 1 min with vigorous shaking.

10. After wash II, transfer and sequentially wash the slides for 30 s as in step 9 in the two troughs containing wash buffer III.

11. Submerge the slides briefly in water (approx. 20 dips) and then in ethanol (approx. 10 dips) and shake off excess ethanol. Inspect the slide for any dust particles; if any are present, submerge the slide in ethanol until you are satisfied with the slide.

12. Dry the slides by placing the **barcode down** individually into 50 mL centrifuge tubes and centrifuge at 1500 rpm for 5 min and use the slides within 1 h.

C. HYBRIDIZATION

Remember to minimize the natural light falling on the CyDye-labeled probe and keep the slides in the dark.

13. Prepare the hybridization probe mixture (30% formamide, 5× SSC, 0.1% SDS, 0.1 mg μL^{-1} salmon sperm DNA, 1× Denhardt's solution and Cy3 and Cy5 probe) as below for lifter coverslip:

100% formamide	24 μL
20× SSC	20 μL
10% SDS	0.8 μL

Salmon sperm DNA 5 mg mL^{-1}	1.6 μL
100× Denhardt's solution	0.8 μL
Cy3- + Cy5-labeled probe	33 μL
Final volume	approx. 80 μL

The final volume of hybridization probe mixture should be approx. 80 μL; if not, make up to the desired volume (80 μL) with filtered HPLC-grade water

14. Gently mix the hybridization probe mixture by tapping the tube and then spin briefly.

15. Heat hybridization probes to 95°C for 5 min and **do not place on ice.**

16. Centrifuge the probe briefly to collect condensation. Apply to the slides immediately.

17. While denaturing the probes, place the slide in the hybridization chamber and add 10 μL water to each end well in the hybridization chamber to keep the chamber humid.

18. Place the Lifterslip onto the array and gently pipette the probe onto the edge of the Lifterslip, allowing capillary action to draw the probe across the whole area. Avoid moving the coverslip as this may cause spots to streak.

19. Assemble the hybridization chambers as stated in the supplier's instructions.

20. Place the chambers in a moist box – an airtight container lined with moist paper towels – seal with parafilm and cover with foil. Place in a 42°C incubator overnight or for 16–20 h.

D. POST-HYBRIDIZATION WASHING

Do not allow arrays to dry out between washes. Perform steps 22 and 23 separately for arrays hybridized with different probes.

21. Pre-warm overnight at 42°C wash buffer I in a black plastic trough, with lid.

22. Next morning, place wash buffer II into a separate clean black plastic trough, with a lid.

23. Place wash buffer III into two separate clean black plastic troughs, each with a lid.

Remember to minimize the amount of natural light falling on the slides: keep the slides in the dark.

24. Disassemble hybridization chambers and dip the slides into the pre-warmed wash buffer I until the cover slips fall off and wash the slides for 2 min with vigorous shaking.

25. After wash I, transfer the slides from slide holder in wash I to wash II and place into black plastic trough with lid containing

wash buffer II and wash the slides in wash buffer II for 2 min with vigorous shaking.

26. After wash II, transfer and sequentially wash the slides as in step 25 in the two troughs containing wash buffer III. Inspect the slide for any dust particles. If any are present submerge into wash buffer III until you are satisfied with the slide.

27. Dry the slides by placing individually into 50 mL centrifuge tubes and centrifuge at 1500 rpm for 5 min and immediately place the slides in a dark and dry container.

28. Immediately, scan the slides inverted in the Axon scanner.

- Check the PMT voltage of the two lasers is equal approx: 600 mV

- Do a preview scan

  Check the histogram while scanning

  Adjust the PMT voltage until near saturation is achieved; that is, the signal from both dyes is more or less uniform across the x-axes of the histogram and the two dyes are balanced

  Rescan

  Save image as multi-image TIF file and make your own copy!

Protocol 5: ChIP procedure

A. MAKING CROSSLINKED CELL EXTRACTS

Methods for growing cells, formaldehyde crosslinking, cell lysis, and DNA fragmentation by sonication are organism-dependent. Published protocols exist for many organisms, including *E. coli*, *S. cerevisiae*, rodents, and cultured human cells.

B. IMMUNOPRECIPITATION

A fraction of the crosslinked cell extract is used for the immunoprecipitation (IP) step. Typically, extract from 1×10^8 cells is used for a single immunoprecipitation. The following solutions are required.

- IP buffer
  50 mM Hepes–KOH (pH 7.5)
  150 mM NaCl
  1 mM EDTA
  1% Triton X-100
  0.1% sodium deoxycholate
  0.1% SDS
  2 mM PMSF (add fresh)

- High salt IP buffer
  50 mM Hepes–KOH (pH 7.5)
  500 mM NaCl
  1 mM EDTA
  1% Triton X-100
  0.1% sodium deoxycholate
  0.1% SDS

- Wash buffer
  10 mM Tris-HCl (pH 8.0)
  250 mM LiCl
  1 mM EDTA
  0.5% Nonidet P-40
  0.1% sodium deoxycholate

- Elution buffer
  50 mM Tris-HCl (pH 7.5)
  10 mM EDTA
  1% SDS

- TE
  10 mM Tris-HCl (pH 7.5)
  1 mM EDTA

1. Dilute crosslinked cell extract to 800 μL in IP buffer.

2. Add 20 μL 50% (v:v) protein A-sepharose (or other suitable beads) and an appropriate amount of antibody.

3. Incubate at room temperature for 90 min (for some antibodies the immunoprecipitation is better performed overnight at 4°C).

4. Wash twice with IP buffer, once with high salt IP buffer, once with wash buffer, and once with TE.

5. Elute by incubation with elution buffer for 10 min at 65°C. Centrifuge at low speed in a microfuge and keep the supernatant.

6. Add an equal volume of TE and 400 µg proteinase K. Incubate at 42°C for 2 h and at 65°C for 6 h.

7. Purify immunoprecipitated DNA by phenol extraction or use an affinity column.

8. To generate 'input' DNA take a fraction (1/10th to 1/100th of the material used for immunoprecipitation) of the crosslinked cell extract (before immunoprecipitation) and proceed to step 6.

C. AMPLIFICATION

Protocols for amplification by random priming and PCR can be found at www.microarrays.org. Protocols for amplification by ligation-mediated PCR are published (Lee *et al.*, 2002; Harbison *et al.*, 2004), as are protocols for amplification using T7 RNA polymerase (Liu *et al.*, 2003; Bernstein *et al.*, 2004, 2005).

REFERENCES

Bernstein BE, Liu CL, Humphrey EL, Perlstein EO, Schreiber SL (2004) Global nucleosome occupancy in yeast. *Genome Biol* **5**: R62.

Bernstein BE, Karnal M, Lindblad-Toh K *et al.* (2005) Genomic maps and comparative analysis of histone modification in human and mouse. *Cell* **120**: 169–181.

Harbison CT, Gordon DB, Lee TI *et al.* (2004) Transcriptional regulatory code of a eukaryotic genome. *Nature* **431**: 99–104.

Lee TI, Rinaldi NJ, Robert F *et al.* (2002) Transcriptional regulatory networks in *Saccharomyces cerevisiae*. *Science* **298**: 799–804.

Liu CL, Schreiber SL, Bernstein BE (2003) Development and validation of a T7 based linear amplification for genomic DNA. *B.M.C. Genomics* **4**: 19.

Protocol 6: Array comparative genomic hybridization (CGH)

A. DNA LABELING

This protocol is intended for genomic BAC arrays printed onto Corning UltraGAPS™ Slides (Corning Incorporated, NY, USA).

Hybridization solution

1. Add together the following components:
   200 μL of human Cot1 DNA at 1 mg mL^{-1}
   50 μL of yeast tRNA at 20 mg mL^{-1}
   1 μL of glycogen at 20 mg mL^{-1}
   50 μL of NH4Ac 5 M
   950 μL of 100% ethanol

2. Mix and store in the freezer at –80°C for 15 min.

3. Spin at 16 000 × g for 5 min.

4. Discard the supernatant and wash the pellet with 1 mL 70% ethanol.

5. Spin at 16 000 × g for 5 min.

6. Discard the supernatant, dry the pellet and dissolve it at 37°C in 100 μL (enough volume for two arrays under a 24 × 50 mm coverslip) of the following solution:
   50% of deionized formamide
   10% of dextran sulfate
   2 × SSC (saline-sodium citrate buffer)
   2% SDS (sodium dodecyl sulfate)

Labeling of the DNA

Prior to the hybridization, the genomic test and reference DNA are labeled with Cy3-dUTP and Cy5-dUTP fluorescent dyes respectively. The random primed labeling procedure is carried out with components from the BioPrime DNA Labelling System (Invitrogen, Carlsbad, CA, USA). This procedure should be performed protecting the dyes from direct light.

7. Mix 800 ng to 1 μg of genomic DNA with 20 μL of 2.5 × random primer solution in a 0.2 mL vial and make the volume up to 42 μL with pure water.

8. Denature the DNA by heating the mixture at 99°C for 10 min in a thermal cycler.

9. Place the tubes on ice for 10 min.

10. Add the following components:
    5 μL of a 10 × dNTP mixture (2 mM dATP, GTP and CTP; 0.5 mM dTTP, 1 mM EDTA, 1 mM Tris-Cl pH: 7.5)
    2 μL of 1 mM Cy3-dUTP (for the test DNA) or 1 mM Cy5-dUTP (for the reference DNA)
    1 μL Klenow DNA polymerase 40 U mL^{-1} supplied in the BioPrime kit

11. Mix well and incubate the vials for 24 h at 37°C.

Purification of the labeled DNA and preparation of the probe

12. Purify each DNA probe through a ProbeQuant™ G-50 Micro Column according to the user manual (Amersham Biosciences UK Limited, Little Chalfont, UK).

13. Pool the labeled test and reference DNA samples in a single tube and mix well.

14. Apply the sample to a Microcon YM-30 column (Millipore Corporation, Bedford, MA, USA) and centrifuge at 16 000 × g for 10 min at 22°C.

15. Wash the DNA with 200 μL of 10 mM Tris-HCl pH: 8.0 and centrifuge at 16 000 × g for 10 min at 22°C.

16. Add 50 μL of the hybridization solution and incubate for 5 min at 37°C.

17. Collect the probe by inverting the column into a fresh 1.5 mL Microcon tube and centrifuge for 3 min at 1000 × g.

The eluate, the labeled test, and reference DNAs ready for hybridization, should show a purple color indicative of Cydye incorporation.

B. HYBRIDIZATION AND STRINGENCY WASHES

This protocol is intended for genomic BAC arrays printed onto Corning UltraGAPSTM Slides (Corning Incorporated, NY, USA).

Reduction of slides with sodium borohydride

18. After being printed the slides must be treated with sodium borohydride by washing them in 100 mL of 66 mM of BH$_4$Na in 2 × SSC, 0.05% SDS at 42°C for 30 min.

19. The slides are then washed twice in 1 × SSC at room temperature (RT) for 5 min and three times in 0.2 × SSC at RT for 2 min.

20. Spin-dry the slides in a centrifuge at 800 × g for 3 min.

Pre-hybridization of the slides

21. Wash the slides in 300 mL of 0.1% SDS for 5 min and rinse them in ddH$_2$O.

22. DNA on the array is denatured incubating the slides for 1 min and 30 s in 500 mL of boiling ddH$_2$O.

23. Slides are then transfer to ddH$_2$O at RT for 1 min

24. Pre-heat the pre-hybridization solution at 45°C:
    25% deionized formamide.
    5 × SSC.
    0.1% SDS.
    0.01% BSA (bovine serum albumin).

25. Block the slides for 1 h at 45°C in the pre-hybridization solution.

26. Wash the slides twice in H$_2$O for 5 min at RT.

27. Spin-dry the slides in a centrifuge at 800 × g for 3 min.

Hybridization

28. Denature the DNA probe at 72°C for 1 min. Keep it at 42°C.

29. Place the slide and a clean coverslip on a heated block at 42°C.

30. Apply the denatured probe onto the coverslip in a line along the major axis.

31. Place the coverslip with the probe on top of the array without introducing air bubbles.

32. Put the array into a hybridization chamber and hybridize for 40 h in an oven or waterbath at 37°C.

Post-hybridization stringency washes

Proceed with the following steps in the dark.

33. Carefully wash off the coverlip in 2 × SSC, 0.1% SDS.

34. Wash the slide through the following solutions:
    15 min at RT in 2 × SSC, 0.1% SDS
    2 min at RT in 2 × SSC
    15 min at 45°C in 50% deionized formamide in 2 × SSC, pH: 7.0
    30 min at 45°C in 2 × SSC, 0.1% SDS
    15 min at RT in 0.2 × SSC
    1 min at RT in ddH$_2$O

35. Spin-dry the slides in a centrifuge at 800 × g for 3 min.

36. Scan the arrays.

Protocol 7: Run-off microarray analysis of gene expression (ROMA)

A. *ESCHERICHIA COLI* OLIGONUCLEOTIDE ARRAY

E. coli oligonucleotide arrays were produced by the UBEC group (University of Birmingham *E. coli* group). The oligonucleotides were designed and synthesized by Qiagen Operon Ltd as the *E. coli* Array ready oligo set v 1.0 (Qiagen, Crawley, UK) and represented 4289 *Escherichia coli* K12 strain (MG1655) ORFs; 1416 ORFs designed to the O157:H7 (EDL933) strain; and 273 ORFs unique to the O157:H7 (Sakai) strain (http://oligos.qiagen.com/arrays/oligosets_ecoli.php). In addition, 110 oligonucleotides representing ORFs from the EDL933 and Sakai plasmids were added to the oligonucleotide array set. Also included within the array set were 12 positive and 12 negative control oligonucleotides, which were printed within each subarray. The average size of each oligonucleotide used in the array was a 70-mer, with a $T_m = 75 \pm 5°C$. The position of each oligonucleotide within the ORF was more than 40 bases away from the 3' end. The oligonucleotides were arrayed on Corning CMT-GAPS II slides in 48 blocks, each containing 324 spots (18 rows by 18 columns) using a MicroGrid II robot (BioRobotics, UK). Each oligonucleotide was printed in duplicate on the array, and Amersham Lucidea™ Universal Scorecard™ (Amersham, UK) controls were printed within each subarray.

B. GENOMIC TEMPLATE PREPARATION

Genomic DNA was isolated from the *E. coli* MG1655 (CGSC 7740) wild-type strain by phenol/chloroform extraction and fragmented by EcoRI digestion as follows.

1. Cells were collected from 5 mL of overnight culture and resuspended in 453 µL TE buffer.

2. Cells were lysed by the addition of 10 µL of 100 mg mL⁻¹ lysozyme, and incubated at 80°C for 10 min.

3. To remove RNA, 4 µL of 25 mg mL⁻¹ RNase A was added to the cell lysate and incubated at 37°C for 1 h. Complete removal of RNA is crucial.

4. Proteins including RNase A were digested by incubation with 3 µL 20 mg mL⁻¹ proteinase K and 30 µL 10% SDS at 37°C for 1 h.

5. To remove cell debris and proteins, the cell lysate was extracted with 500 µL phenol/chloroform and centrifuged for 10 min at room temperature. The aqueous layer was carefully transferred to a fresh tube and extracted with phenol/chloroform twice more. Eppendorf Phase Lock Gel™ works better to separate the two layers.

6.   The final upper aqueous layer was transferred to a fresh tube and DNA precipitated by adding 50 µL of 3 M NaOAc and 360 µL of isopropanol. The tube was gently inverted until hair-like DNA precipitated.

7.   The hairy DNA pellet was recovered using a pipette tip and washed in 1 mL 70% ethanol twice. Spinning is not recommended as this may cause DNA fragmentation and also difficulty in resuspending the pellet.

8.   The DNA pellet was air-dried at room temperature for 10 min and resuspended in 100 µL sterilized water. Incubation at 65°C for 30 min may be necessary.

9.   Genomic DNA was fragmented by digestion with restriction enzyme EcoRI at 37°C overnight. For complete DNA fragmentation, excess EcoRI was added.

10.   DNA fragments were extracted by phenol/chloroform followed by ethanol precipitation. The DNA was checked on 1% agarose gel and the concentration determined by the absorbance at 260 nm and 280 nm.

C. RECONSTITUTION OF RNA POLYMERASE

The holo RNA polymerase (RNAP) used in this study was reconstituted from purified His-tagged α subunit and stocks of washed β, β' and σ^{70} subunits inclusion bodies (Tang *et al.* 1995). It is also commercially available, e.g. Epicentre Technologies. Wild-type CRP, AR1-mutated CRP (HL159) and AR2-mutated CRP (KE101) were purified as described previously by Rhodius *et al.* (1997).

D. *IN VITRO* RUN-OFF TRANSCRIPTION WITH GENOMIC DNA

Two run-off transcription reactions were set up in parallel: the control reaction contained 4 µg of EcoRI digested genomic DNA, 1 mM each of ATP, GTP, CTP and UTP, 50 units RNase Out (Invitrogen, UK), in transcription buffer (40 mM Tris/acetate pH 7.9, 10 mM $MgCl_2$, 1 mM DTT, 100 mM KCl, 0.1 mg mL^{-1} BSA); the test reaction contained 2 µL of 5 mM cAMP (final concentration 200 µM), and 4 µL of 10 µM CRP (40 pmol of monomer, at a final concentration of 800 nM) in addition to the components of the control mixture.

11.   The mixtures were incubated for 10 min at 37°C and the transcription reactions started by the addition of 20 pmol of RNA holoenzyme to each tube.

12.   The reactions were incubated for 30 min at 37°C and then stopped by the addition of 5 µL 250 mM EDTA and placed on ice.

13.   The RNA transcripts were purified using the Qiagen RNeasy mini kit (Qiagen Ltd, Crawley, UK). On-column DNase I digestion (Qiagen) is necessary to remove contaminating genomic DNA templates.

14. The concentrations of RNA were determined by the absorbance at 260 nm and 280 nm. Typically 3–6 µg of RNA can be obtained.

E. AMINO-ALLYL dUTP LABELING OF RNA AND DYE INCORPORATION

An amino-allyl dUTP-based labeling method was used to obtain Cy-labeled cDNA from RNA transcripts purified.

15. The RNA transcripts obtained from the last step were vacuum-dried and resuspended in 16.4 µL of RNase-free water. In this experiment, Lucidea™ Scorecard (Amersham) mRNA spike mixes were added to the RNA samples.

16. RNA samples were mixed with 2 µL of 3 mg mL^{-1} random hexamers (Amersham Biosciences, Little Chalfont, UK) and incubated at 70°C for 10 min.

17. The mixtures were snap-cooled and then incubated at room temperature for 10 min.

18. Reverse transcription labeling mixture (11.6 µL) was then added to obtain 0.5 mM dATP, dCTP, dGTP, 0.2 mM dTTP, 0.3 mM aminoallyl-dUTP (aa-dUTP), RNase inhibitor (30 U), 400 U SuperScript II (Invitrogen, Paisley, UK), 10 mM DTT, and 1× first strand buffer in the final reaction.

19. The reactions were carried at 42°C for at least 3 h to generate aminoallyl-labeled cDNA. The reaction was stopped and the RNA template hydrolyzed by the addition of 10 µL 0.5 M EDTA and 10 µL of 1 M NaOH and incubation at 65°C for 15 min. The solution was then neutralized by the addition of 10 µL 1 M HCl.

20. The unincorporated aa-dUTP and free amines were removed by washing through a microconcentrator (Microcon YM-30, Millipore) three times with water. It is important to remove Tris buffer from the reaction as Tris contains amine group which can bind to monoreactive CyDye™ (Amersham).

21. The samples were dried and resuspended in 4.5 µL 0.1 M sodium carbonate buffer (pH 9.0) and 4.5 µL aliquot of Cy3 or Cy5 monoreactive dye (Amersham), prepared in DMSO. The aa-cDNA from the control reactions was coupled with Cy3 and aa-cDNA from the CRP reactions coupled with Cy5.

22. The uncoupled dyes were removed using a QIAquick PCR purification kit (Qiagen).

F. PREHYBRIDIZATION AND HYBRIDIZATION

23. Before hybridization, the arrayed slides were prehybridized in a buffer containing 25% formamide, 5× SSC, 0.1% SDS and 10 mg mL^{-1} BSA at 42°C for 2 h.

24. The slides were washed by dipping twice in distilled water and once in 95% ethanol and dried in a clean 50 mL centrifuge tube by centrifugation at 1500 × g for 10 min. The slides can also be dried by compressed air (filtered) if available.

25. The Cy3- and Cy5-labeled cDNAs were mixed, vacuum dried, and resuspended in 70 µL hybridization buffer containing 25% formamide, 5× SSC, 2 µL 50× Denhardt's, 2 µL yeast tRNA (20 µg mL^{-1}) and 0.1% SDS.

26. The labeled cDNAs were denatured by heating at 95°C for 5 min, and applied to the prehybridized slide in a CMT-Hybridization chamber (Corning Inc., Corning, NY). A HybriSlip (Sigma) was carefully lowered onto the slide. Air bubbles were removed by gently pressing the coverslip. To maintain humidity inside the chamber, 10 µL of distilled water was added to the two reservoir wells.

27. The chamber was tightly sealed and incubated at 42°C for 16–20 h in the dark.

Slide washing and scanning

28. Slides were removed from the chamber, washed for 5 min sequentially in 2× SSC/0.1% SDS buffer, 0.1× SSC/0.1% SDS buffer and 0.1× SSC buffer, rinsed in distilled water for 5 s, and dried by centrifugation at 2000 rpm for 10 min.

29. The hybridized slides were scanned with a confocal laser scanner (Axon GenePix 4000A) using appropriate gains on the photo-multiplier tube (PMT) to obtain the highest intensity without saturation.

G. *IN VIVO* TRANSCRIPTIONAL PROFILING

30. For *in vivo* microarray experiments, both wt and Δ*crp* strains were grown in M9 minimal media (Sambrook, 1989) containing 0.2% fructose at 37°C to OD$_{600}$ 0.8. Glucose (0.2%) was then added to the cultures, which were grown for a further 15 min, prior to harvesting.

31. Cell samples for RNA preparations were collected before and after the addition of glucose. Two volumes of RNAProtect (Qiagen Ltd, Crawley, UK) were added per volume of bacterial culture, prior to centrifugation.

32. The total RNA from cell cultures was extracted using the Qiagen RNeasy mini kit and on-column DNase I digestion (Qiagen Ltd, Crawley, UK) using the manufacturers instructions.

33. The quality and concentration of the RNA prepared was assessed using the Agilent 2100 Bioanalyser.

34. Total RNA (10–20 µg) was labeled using the CyScribe Post-Labelling Kit (Amersham Biosciences) as described by the manufacturer.

35. The slides were hybridized, washed, and scanned as described in the ROMA protocol above.

Image extraction and data normalization

Scanned images for Cy3 and Cy5 were then overlaid with GenePix Pro 3.0 software. Only data generated from spots representing *E. coli* MG1655 genes were analyzed. Spots with background-subtracted intensity lower than 100 in both Cy3 and Cy5 channels were filtered out. GeneSpring software (SiliconGenetics) was used for global normalization and density-dependent normalization (Lowess) to correct artifacts dependent on density or caused by different dye incorporation rates for the two dyes. The duplicate spots for each gene on a single slide were taken as two individual spots. Three independent experiments were performed for each comparison and therefore, six replicate data sets were obtained for each gene. The normalized data from GeneSpring software were exported to a Microsoft Excel spreadsheet.

Data reproducibility

To assess data reproducibility, the spot-to-spot variation was calculated as described by Loos *et al.* (2001) and correlation between data sets analyzed by the Pearson correlation coefficient (*r*). The spot-to-spot variation was represented by the percent error calculated by dividing the standard deviation by the average ratio for each gene (% = σ/μ). Average variation within a slide was calculated as the average of the gene spot-to-spot variations on that slide (two spots), and average variation across slides was calculated as the average of gene spot-to-spot variations based on the six spots of the data set. The latter variation included spot, slide, and cDNA variability. The Pearson correlation coefficient (*r*) was calculated for Cy3 data sets from any two slides and for Cy5/Cy3 from any two replicate slides.

REFERENCES

Loos A, Glanemann C, Willis LB, O'Brien XM, Lessard PA, Gerstmeir R, Guillouet S, Sinskey AJ (2001) Development and validation of *Corynebacterium* DNA microarrays. *Appl Environ Microbiol* **67**: 2310–2318.

Rhodius VA, West DM, Webster CL, Busby SJW, Savery NJ (1997) Transcription activation at class II CRP-dependent promoters: the role of different activating regions. *Nucl Acids Res* **25**: 326–332.

Sambrook J, Fritsch E, Maniatis T (1989) Molecular cloning: a laboratory manual, 2nd edn, Cold Spring Harbor Laboratory, Cold Spring Harbor, NY.

Tang *et al.* (1995) Rapid RNA polymerase genetics: one-day, no-column preparation of reconstituted recombinant *Escherichia coli* RNA polymerase. *Proc Natl Acad Sci USA* **92**: 4902–4906.

Protocol 8: Selection and genotyping of single nucleotide polymorphisms

A. SNP SELECTION

1. Identify SNPs across the candidate gene (PupaSNP tool).

2. Refine the selection of SNPs using Pupas View, taking into account the potential functional effect, LD patterns, phylogenetic conservation, validation status and minor allele frequency.

3. Further refine this list of SNPs considering the probability of successful genotyping in the chosen platform.

B. SNP GENOTYPING (SEE *FIGURE P8.1*).

4. DNA (250 ng at 50 ng μL^{-1}) is activated to bind to paramagnetic particles.

5. Three oligonucleotides are designed for each SNP locus: two allele-specific oligos (ASOs) and a locus-specific oligo (LSO). All three oligonucleotides contain complementary genomic sequences and universal PCR primer sites. ASOs are specific to each allele of the SNP site and LSO hybridizes between 1 and 20 bases downstream from the ASO site. In addition, the LSO also contains a unique address sequence for each hybridization in the array.

6. The oligo pool, hybridization buffer, and paramagnetic particles are combined with the activated DNA in the assay hybridization step, when the oligonucleotides hybridize to the genomic DNA sample which is bound to paramagnetic particles.

7. When DNA ligase is added, the successfully extended ASOs are ligated to the LSOs. The ASO–LSO product serves as a template for amplification by PCR, where three universal PCR primers are used: two corresponding to the ASOs (each fluorescently labeled with Cy3 and Cy5 dyes) and the third to the LSO (biotinylated).

8. The PCR product is captured on a solid support and single stranded, fluorescently labeled material is eluted and hybridized to the *SentrixTM* array matrix.

9. After the hybridization, the *Illumina BeadArray Reader* simultaneously scans *Sentrix Arrays* at two different wavelengths with sub-micron resolution.

10. The automatic calling of genotypes is performed by the software package *GenCall* (see also Section 4.4.2).

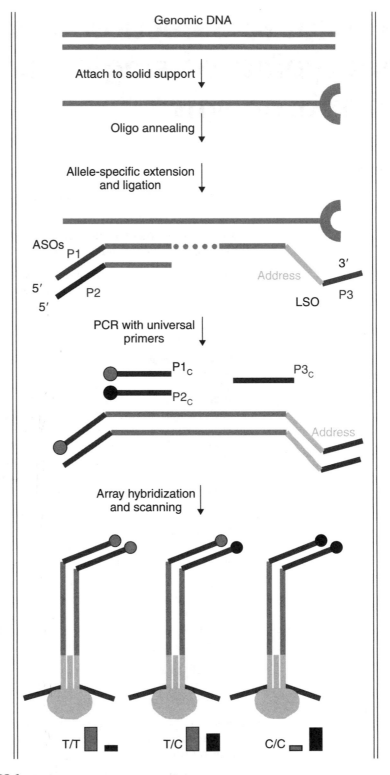

Figure P8.1

Illustration of the GoldenGate genotyping assay process. (A color version of this figure is available on the book's website, www.garlandscience.com/9780415378536)

Appendix 1: Notes on printing glass DNA microarray slides

Antony Jones

A1.1 FOUNDATIONS OF THE MICROARRAY EXPERIMENT

Glass slides are one of the most commonly used surfaces for building an array of DNA or oligonucleotide samples. There are several slide surfaces available from commercial companies which are all coated with a variety of different binding substrates. The glass slide has unique advantages by allowing DNA samples to be covalently attached to the slide surface and is a durable material that sustains high-temperature washes. Choosing the best slides for your samples is probably the most important part of any microarray experiment. A high quality foundation, i.e. the glass slide, will increase immobilization efficiency, improve spot morphology, and enhance array reproducibility.

The slide surface must be flat and the chemical coating consistent to produce robust hybridization of the target with the probe and have low fluorescence, thus not contributing to background 'noise'. Inspection of the quality of any glass slides that are being considered for use in a microarray slide printing program is recommended. The slides should be checked for a uniform coating and no evidence of surface damage which can affect the spotting process and result in difficulties in the data collection stage of the experiment. Simple ways of checking the slide surface is to perform a scan using any microarray slide scanning system on a few random slides before they are printed. The uniformity of the coating and surface should be observed whilst looking for evidence of 'pitting' or scratches on the slide.

There are several different approaches to binding DNA onto microscope slides, but they can be divided into two main groups: those which involve hydrogen bonding or those that use covalent linkage. Hydrogen bonding occurs between amine groups coated onto the surface of the slide and the phosphate groups in the backbone of the DNA. These types of slides are usually coated with either poly-L-lysine or aminosilane to give an amine group coated slide. After printing the slides with samples of DNA they are UV cross-linked to allow covalent bonds to form between the amine groups and the DNA. Covalent bonding occurs between active groups such as aldehyde or epoxy coating of the slides and amine modified DNA. To obtain excellent spot morphology the slides need to have a good hydrophobic surface which results in smaller spots and allows for higher density arrays. The most common slides used are the Corning UltraGaps and Gaps II slides. The Gaps II slides are mostly used for the printing of cDNA samples and Ultragaps are commonly used by the oligonucleotide community.

The glass slides are manufactured under controlled conditions in a clean environment and any particles of dust falling onto the slide surface can cause problems during the printing of samples.

During the loading of slides onto the arraying system it is important that checks should be made to observe any particles of dust on the slide surface and remove poor slides slides from the run if necessary. Particles of dust, glass or fibers which are sitting on the slide surface will block the pin reservoir by sticking to the microscopic tip of the pin and therefore prevent the sample loaded into the pin channel being deposited onto the slide. Such an event occurring will result in irregular sample deposition producing arrays with the possibility of hundreds of samples missing. This can be extremely time consuming and a very costly mistake for research programmes.

Fibers from clothing and the atmosphere can stick to the pin head to create smears on any spots of samples that have already been deposited and result in a whole array consisting of up to 1000 samples being destroyed. Once the slides have been printed it is recommended that they are stored in their original containers from the supplier, preferably in a temperature-controlled environment and desiccated.

A1.2 PRINTING GLASS MICROARRAY SLIDES

There are many different shapes and designs of microarray printing robots on the scientific market, and one of the most commonly used models of arrayer appears to be the MicroGrid Total Arraying Suite - TAS, from Genomic solutions® (http://www.genomicsolutions.com/showPage.php – this machine was manufactured by Biorobotics, which are now part of the Genomic Solutions portfolio of companies). This particular model will be the focal point of this chapter as it is used in the Birmingham Functional Genomics laboratory, but many of the general points made in this section are pertinent to all microarray spotting robots.

The MicroGrid TAS has the ability to print over 100 slides per run and can also hold up to twenty-four 96-well or 384-well plates at a time in a storage unit called a Biobank. The oligonucleotide or DNA samples for spotting are stored inside the robot's Biobank which should be chilled to 11°C. This reduces the evaporation rate of precious samples, and keeps the samples at a suitable printing temperature, as arrays that contain tens of thousands of spots can take over 24 h to print. Although the Biobank holds a maximum of 24 plates during large printing sessions it may be reloaded to allow additional samples to be printed to complete the desired custom array size. The MicroGrid TAS system has the additional feature of two vacuum ports linked to a sample plate lid lifting system which removes the sample plate lid to allow the microspot pins access to the samples for filling of their microscopic reservoir. After filling the pin head with the first source visit of samples, the robotic lid lifting system replaces the sample plate lid to minimize evaporation of the samples.

During the printing process it is vital that the glass slides do not experience any external vibration or movement while the printing trays are in motion. Movement of slides due to vibrations or a leak in the vacuum system can result in the sample alignment on the glass slides moving, thus resulting in 'smiling' or 'rollercoaster' arrays which are often difficult to analyze using standard microarray scanners and gridding programs. Microarray data analysis software works on the principle of excellent alignment of the deposited samples and requires the spots to sit within an analysis grid. Serious problems occur with alignment of the spots within the grids if the slides have moved creating difficulties with data analysis, which has been discussed in Chapter 1.

The problem of holding the slides tightly in position during printing is overcome by the robot's effective vacuum system which requires regular checking and should be maintained by regular cleaning of the vacuum ports responsible for holding slides tightly to the printing platform.

The vacuum's efficiency can be checked before conducting an expensive print run by physically trying to move a slide by hand. If a slide can be moved by hand then the vacuum is insufficient and the print run should be delayed and evidence of a vacuum leak should be explored by listening for hissing from a leaking port. One common cause of vacuum leakage is from fragments of glass or dust sitting under one slide in a group of 100 slides which can break the vacuum seal, thereby affecting all the other slides that have been loaded. Each slide loaded onto the printing tray is positioned over a vacuum port which is responsible for pulling the slide downwards producing a slight cracking noise as the glass slides are held fast into their positions. A cracking noise from the slides is a good indication that the vacuum is working efficiently and should not be any cause for alarm.

The loading of slides onto the printing shelves is an important part of the printing process to achieve reproducibility of slide quality for all the slides loaded. Each slide needs to be an exact duplicate within the completed set of slides to ensure good comparative data for statistical analysis.

Each glass slide is carefully positioned between four raised nodules on the printing loading tray so that the bottom and the left hand side of each slide are supported and the risk of movement is minimal.

Once the slides are loaded they are checked for signs of dust or fibers that may have floated or dropped onto the slide surface from the operator. Fragments of dust or hair can block the pin reservoir and result in the loss of sample deposition to the slides and result in the failure to print the whole set of samples, or to completely ruin the whole set of slides. To help prevent this occurrence glass slides are usually loaded from the back of the printing tray towards the front, until the robot is full to capacity. This means that the operator's arms do not pass over the loaded slides in order to load new slides, minimizing the chance of fibers, dust, and other materials dropping from the operator on to the slides.

During the loading of sample plates into the Biobank it is important to check each plate lid for any signs of damage or difficulty in removing them from the sample plate. The lid lifting system is dependent on the vacuum system to remove the plate lid before a source visit can proceed. Failure to lift a lid due to 'sticky' sample plates will cause the robot to abort the run. Once a run has been aborted by the software it is not possible to continue the run and the whole print run has to be started again and the unfinished slides thrown away, because it is not possible to programme this particular robot to continue a print run from a selected plate and appropriate source visit. Other robots may have this capacity. One major source of lid sticking problems we have found is that some commercial foils used for covering plates can leave traces of glue on the surface of the sample plate and evidence of this should be taken seriously to avoid loss of slides caused by a lid lifting failure.

Before any print run begins it is always very important to check the orientation of the sample plates in the Biobank as errors can occur if the operator does not check the plates' orientation, which results in the planned array design being scattered. The software for the TAS (total arraying suite) system works on the principle of what samples the bottom left hand microspot pin is going to deposit and their positions. The application software which is responsible for designing the arrays allows the operator to produce a layout map indicating each source visit and where each sample from the source plates will be positioned on the slide. A diagrammatic representation of what the final slide will look like can be produced and exported into an Excel spreadsheet which is useful to check the array design. An array is designed on this principle, and the sample plates must be loaded with well A1 facing the front of the robot. However, if the plates are positioned incorrectly in the Biobank this will produce slides with a totally randomized pattern and clone tracking and producing a GAL file for identification purposes will prove to be an incredibly time consuming task.

Once the slides are in position and the sample plates have been loaded the print run can be started.

A1.3 PRIMING THE ROBOT BEFORE PRINTING GLASS MICROARRAYS

Before printing a set of expensive glass slides the robot's internal environment needs to be primed in order to reach optimum specific temperatures and humidity conditions that enhance the spot morphology. The MicroGrid TAS has two re-circulating water baths and a main wash-station that work together during the printing process to wash and dry the microspot pins before filling with the next set of samples. The main wash-station for the microspot pins is fed by a large water tank concealed within the robots pump mechanism and produces a high speed jet of water against the microspot pins sample reservoir during a print run. This process flushes out any remaining sample from the pin reservoir within the pin's head to avoid cross-contamination of samples, and the vacuum pump complements this action by creating a strong drying vortex in the main wash area which dries the pins before they move to the next group of samples.

Two glass Duran bottles are located outside the robot, which contain fresh 18 MΩ water and are responsible for re-circulating water to two re-circulating baths inside the robot. The purpose of the re-circulating baths is to reduce the carry over of samples during printing and they are also effective at removing traces of $3 \times$ SSC that may have been used as a spotting buffer. Samples resuspended in $3 \times$ SSC are susceptible to blocking the microspot pin head due to crystals forming on the pin tip during the printing process. Crystals forming on the pin tip have resulted in the failure of samples being deposited across a full set of slides. To avoid this problem it is recommended that the pins are washed twice in each re-circulating bath for 7 s and the 'wiggle' effect set to 3 mm within the robot settings. The 'wiggle' setting on the robot allows the pins to be agitated in the re-circulating baths from left to right and this ensures that all traces of SSC are removed. This process is excellent at keeping the microspot pin heads clean and therefore producing a successful print run for 'challenging spotting buffers'.

A humidifier located to the side of the printing robot is used to maintain a desired and controlled humidity inside the printing environment of the robot, and filters the air entering the robot to remove particles of dust, which can inter-fere with print quality. The humidifier helps prevent the spots drying out too soon upon deposition onto the slide surface and therefore prevents production of spots with a dark center or hole (otherwise known as the 'donut effect'). To prepare for a print run the robot is filled with fresh 18 MΩ water and all the feeder pipes are flushed through to ensure the removal of air-locks to enhance pin washing.

Slides are loaded from the top shelf downwards until all the shelves are full which is controlled by the printing software which prompts the operator.

Compressed air can be used to clean off slides and shelves but can sometimes spray droplets of oil across the slides and this method of removing traces of dust should be avoided. The best method for cleaning the shelves on the robot is to use an ordinary lint-free cloth slightly dampened with isopropanol, which helps remove grease from the slide trays.

The robot should be allowed to adjust to the desired humidity before the print run is started to ensure that the humidity reading is stable at the start of the process rather than increasing during the run time.

Usually a humidity of 50–60% produces a good quality spot with most samples and slides but it may require a few test runs to optimize the humidity and the temperature of the robot to suit the spotting buffer in which the samples have been resuspended, and the type of glass slide used.

The loading of the microspot pins onto the robot should be done just before printing is due to start and during the initial set up they are left in the printing tool hanging in MilliQ water to soak.

Allowing the microspot pins to sit in MilliQ water helps remove any crystals from a previous printing exercise. The pins are sonicated for 15–20 s in MilliQ water and then loaded onto the robot.

Sonication of the pins helps prime the microspot pins before a run and hopefully removes any blockages in the microspot pin reservoir.

A1.4 PRESPOTTING AND SPOTTING BUFFERS

In most microarray experiments a group of slides are commonly used for prespotting at the beginning of the main print run which is a valuable technique that is used to remove any excess of a sample from around the outer surface of the microspot pin to avoid the 'rush' effect.

The prespotting slides we use are inexpensive aminosilane-coated slides which can be used as a blotting mechanism to remove excess sample from the pin before printing onto the more expensive experimental slides. This process helps produce spots having the same diameter throughout the whole set of slides. Prespotting can also be used to check that every microsopt pin is deliv-ering samples before commencement of printing of the experimental slides. The printing process can be aborted if there is evidence that a microspot pin is not delivering sample and the cheap slides are wasted rather than a set of expensive experimental slides. The experimental expensive microarray slides cannot be re-used on the robot once sample has been delivered as the alignment of the spots will be affected and 'double' printing occurs. Most probe DNA samples were originally dissolved in a high salt spotting buffer such as $3 \times SSC$, but new deter-gent additives and spotting buffer chemistries have been produced which have improved the spot morphology. Further improvements have been made and the introduction of hygroscopic universal spotting buffers have resulted in improved spot structures thus giving higher signal intensities. Sometimes additives can increase the spot size but this can usually be overcome by reducing the contact angle of the pin and the speed at which the pin strikes the slide. There are currently a variety of spotting buffers on the market which seems to suggest that no spotting buffer is ideal. High salt buffers evaporate faster and can cause printing problems by blocking pins, but they allow easy desiccation and rehydration of probes which reduces variability between print runs due to changes in DNA concentration. Hygroscopic buffers evaporate at a much slower rate but cannot be easily desiccated and rehydrated and tend to be used in applications where probe DNA is not limiting. Corning Life Sciences currently produce a universal spotting buffer in which samples can be resuspended and then stored in the freezer.

The beauty of this buffer is that the samples do not freeze while they are in storage and can be directly removed and loaded onto the printer without waiting for the samples to thaw. This removes the chance of ice crystals forming in the samples during storage which can damage DNA samples (particularly oligonucleotides) and crystals of salts are not precipitated during any freeze/thaw cycles which could block the microspot pins; this is extremely reassuring to any microarray printing exercise.

A1.5 SPOTTING PINS

Most microspot pins on microarray instruments are made from stainless steel or titanium with a capillary cut into the tip using laser cutting technology. The process of laser cutting can lead to some inconsistencies in the pin tips which can result in some pins depositing samples faster than others within a batch of pins and therefore spotting variations can be observed. Running several test print runs is a valuable exercise to check for signs of variation and the removal of pins and replacement of them with 'cherry-picked pins' that are batch-tested is something that operators will practise to eventually obtain a perfect set of microspot pins for their printing service.

New ceramic and foil pins have been introduced onto the market recently and are considerably cheaper to manufacture and can print up to 250 000 spots on a single microarray slide. It is doubtful, however, that anyone would require such an array size and anyone requiring an array of this size would have to upgrade their current spotter or buy one of the new generation of spotters.

The most commonly used pins are the 2500 (100 µm) and the 10K (50 µm) which are selected depending on the density of the array required. The 10K pins can allow up to 1600 spots per sub-array and the 2500 are for the smaller less dense arrays allowing up to 500 spots per sub-array.

The 10K pins are the most commonly used for high density oligo arrays and produce excellent spot morphology if care and attention is paid to the soft touch settings on the robot. The MicroGrid Software has many features which allow the operator to change the behavior of the pins and how contact is made with the slide surface. These settings are crucial, depending on the type of spotting buffer used and the density of the array required. Generally, the pins are set using the software to slowly lower into the samples in the source place and then rise out of the samples at an extremely slow rate. This action makes sure that any excess sample sitting on the outside of the pin is removed which then allows for smaller spots to be produced. Contact with the slides is also an important factor affecting the quality of arrays which should be tailored according to the array required. For the high density arrays the pins are simply allowed to 'kiss' the surface of the slide to deposit a fraction of the sample onto the slide surface. To obtain the ideal setting requires experimenting with different slide surface coatings, the robot settings and the sample buffer before proceeding to print the experimental slides. Although this can take time and a great deal of patience the results obtained are always rewarding for facilities and their operators. Recently, it has been evident that those wishing to work with microarray slides tend to ask those facilities that are set up to print their slides for them rather than having to spend lots of valuable research time and resources optimizing a microarray printing robot.

Appendix 2: Comparative genomics. The nature of CGH analysis and data interpretation

Lori A.S. Snyder, Graham Snudden, Nick Haan, and Nigel J. Saunders

A2.1 STRENGTHS, WEAKNESSES, AND THE NATURE OF CGH

Comparative genome hybridization (CGH), or genomotyping, is a powerful means to rapidly identify the presence of genetic sequences within a genome. It is a development of the Southern blot, conducted in a format in which thousands of probes can be simultaneously assessed. Whereas, with Southern blotting, a collection of bacterial strains can be surveyed through digestion of their DNA, resolution on a gel, blotting to a membrane, and hybridization with a single-labeled probe, now the experiment can be turned upside-down. In this way, the strain's DNA can be simultaneously probed for as many genes as are printed onto a microarray slide surface, potentially giving the researcher an overview of the complete gene complement of a strain.

To meaningfully assess gene complements, the same principles established in Southern blotting apply. The likelihood of obtaining a Southern blot hybridization band is dependent upon both the sequence length and homology of the probe to the DNA target, and the stringency of the hybridization conditions. Lowering the stringency of the hybridization buffers could reveal less homologous sequences, but possibly at the cost of generating less reliable/specific results. It is therefore important to have designed the probe properly, taking into account the most homologous regions of the sequence in question and any potential cross-hybridizations that might occur.

It is not a trivial aspect of microarray design to obtain optimal probes that combine high specificity with sufficient lengths to facilitate stringent processing. As the length of the probe increases, the diversity of sequences available for cross-hybridization increases, and the opportunity for false hybridization increases with the length of the genome and the size of paralogous families. A recent study has investigated how differences in probe length influence the experimental data, through differences in reproducibility of probe binding and in cross-hybridization with other sequences (Chou *et al.*, 2004). The key points, in relation to probe design and selection, are shown in *Figure A2.1*. However, the degree of probe cross-hybridization can be reduced substantially through

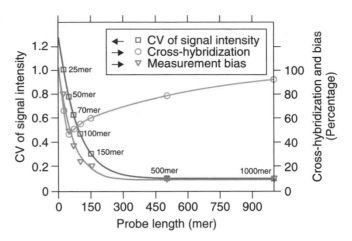

Figure A2.1

Determination of the optimal probe length. Plots of cross-hybridization, measurement bias for one probe per gene, and the CV of the signal intensity versus probe length. All the solid lines were fitted by regression analysis of the data points. The scale for cross-hybridization and measurement bias is on the right of the figure. Taken from Chou *et al.*, 2004.

careful selection of probe regions. This approach does normally require significant additional time invested in probe design and selection and expert input from the operator, rather than reliance on completely automated systems. The suitability of the microarray probes that will be used for CGH studies may be qualitatively different than for those used for expression studies. In expression studies, the RNA is normally extracted from a strain with the same or very similar sequence to that used for the design of the microarray probes. In CGH studies, diverse samples are explored which frequently differ substantially between strains, therefore if poorly conserved regions of genes are selected as probes this can lead to false negatives, while using regions of low specificity between paralogues can lead to cross-reactivity and false positives. Probes specifically designed for CGH studies, taking into account all of the aspects of design discussed here, are very robust and also make excellent expression study microarray probes. The reverse is not necessarily true. It cannot be assumed that one can take a microarray developed for expression studies 'off the shelf' and apply it as successfully to a CGH study, and the suitability of the probes are likely to vary on a probe-by-probe basis.

Sometimes it is not possible to obtain probes that will differentiate between different allelic copies of a gene, or very similar paralogous genes, and of course it is never possible to distinguish between the different copies of repeated genes. These, sadly, are unavoidable aspects of using this (or most alternative) methodological approaches, other than genome re-sequencing, but the currently available and developing technologies do not make re-sequencing a viable option for genomes as divergent as those of many bacterial species. It is important, however, for the interpretation of results that full information on potential cross-hybridization of microarray probes is made available for studies of all types.

A2.2 THE EXTENSION OF CGH BEYOND THE GENOME 'INDEX' SEQUENCE

The nature of both the microarray and the populations to be studied determine the applicability of CGH and the nature of the probes that must be used. Within eukaryotic systems there is typically a very high degree of conservation of sequence identity between related genes within species, and in some instances even between species. Examples of this include studies of chimpanzees using human microarrays, and of radishes using *Arabidopsis* microarrays. In some cases such comparisons have been performed using oligonucleotide-based probes. Hybridization with oligonucleotide probes is highly susceptible to mismatches within the target, a fact that is exploited in probes designed to identify local sequence polymorphisms, and the influence of mismatches on hybridization is relatively unpredictable with regard to the number of mismatches that will be tolerated and their location within the probe. Depending upon their location and the stringency of the hybridization, as few as three mismatches can significantly disrupt detection using long oligo probes (e.g. 70 bp – and the problem is greater if shorter probes are used). For an expression study, in which one is comparing the responses to different conditions or mutations in the same strain, the use of a microarray designed against something sufficiently similar but not identical can be tolerated because mismatches influencing the signal of one channel will also influence the intensity of the other, and thus not greatly affect the fold-ratio which is measured. However, for CGH applications oligonucleotide-based microarrays have a tendency to be highly strain-specific in most bacterial systems, and are frequently unsuitable even for the purposes of CGH comparisons of unrelated strains within the same species. There are some exceptional bacterial systems in which sequences are sufficiently conserved between strains for the use of oligonucleotide probes, especially if more than one probe is used per gene, but usually this is not the case.

For these reasons, PCR probes are often superior to oligonucleotide probes for microarrays intended for CGH studies, and also for comparative transcriptional studies in bacterial systems when the response of different strains to the same conditions will be considered, for example. This is not due to inherent differences, such as the presence of both strands, but simply reflects the fact that currently it is not possible to cost-effectively synthesize oligonucleotides accurately within the length range needed for the most stringent hybridizations and that will be resistant to moderate inter-strain sequence divergence.

The conservation of the target sequences determines how functionally specific a microarray designed against a single genome sequence is with respect to the species and strains to which it can meaningfully be applied. Regardless of the context, properly designed and selected probes, whether oligonucleotide or PCR product, will work well for the strain against which it was designed, but its application to other strains cannot be guaranteed to perform as well. With only one genome sequence represented by the probes, any CGH data generated can also only identify those genes that are absent from the strains that are being tested, compared with what is in the genome sequence. No genes that are not in that single genome sequence can be identified, although it is possible to identify regions for further investigation through targeted sequencing (Wolfgang *et al.*, 2003; Snyder *et al.*, 2004). Absence of signal cannot differentiate between either sequence absence, or of divergence of sequence below that permitted by the stringency of the hybridization.

A single genome-based microarray is a perfectly good platform for performing transcription studies with the strain used for its design, but for CGH applications a microarray designed on the basis of comparative analysis of multiple genomes has significant advantages. By utilizing multiple genome sequences, the probes selected for each of the genes that are present in more than one genome sequence can be targeted to the most conserved regions of each gene. For genes under strong diversifying selection, this means that the variant regions of the sequence can be avoided and the probes can be placed within the conserved regions. This is greatly facilitated if people experienced in comparative genomics and familiar with the species being addressed are involved in this process. Also, genes that are known to be likely to be problematic can be aligned and the best available probe regions can be selected manually.

The microarrays that we have used to date for CGH applications have been designed along these lines. The main microarray that we have used in this field is the pan-*Neisseria* microarray (Snyder *et al.*, 2004), which in version 1 was designed using the genome sequences of *N. gonorrheae* strain FA1090, *N. meningitidis* strains MC58 and Z2491, and genes specific to *N. gonorrheae* strain MS11. Version 2 of this microarray contains additional probes for genes unique to *N. meningitidis* strain FAM18, plus a number of other published and unpublished neisserial genes. The other microarray we have used in this area of work is a targeted chip focussed upon the surface proteins, regulators, and strain-divergent genes of group B streptococcus. These provided very different starting points, that presented very different problems, and the fact that the group B streptococci show less sequence variation between strains meant that the design was far simpler and required fewer genome sequences than were needed for the *Neisseria*.

A2.3 PROBE SELECTION CRITERIA AND SLIDE DESIGN FOR CGH MICROARRAYS

There are no firmly established guidelines for probe selection criteria, nor for the design of the final layout of microarrays, and there are a wide range of different solutions that have been implemented by different groups.

In a eukaryotic context, there are three broad models for microarrays for CGH applications (in addition to the use of gene-directed microarrays that are also applicable to expression profiling): two addressing chromosomal structure based upon 'tiling', and one specifically addressing known polymorphic sites. In the tiling arrays probes are either prepared from existing clone banks – usually amplified using common primers to conserved flanking regions in the vector into which they are cloned, or using oligonucleotides to regions at intervals along the chromosome. The former approach has the advantage of lower costs if a pre-existing and positionally defined library is available, and of having long probes that can be hybridized at high stringency. However, it also has the disadvantage that the chances of cross-hybridization with other locations also increase with the probe length. The oligonucleotide approach has the disadvantages of synthesis costs (which are falling all the time), but has the advantage that it can be more easily focused upon regions of specific interest, and sequence specificity. The size of the chromosomal changes that are often being sought, either as locations of chromosomal deletions or increases in copy number, in experiments using these 'tiling arrays' are usually quite large, for example in studies of changes associated with malignant transformation. Therefore interpretation is often not based upon the presence or absence, or change in sample-associated signal intensity, for a single probe. Instead (if the microarray contains probes at a sufficient density) a number of consecutive probes will show similar changes, which is helpful to data interpretation, and overcomes some issues of probe specificity.

In a prokaryotic context, one considerable advantage over eukaryotic investigations is that the chromosomal size is relatively small – reducing the source of cross-reactive nonspecific sequences. In addition, paralogous gene families in eukaryotes can frequently be so similar that designing specific probes can be highly problematic, whereas bacterial gene families tend to be more diverse. This is where the advantages end. Not only are there differences in gene presence and absence, but there can also be significant sequence differences in the genes that are commonly present between strains. The oligonucleotide probes currently in use are too short to be helpful in determining gene presence and absence because they are too susceptible to local sequence variation. Whole gene-based PCR probes (or, even worse, sequencing library-derived probes) are completely incapable of differentiation between the presence and absence of different members of moderately similar paralogous families. Partial gene PCR probes are the only remaining option, and not only are these labor intensive (and hence expensive) to produce, but the design path (as discussed above) is also not trivial.

What can be achieved for specificity and sensitivity will vary when working with different species, with differing base compositions, and inter-strain sequence conservations. There is no universally applicable model. However, the 'ball-park' parameters that we have settled on for one bacterial probe set, for the pan-*Neisseria* microarray, will illustrate the window that can be obtained. For this slide a probe range of 150 bp to 450 bp was selected, unless the length of a short predicted coding sequence required less than this. In each case the region selected for primer design was conserved between orthologues between the genomes, and avoided 'low complexity' regions. Using this probe length, and four rounds of sequential design, *in silico* hybridization, and probe refinement, it was possible to design probes that addressed orthologues with greater than 90% identity over at least 150 bp in the vast majority of cases, and which shared no more than 80% identity over any region of 20 bp or more between the intended and other target sequences. Obviously, it was not possible to design probes that differentiated between repeated genes, nor of a very small number of near identical members of paralogous families, but this illustrates that very high degrees of specificity are obtainable, even with PCR product-based probes. For extensions to this microarray, and in other projects, since there is no particular advantage in using the longer probes on this array for sensitivity, a lower upper limit for probes is now used, with an initial probe length of 150 to 300 bp being selected, which helps in obtaining specific probes. This probe size range has been independently determined to be optimal (*Figure A2.1*), once cross-hybridizations have been excluded through careful design (Chou *et al.*, 2004).

If all microarrays were perfectly printed, and no hybridization artifacts ever occurred due to positional effects or processing, then spot replication would not be necessary. However, this is not the case. Slide design is the final preliminary issue to be addressed. If at all possible all microarrays should include spot replication or, if this is not practical due to the use of a very large number of probes that will not fit onto a reasonable number of slides for processing, the experiments should be technically replicated. On-slide spot replicates should never be adjacent, since this simply reproduces local slide artifacts across each of the replicated spots. For similar reasons, in 'tiling arrays' the adjacent probes on the chromosome should not be physically adjacent on the microarray.

A2.4 PRINTING QUALITY CONTROL FOR CGH EXPERIMENTS

CGH studies are particularly prone to the effects on the results of missing spots, i.e. ones that were not actually printed onto the microarray surface. While a comparative expression study will simply fail to detect a change in a gene for which the probe is not present on the microarray, depending upon the analysis tools used, in a CGH experiment this gene is likely to be called as either a false positive or a false negative (discussed below). The cumulative effect of this sort of error when addressing a large number of probed regions in an extensive collection of samples can be catastrophic, even when the number of such failures in any one dataset is fairly small. It is therefore, important to perform as full a quality control of the slides post-printing as possible. All slides should be scanned prior to pre-hybridization, to identify any spots that have not been printed. Even then, there are (rare) occasions even when material has been arrayed, that a spot will simply fail to generate a hybrization signal. This is illustrated in *Figure A2.2*, and further highlights the need for adequate spot replication in this type of study. Using non-adjacent triplicate spots, we have observed strong hybridization to one or two out of three spots with absolutely no detectable hybridization to the other replicate(s). On a slide with no replicates, especially if only one or two slides provide the source data, the lack of a hybridization signal cannot always be attributed to gene absence or sequence divergence.

A2.5 CGH DATA ANALYSIS: GENERAL PRINCIPLES

The fairly lengthy pre-amble to this point in this Appendix has been to describe the context in which CGH interpretation has to be performed, and to this extent 'defines' the analysis problem. This discussion has sought specifically to introduce the following concepts: that the type of probe selected, the probe design strategy, the nature of the biological population, and the printed format of the microarray, all contribute to the nature and interpretability of CGH data. Furthermore, it will frequently be the case that simple yes/no answers will not be obtainable, and that the analysis tools available need to facilitate interpretation in the context of the real data that can be obtained, rather than on the basis of idealized but unrealizable concepts of the nature and functionality of the tools that are available. Also, that there are considerable additional complications associated with working beyond the 'index strains', particularly in divergent populations.

For all of these reasons (and more) two issues become clear. First, that while interpretation of multiple associated data-points in tiling arrays may be

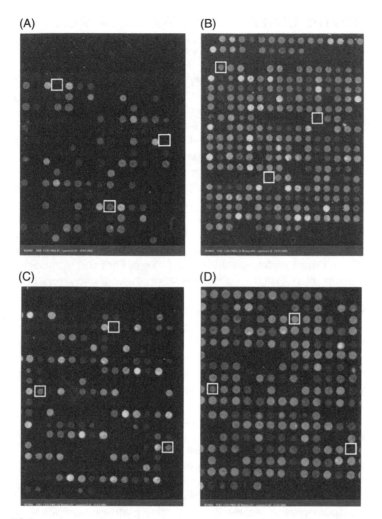

Figure A2.2

Examples of replica spot hybridization differences in CGH. Hybridization has occurred in 1 of 3 replicate spots (panels A and B) and two of three replicate spots (panels C and D). (A color version of this figure is available on the book's website, www.garlandscience.com/9780415378536)

sufficiently mutually supportive to facilitate firm conclusions, analysis of spot intensity information alone is frequently likely to be insufficient to draw firm conclusions without additional analysis and comparisons. Second, that a close link between the extracted data and the images (which are the primary data in microarray experiment – rather than the tables derived from them) is essential if errors are to be avoided. What is really needed in an analysis tool is something that allows data to be sorted into: 'reliable positive', 'reliable negative', and 'needs to be specifically addressed' categories – in which the latter set is of a manageable size and can be readily addressed with reference to the extracted data and the image files.

A2.6 WHY THE GENERAL COMPARATIVE APPROACH CANNOT BE EASILY EXTENDED TO CGH USING MULTI-GENOME-BASED MICROARRAYS

If a microarray based upon one or a very small number of genomes is used, and if one of the channels in an experiment is used as a positive control hybridization using the strain(s) used for design, or there is an appropriate alternate comparison strain for direct comparative studies, and the presence of all spot features have been confirmed in prehybridization checks, then there are tools that have been available for several years that can effectively analyze the data. These use the 'fold ratio' of signal intensities in a somewhat similar fashion to the approach used in comparative transcription analysis. The principle is that the genome sequence strain will hybridize to all the probes, generating a baseline hybridization. The 'test' strain will then either generate a similar signal intensity, making a fold-ratio value close to or greater than 1 and be reported as 'present', or it will generate lesser signal intensity and be reported as 'absent' or 'divergent'. The most widely used tool that exemplifies this approach is GACK available from: http://falkow.stanford.edu/whatwedo/software/software.html (Kim *et al.*, 2002).

The limitation of working with tools that require a control channel that will hybridize reasonably uniformly with all of the probes is that they are not well suited to use with multi-genome based microarrays and therefore in broader population-based studies. As described above, for a comparative ratio-metric analysis method to be used efficiently, one channel must always be hybridized to the reference strain against which the microarray was designed. In microarrays containing probes for genes from multiple genome sequences, such as the pan-*Neisseria* microarray, no single strain can serve as a reference that will hybridize to all spots. Using a pool of template DNAs in the control channel is not a solution to this problem, because it will contain a wide range of 'gene doses' that will influence the hybridization signal intensity ratios, and thus influence the presence/absence results based upon them. To construct the pan-*Neisseria* microarray, 18 different templates were used in PCR amplification of the probes: four genome sequence strains, five plasmids containing antibiotic resistance markers, and nine additional neisserial strains. Using methods based upon signal ratios, a spot which has hybridized in both channels is called 'present', or just in the test strain, is called 'present' in both strains because the test is similar or greater than the reference signal, unless additional analysis and manual spot-checking is also used. Equally, on a comparative basis alone, spots in which there is no hybridization in either channel are also called 'present' in both strains. Therefore, in experiments such as those described here, this approach only accurately reports genes that are present in both strains, or present in a strain assigned as the 'control' strain and absent in the 'test' strain when used to interpret multi-genome microarray data.

A2.7 CGH INTERPRETATION BASED UPON INDEPENDENT SCORING OF EACH SPOT IN EACH CHANNEL

The pON score within the microarray data extraction program BlueFuse for Microarrays (BlueGnome) has been developed to noncomparatively evaluate microarray spot data to determine the presence or absence of signal (reported as a probability of spot hybridization – pON), which is not dependent upon the signal in the other channel. The pON score is expressed in a range from zero (there is no evidence for a spot) to one (there is strong evidence for a spot). More quantitatively, the pON measure is calculated by combining two features calculated from the data: the proportion of pixels in the region that are considered to be inconsistent with the statistical characteristics of the background noise, and the evidence for an approximately circular and uniform form in the region. In the simplest case, where the background noise is normally distributed, the first measure finds the proportion of pixels that are greater than a certain number of standard deviations greater than the background. However, this measure alone can be overly sensitive to high intensity noise. The second feature therefore provides a probabilistic measure of circularity and uniformity to discriminate between signals which are simply large as opposed to those which are consistent with a spot. In CGH applications, it therefore provides an independent assessment of gene presence or absence in each channel. This measure is not primarily a measure of intensity, although obviously the intensity of the feature must be above that of the background for it to be detected and quantified, and the pON score as a result is largely independent of the PMT settings and mean spot intensity. By combining this analysis with replicate comparative hybridizations of each strain against two different unrelated strains, it is possible to obtain gene presence and absence data for diverse strains, using a microarray based upon multiple genomes and in which each channel of the hybridization is being used to investigate experimental strains.

Using this approach does not eliminate, but significantly reduces, the operator time needed to assess the data. Initially control hybridizations, preferably ones using the strains used for microarray design, are performed and the thresholds for pON values associated with gene presence are determined for an experimental system. These are then used to define the range of pON values that must be checked by the operator, although in some experimental designs it may be possible to rely upon the pON value alone.

By using pON, a slide that would otherwise take many hours of manual analysis, can be analyzed more quickly and efficiently. It should be repeated that variations in hybridization and slide artifacts can never be completely avoided and no software program can replace the need for thorough flagging of the slide before final data interpretation. In CGH, unhybridized spots, as in *Figure A2.2*, can influence the average pON scores and thus the results in the final 'present' and 'absence/divergent' tables. Therefore, in addition to excluding all of the unintentional features from the slide (dust, smears, etc.), whether all of the replicate spots are producing equivalent hybridizations and equivalent pON scores should be assessed.

A2.8 SELECTION OF APPROPRIATE THRESHOLDS FOR USING pON SCORING

Certain preliminary steps are necessary for studies of this kind to determine the optimal parameters for the analysis of particular experiments, because the optimal settings are dependent upon the signal, background, and probe and target characteristics. Initially, control CGH comparisons should be performed using combinations of the strains used to design the probes present on the microarray. These hybridizations should be used to correlate the pON scores associated with genes that are present and highly conserved, and absent. In our studies using the pan-*Neisseria* microarray, we have found a pON score of 0.68 (using BlueFuse version 2) in two or more of three spot replicates to be a robust indicator of gene presence, but this should be determined for each new experimental system used, and will also potentially vary between versions of BlueFuse if this algorithm is refined.

It should be noted that this approach cannot be used (and it is problematic using other methods also) with slides that generate a consistent false positive signal from the spotted features. This is not a common problem, but is associated with some types of microarray either due to the chemistry used for linkage/fabrication, or due to the use of marker dyes in printing quality control. In these cases pON scores of 1 will be attributed to all spots in the affected channel.

The pON score can be used for CGH in a number of ways. It can be used as a guideline for manual investigation of spot presence. It can be used as a spot-by-spot probability of presence. The pON scores for replicate spots can be combined to give a single score for each probe on the microarray. The pON scores across several slides can be combined to generate a score for each experiment. The optimal use of the pON score are up to the end-user to determine and the goals of the experiment. The pON score can also be used in transcription studies to determine inclusion and exclusion criteria for features not generating anything other than background signal, which generate unnecessary noise and fluctuations in down-stream analysis. This is not the subject of this Appendix, but it should be noted that the thresholds best suited to transcription analysis are normally less stringent than those applied to CGH – but again the values used should be titrated in pilot experiments.

A2.9 EXAMPLE: ANALYSIS OF CGH DATA IN A BACTERIAL STRAIN COMPARISON STUDY

Each strain of *N. lactamica* was hybridized to two slides, was paired with different strains in each hybridization, and was labeled with Cy3 for use on one slide and Cy5 for the other. This generated triads of strains and slides; for example strain 'A'-Cy3 was hybridized with strain 'B'-Cy5, strain 'B'-Cy3 with strain 'C'-Cy5, and strain 'C'-Cy3 with strain 'A'-Cy5. Microarray spot features were identified, quantified, and assessed using BlueFuse for Microarrays v2 (BlueGnome). Once the spots were flagged, the pON values of the spots in each channel were assessed. Each probe was printed onto the microarray slide in three non-adjacent locations, therefore three pON scores were available for each probe. Unless the spot(s) had been manually excluded during flagging, if all three probe spots had pON scores > 0.67 then the presence of that gene within the *N. lactamica* genome was accepted without review of the slide image. For any other probes with pON scores > 0.67 for one or more spots, the slide image was reviewed to ascertain whether hybridization was present, indicating gene presence in the strain. In all, with 8535 probes on each microarray (2845 probes printed in triplicate), 17 070 probes were assessed per strain, for a total of 221 910 features that were evaluated, which was only possible in an experiment of this type by using this approach.

A2.10 EXAMPLE: ANALYSIS OF CGH DATA TO IDENTIFY HUMAN CHROMOSOMAL ABNORMALITIES

As mentioned previously, human chromosomal abnormalities can result in certain medical disorders and large changes are strongly associated with changes in the development of cancer. Regions of the chromosome may contain additional copies of genes or may be missing genes. By comparing the microarray hybridization of a patient sample of DNA with a pooled control, it is possible to detect chromosomal abnormalities that may lead to an understanding of the nature and possible cause of the medical disorder. Because of the much higher degree of sequence conservation, the comparative approach is currently the most robust for analysis of CGH experiments of this type.

This is a qualitatively different process to the one described above for the prokaryotic study, because a robust highly similar comparison can be used in a systematic dual channel fashion, and as such analysis can be based more upon the relative intensities of a test and a universally applicable control which can be expected normally to hybridize to all of the probes within the experiment, once the data has been appropriately assessed and triaged. To begin, signal intensity information from the scanned microarray images of the sample and control DNA hybridizations is extracted. This produces the raw data for the hybridization of each probe and generates a signal intensity ratio between the data from the sample and control probe hybridizations. The location of each probe is also mapped to locations on the chromosome, which facilitates the analysis of the final results relative to the chromosomal architecture. This primary raw data is then normalized to correct for nonspatial intensity variation between the two scanned images, spatial variations within the images, and potentially any intensity related variation within the images. Any of the probe spots that have produced poor quality results or results that are inconsistent between replicates must then be removed from the normalized results, and the pON analysis can be used at this stage in this type of study, as can other indicators or spot quality. The remaining, unflagged, replicate data spots are then combined to generate a single data point for each probe on the microarray.

The detection of copy number variations is then assessed through investigation of the ratios of signal intensities between the two channels of data. The results generated must be properly assessed and analyzed in the context of the experiment to then be able to classify the hybridizations on the microarray slide as indications of changes in the chromosome, either copy number amplifications or deletions. Regions of the chromosome containing abnormalities can be assessed using chromosomal mapping functions, which facilitates the interpretation of the data.

A2.11 DATA NORMALIZATION FOR EUKARYOTIC AND DIAGNOSTIC CGH

Normalization of microarray data is a complex area that excites great debate. At its heart, however, is the simple requirement to correct the primary results for unwanted effects observed in the processing of the sample. Normalization involves a mathematical manipulation of the results; therefore it is important to ensure that the effects are indeed present before any form of correction is applied. In eukaryotic microarray CGH experiments the three most likely effects

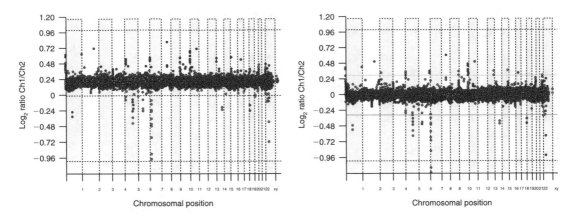

Figure A2.3

An array subject to a global median normalization to correct for intensity variation of approximately 0.24 between images. Un-normalized results: left; normalized: right. (A color version of this figure is available on the book's website, www.garlandscience.com/9780415378536)

are intensity variation between images, spatial variations within images, and potentially intensity related variation within images. As a bottom line, all data transformations involve a loss of information. Therefore the minimum necessary, and least complex, applicable normalization is always the most appropriate way to process the data, and in an ideal situation no normalization would be used at all.

The different fluorescent emission properties of the dyes used to generate the images for the sample and control make it unlikely that a 1:1 ratio will be achieved for genes which are present to the same degree in both channels. A simple global median normalization, equivalent to the dye bias, may be all that is required to correct for this effect (*Figure A2.3*).

Spatial variations in signals within images may result from a combination of factors relating to the printing of the microarray or its subsequent hybridization and washing. If there has been a tip failure or other tip effect, then one of the subgrids of the microarray may not have hybridized as well as the other blocks. Where a clear printing effect is evident in the data, a block median normalization may be more appropriate than a global median. If there is a washing effect that influences only a portion of the slide, then a more localized normalization may also be needed. A large number of probes must be present in each block for this normalization to be robust.

Because CGH experiments are labeling chromosomal DNA, which is present in similar amounts, the hybridization signals do not usually exhibit the very large dynamic range found in expression based microarray experiments. It is therefore less common to have to correct for the nonlinear response of fluorescent dyes and background correction effects. Particular care should be taken when applying the common Lowess intensity based normalization to CGH data as the major assumption of Lowess that the data is consistent and is centering around the 1:1 ratio may be inappropriate for experiments that exhibit large regions of copy number change. Lowess is primarily needed in transcription studies to address the different incorporation effects of dyes into each labeled extract, which when using some labeling methodologies show both sequence and length dependence. In expression studies a significant component of these differences is the substrate specificity and modified base incorporation efficiency of reverse transcriptase used to generate the cDNA. In CGH studies Klenow is normally used for labeling, which exhibits far fewer template-specific effects, this is another reason why Lowess normalization is not appropriate or necessary in this setting (also see below).

A2.11 REPORTING 'NEGATIVE DATA'

Choosing the right controls for any experiment is critical to being able to interpret the results. Properly conducted controls determine whether the experimental data produced is reliable and interpretable. The same is true in CGH microarray experiments. In a dual channel CGH the 'control' can often be chromosomal DNA prepared from the strain used for microarray design, and this can work efficiently for microarrays based upon one or two strains if a pool of DNA is used. This will serve as a positive control and it will hybridize to all of the spots upon the microarray that have been processed correctly. Hybridization of the control channel will validate the presence of printed material in each spot and confirm that the localized hybridization conditions were appropriate for hybridization and produce good signals above background. This positive control becomes more problematic to implement if there is no one DNA source that can be used and if a pool of DNA cannot be suitably constructed, as discussed above for multi-genome bacterial microarrays. If there is hybridization in one channel, then the lack of hybridization in the other channel can be reliably interpreted as either a complete absence of the gene or sequence divergence, such as is encountered in genes encoding antigenically variable proteins. Therefore, even when there is hybridization in the 'control' channel, in some settings it should not be assumed that an equivalent variant allele is absent. If both channels are unknowns, the second experimental channel can also be used as a control on a spot-by-spot basis when there is hybridization in one or other of the channels. However, if neither channel produces a hybridization, this spot data is of lower informational value. The lack of hybridization to an individual microarray spot can arise as a result of technical issues related to the printing and localized hybridization conditions on a microarray. Even a lack of hybridization from non-adjacent replicate spots must be interpreted cautiously in the final analysis. While obtaining consistent results of this type from technical replicates, or the consistent observation of a particular gene as present or absent in different strains, this must be considered to be corroborative, rather than definitive, evidence of absence, and the qualitative difference between a positive hybridization, and the failure to detect a hybridization must be remembered.

It must therefore be stressed that negative hybridization on CGH is non-data and should at best not be reported and at worst be presented for the technically unreliable data that it is, and not as a conclusive indicator of gene absence. For example, in our own recent publication of the gene complement of *N. lactamica*, 11 gene probes to reported virulence genes produced no hybridizations for any of the 13 *N. lactamica* strains tested, but five of these genes were present within the genome sequence of a different *N. lactamica* strain available from The

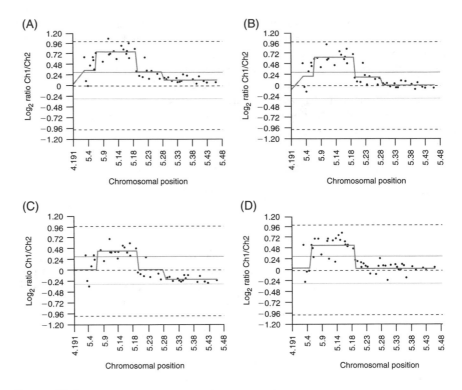

Figure A2.4

Diagnostic array subject to: (A) global median normalization using probes from a stable region on chromosome 1; (B) global median normalization; (C) block median normalization; and (D) global Lowess normalization. (A color version of this figure is available on the book's website, www.garlandscience.com/9780415378536)

Wellcome Trust Sanger Institute (Snyder, *et al.*, 2006). While it is possible that all 11 genes are truly absent from the 13 strains tested, it is far more likely that the sequence divergences observed in the genome sequence or technical issues with the microarray are the factors which produced no hybridization signals for the five genes in question.

Particular care should be exercised when deciding which results are used to calculate the normalization factor. In eukaryotic CGH it is typical to exclude the sex chromosomes, X and Y, and other regions which might exhibit significant copy number instability. In the case of a diagnostic microarray in which the probes are heavily tiled across an unstable region, it is good practice to include a series of probes at locations that are known to be stable. *Figure A2.4(A)* illustrates this approach on a diagnostic microarray that is heavily tiled on part of chromosome 5. The results have been normalized using probes on a stable portion of chromosome 1 included on the microarray for this purpose. *Figure A2.4(B to D)* illustrates results obtained using other approaches. The results have been subjected to an algorithm for detecting regions of copy number change, which further illustrates that different regions of change might result from differing, and possibly inappropriate, approaches to normalization.

A2.12 FUSION OF REPLICATE SPOT RESULTS

Once the raw data has been normalized and any unreliable results have been excluded, then the results from the replicate hybridized probe spots must be combined, to generate a single result for each probe. This is usually achieved by a simple mean or a median of the replicate probe data. Alternately, the replicates can be combined using the data fusion approach from BlueFuse for Microarrays, which uses the confidence in each result, calculated based on the hybridization characteristics, and the agreement between replicates to generate a single fused result.

A2.13 COPY NUMBER DETECTION

Eukaryotic CGH is typically used to answer two well defined questions: what region of the chromosome is aberrant; and how many copies have been lost or gained? In contrast to microarray expression analysis there is frequently a theoretical 'truth' against which actual results may be compared; a loss of a region of one chromosome should show a ratio of 1:2 between the sample and the control, an equivalent gain a ratio of 3:2. Likewise, by visualizing and analyzing probes on the basis of their genomic location it should be possible to identify the start and end points of each region of copy number change. It should therefore be possible to automate the identification of regions of copy number change for routine clinical testing.

A2.14 SUMMARY

Different CGH experiments require specific types of microarray and specific types of microarray analysis to be performed optimally. The selection of the best tools for the job, and ones that facilitate ready access to the primary data (the slide images) and the final results, are fundamental to this area of work, and should be carefully selected. Significant errors can be generated by the use of suboptimal tools or a lack of understanding of when and how different methods should be used, but it is increasingly possible to perform complex and largely automated analysis of CGH data, although it will always be necessary for careful checks to be made of the quality of the tools, hybridizations, and final data generated.

REFERENCES

Chou CC *et al.* (2004) Optimization of probe length and the number of probes per gene for optimal microarray analysis of gene expression. *Nucleic Acids Res* **32**: e99.

Kim CC, Joyce EA, Chan K, Falkow S (2002) Improved analytical methods for microarray-based genome-composition analysis. *Genome Biol* **3**: RESEARCH0065.

Snyder LAS, Davies JK, Saunders NJ (2004) Microarray genomotyping of key experimental strains of *Neisseria gonorrhoeae* reveals gene complement diversity and five new neisserial genes associated with Minimal Mobile Elements. *BMC Genomics* **5**: 23.

Snyder LAS, Saunders NJ (2006) The majority of genes in the pathogenic *Neisseria* species are present in non-pathogenic *Neisseria lactamica*, including those designated as 'virulence genes'. *BMC Genomics* **7**: article 128.

Wolfgang MC, Kulasekara BR, Liang X, Boyd D, Wu K, Yang Q, Miyada CG, Lory S (2003) Conservation of genome content and virulence determinants among clinical and environmental isolates of *Pseudomonas aeruginosa*. *Proc Natl Acad Sci USA* **100**: 8484–8489.

Appendix 3: Useful web links to microarray resources

Resource	Site	Remarks
Construction of cDNA library		
Lambda ZAP	www.stratagene.com	Commercial Product
CloneMiner	www.invitrogen.com	Commercial Product
SMART	www.clontech.com	Commercial Product
Clone LIMS and tracking		
AlmaZen	almazen.bioalma.com	Commercial Product
CloneTracker	www.biodiscovery.com/index/clonetracker	Commercial Product
B.A.S.E.	base.thep.lu.se	OpenSource Product
EST trimming, assembly, and processing pipelines		
phred, phrap, cross-match	www.phrap.org	Trimming algorithms
lucy	www.tigr.org/software/	Trimming algorithms
CAP3	genome.cs.mtu.edu/cap/cap3.html	Trimming algorithms
CLU	compbio.pbrc.edu/pti	Trimming algorithms
d2-cluster	www.ccb.sickkids.ca/dnaClustering.html	Clustering algorithm
TGICL	www.tigr.org/tdb/tgi/software/	Clustering algorithm
ESTIMA	titan.biotec.uiuc.edu/ESTIMA/	OpenSource Product
PHOREST	www.biol.lu.se/phorest	OpenSource Product
ESTWeb	bioinfo.iq.usp.br/estweb	OpenSource Product
ESTAnnotator	genome.dkfz-heidelberg.de	OpenSource Product
PipeOnline	bioinfo.okstate.edu/pipeonline/	OpenSource Product
ESTree	www.itb.cnr.it/estree/process.php	OpenSource Product
Microarray printing		
Genomic Solutions	www.genomicsolutions.com	Commercial Product
Corning	www.corning.com	Commercial Product
ArrayIt	www.arrayit.com	Commercial Product
Annotation		
Gene Ontology	www.geneontology.org	Gene Ontology Consortium
InterPro	www.ebi.ac.uk/interpro	Protein Family Annotation
Blast2GO	www.blast2go.de	Freeware tool for GO annotation
General database systems		
MySQL	www.mysql.org	OpenSource Product
PostgreSQL	www.postgresql.org/	OpenSource Product
Oracle	www.oracle.com/database/index.html	Commercial Product
Plant genomics databases		
PLANET	mips.gsf.de/projects/plants/PlaNetPortal/databases.html	Network of European Plant Databases
TAIR	arabidopsis.org/	The Arabidopsis Information Resource

Resource	Site	Remarks
MAIZEGDB	www.maizegdb.org/	Maize genome database, homepage
GRAMENE	www.gramene.org/	Resource for Comparative Grass Genomics
SoyBase	soybase.agron.iastate.edu/	Soybean database project
Microarray sites and data repository		
MGED	www.mged.org/	International Microarray Gene Expression Data Society
MIAME	www.mged.org/Workgroups/MIAME/miame.html	Minimum information about a microarray experiment, published by the MGED
SMD	genome-www5.stanford.edu/index.shtml	Homepage of the Stanford microarray database of Stanford University
ArrayExpress	www.ebi.ac.uk/arrayexpress/	EBI repository
RED	red.dna.affrc.go.jp/RED/	Rice Expression Database
NASCArrays	affymetrix.arabidopsis.info/	International Affymetrix transcriptomics service
Microarray protocols		
General information	www.microarrays.org	Site with information on microarrays, University of California at San Francisco
KRL	www.rbhrfcrc.qimr.edu.au/kidney/Pages/Microarray protocol.html	Microarray protocols used by the kidney research laboratory (Australia)
Microarray protocols	research.nhgri.nih.gov/microarray/hybridization.shtml	Microarray protocols used by the National Human Genome Research Institute
Data analysis		
Bioconductor	www.bioconductor.org	Freeware Project for Genomics Data Analysis
GEPAS	gepas.bioinfo.cnio.es	Web resource
CyberT	visitor.ics.uci.edu/genex/cybert	Web resource
Expressionist	www.genedata.com	Commercial Product
Rosetta	www.rosettabio.com/products/resolver/default.htm	Commercial Product
Institutions		
CropNet	flora.life.nottingham.ac.uk/agr/	UK Bioinformatics Resource for Crop Plants
ILSI	www.ilsi.org/	International Life Sciences Institute
NCBI	www.ncbi.nlm.nih.gov/	National Center for Biotechnology Information
EBI	www.ebi.org	European Bioinformatics Institute
TIGR	www.tigr.org/	The Institute for Genomic Research, homepage

Appendix 4: The genes selected for clustering (Chapter 6, Figure 6.4)

Cluster 1		Cluster 2	
INTEGRIN-beta6	TRABID	TNFRSF12	NOL3
PKC alpha	PDGF R beta	TNFRSF17	CXCR-5
PDCD6	GAS2	AATK	TP53
TANK	MYCN	ROBO-2	E2F1
AGRIN	FRIZZLED-5	MAP2K5	HSPA5
DR6	IL-13 R alpha1 Beta-2	ADAM17	TNFRSF11B
PPP2R2A	Microglobulin	PML.1	MCP-2
STAT1	MyD88	VEGFC	RIPK1
RAD50	FRACTALKINE	REF-1	BMP RIB
BMP-4	GMF-beta	NESTIN	MCAM
CASP5	BRAK	TNFRSF10B	HRK
IL-1 RI	GRO-gamma	IGF-II	TNFSF13B
IL-6 R alpha	IRAK2	CUGBP2	AK155
PKC iota	CCR-11	STAT12	BMP-8
E1B-AP5	IL-1 R AcP	WNT-10b	RBBP6
EphrinB 2	NIK	ZK1	PKC beta 1
RADICAL	HNRPA1	WNT-1	IL-11
RAD23B	ICAM-1	API5	GG2-1
BCL2L12	MIP-3 alpha	E2F5	WNT-16 1
JAGGED-2	eNOS	CDC2	BAR
TNFRSF21	IAP-1	IL-1 RII	ERCC2
EphrinB 1	DELTA-LIKE 4	GP130	TERF1
CARD10	EBAF	STK17B	IL-12 R beta1
C-MET	INSULIN R	CLU	ALCAM
HSPCA	CDH3	FRIZZLED-9	p27
ACTIVIN RIIA	STAT3	BIRC1	MMP-9
IL-20	DEDD	TEP1	WNT7b
TNFRSF6	IRS2	PTGES	ARTN
ILF1	Cyclophilin A	IGFBP5	PRKCB1
ASTROTACTIN	HPRT	NME5	INSRR
GRO-beta	IKBKG	TNF	EOTAXIN
BAG4	PTK2	CARD15	TrkB
RPS9	PKC zeta	MIP-1 delta	CASP1
RANTES	NFKBIE	FRIZZLED-7	EOTAXIN-3
INTEGRIN-alpha6	INTEGRIN-alpha2	TNFSF15	AIF1
BMP RIA	ICAM-3	CGR19	CASP10
ACTIVIN A	PD-ECGF	INTEGRIN-alpha5	EMP3
TNFSF13	INTEGRIN-alphaX	GRIN2C	TNFSF12
APG5L	INTEGRIN-beta5	BMP-9	FIGF
COX-1	IFN-gamma R2	WNT-11	TNFSF5
NFATC1	GLB1	G-CSF R CYTOCHROME	LDGF
MAPK8	DELTA-LIKE 1	p450	WNT-4
SURVIVIN	PRKCA	FIL1 epsilon	MUSK
TAJ	MIF	TPO	PRDX5
BIRC3	TDAG8	PML	CFLAR GM-CSF
CDC25A	FRIZZLED-4	TNFSF8	Rbeta

Cluster 3		Cluster 4	
BMP RIIA	BMP-7	AKT1	FRIZZLED-6
CCR-5	CNTF	AKT2	CTNNB1
TNFRSF11A	GPX4	DAPK2	PDCD8
CARD9	IKK-b	WNT-6	INHIBIN A
RbAp48	iNOS	PKC gamma	LTA
IGFBP6	EMP2	BRE	TP53.1
DOCK1	DAXX	INTEGRIN-beta7	EPO
JUNB	EphA 8	TERT	TOLL R 2
p75 NGFR	THROMBIN R	UNC-5	MadCAM-1
TACSTD2	ALG4	WNT-15	STAT4
API5.1	LOC51283	REELIN	BAD
TNFAIP2	PKC eta	WNT-14b	RET
AMH	IL-10	PDGF-B Chain	GDNF
RBL2	BDNF	IL16	
IL-3 R alpha	GALECTIN 3	LOC51275	
INTEGRIN-alphaM	SLC25A4	PLGF	
NSMAF	EphA 1	CXCR-3	
GRO-alpha	SPARC	INTEGRIN-beta2	
MFGE8	JUN	RAB6KIFL	
RELB	PPP1R15A	APC	
BNIP2	TNFSF7	CNTF R alpha	
PQBP1		MIG	

The genes are arranged by their position in the dendrogram.

Index